BIRKHÄUSER

Operator Theory: Advances and Applications

Vol. 206

Founded in 1979 by Israel Gohberg

Convolution Equations and Singular Integral Operators

Selected Papers of
Israel Gohberg and Georg Heinig
Israel Gohberg and Nahum Krupnik

Translated from the Russian by Oleksiy Karlovych

Leonid Lerer
Vadim Olshevsky
Ilya M. Spitkovsky
Editors

Birkhäuser

Editors:

Leonid Lerer
Department of Mathematics
Technion – Israel Institute of Technology
Haifa 32000
Israel
e-mail: llerer@techunix.technion.ac.il

Ilya M. Spitkovsky
Department of Mathematics
College of William & Mary
Williamsburg, VA 23187-8795
USA
e-mail: ilya@math.wm.edu

Vadim Olshevsky
Department of Mathematics
University of Connecticut
196 Auditorium Road, U-9
Storrs, CT 06269
USA
e-mail: olshevsky@uconn.edu

2010 Mathematics Subject Classification: 15A09, 15A29, 15A33, 15B05, 15B33, 33C47, 45E05, 45E10, 45F05, 45Q05, 47A48, 47A56, 47A57, 47A62, 47B35, 47B38, 47G10, 47L15, 65F05, 65F30, 65R20, 65R32, 93A99

Library of Congress Control Number: 2010924098

Bibliographic information published by Die Deutsche Bibliothek.
Die Deutsche Bibliothek lists this publication in the Deutsche Nationalbibliografie; detailed bibliographic data is available in the Internet at http://dnb.ddb.de

ISBN 978-3-7643-8955-0

P.O. Box 133, CH-4010 Basel, Switzerland
Part of Springer Science+Business Media
Printed on acid-free paper produced from chlorine-free pulp. TCF∞
Printed in Germany

ISBN 978-3-7643-8955-0
e-ISBN 978-3-7643-8956-7

9 8 7 6 5 4 3 2 1

www.birkhauser.ch

Contents

Preface

This book consists of translations into English of several pioneering papers in the areas of discrete and continuous convolution operators and on the theory of singular integral operators published originally in Russian. The papers were written more than thirty years ago, but time showed their importance and growing influence in pure and applied mathematics and engineering.

The book is divided into two parts. The first five papers, written by I. Gohberg and G. Heinig, form the first part. They are related to the inversion of finite block Toeplitz matrices and their continuous analogs (direct and inverse problems) and the theory of discrete and continuous resultants. The second part consists of eight papers by I. Gohberg and N. Krupnik. They are devoted to the theory of one dimensional singular integral operators with discontinuous coefficients on various spaces. Special attention is paid to localization theory, structure of the symbol, and equations with shifts.

This book gives an English speaking reader a unique opportunity to get familiarized with groundbreaking work on the theory of Toepliz matrices and singular integral operators which by now have become classical.

In the process of the preparation of the book the translator and the editors took care of several misprints and unessential misstatements. The editors would like to thank the translator A. Karlovich for the thorough job he has done.

Our work on this book was started when Israel Gohberg was still alive. We see this book as our tribute to a great mathematician.

December 10, 2009 The volume editors

Operator Theory:
Advances and Applications, Vol. 206, ix–xxii

Introduction

Leonid Lerer, Vadim Olshevsky and Ilya Spitkovsky

Israel Gohberg has made, over many years, a number of contributions to different branches of mathematics. Speaking about the quantity only, his resume lists more than 25 monographs, as well as more than 500 papers. Among these there are several papers published in Russian which have never been translated into English. The present volume partially removes this omission and contains English translations of 13 of these papers.

The first part of the book comprises a plethora of results related to the paper [GS72]. This paper contains an explicit formula for the inverse of a (non-Hermitian) Toeplitz matrix that is widely cited in many areas especially in the numerical and engineering literature as the Gohberg-Semencul formula. There are at least three reasons for its popularity. One reason lies in the fact that the Gohberg-Semencul formula (that provides an elegant description for the inverses of Toeplitz matrices) leads to efficient (in terms of speed and accuracy) algorithms. Secondly, inversion of Toeplitz matrices is a very important task in a vast number of applications in sciences and engineering. For example, symmetric Toeplitz matrices are the moment matrices corresponding to Szegö polynomials and Krein orthogonal polynomials. The latter play a significant role in many signal processing applications, e.g., [Kai86] in speech processing, e.g., [MG76]. Furthermore, prediction, estimation, signal detection, classification, regression, and communications and information theory are most thoroughly developed under the assumption that the process is weakly stationary, in which case the covariance matrix is Toeplitz [Wie49]. Along with these two examples, there are numerous other applications giving rise to Toeplitz matrices.

Finally, the third reason for the popularity of the Gohberg-Semencul formula is that it has triggered a number of counterparts and generalizations [GK72, BAS86, HR84, LT86, KC89, GO92], as well as theories, e.g., the displacement structure theory was originated in [KKM79] (see also [HR84]).

At the time of publication of [GS72] its authors were unaware of the recursive inversion algorithm that was derived earlier in [Tre64] for the case of positive definite Toeplitz matrices. The paper [Tre64] also presents (without a proof) a generalization to non-Hermitian matrices, but it is stated that all principal minors have to be nonzero. Although the Gohberg-Semencul formula is absent in

[Tre64], it could be derived from the recursions in [Tre64], at least for the special cases considered there. However, in many cases it is useful to have a closed-form formula from which different algorithms can be derived. This is especially true for the case of the Gohberg-Semencul formula, since it represents A^{-1}, the inverse of a Toeplitz matrix, via sum of products of triangular Toeplitz matrices (cf. with a generalization (0.1) below). The latter property has two important consequences. The first is that the matrix-vector product for A^{-1} can be computed in only $O(n \log n)$ operations which is fast as compared to $O(n^2)$ operations of the standard algorithm. Moreover, the second important fact is that it was the form of the Gohberg-Semencul formula that triggered the development of the study of inversion of structured matrices (see the previous paragraph).

We start our systematic account of the papers included in this volume with a description of [2] ([1] is a brief summary of the subsequent papers [2] and [3]).

The original paper [GS72] dealt with Toeplitz matrices $A = \left[a_{j-k} \right]$ with complex entries. Many applications, e.g., in Multi-Input-Multi-Output system theory, give rise to block Toeplitz matrices where the entries are matrices themselves.

In [2] the authors generalized the results of [GS72] to this and even to a more general case of Toeplitz matrices $A = \left[a_{j-k} \right]$ whose entries are taken from some non-commutative algebra with a unit. The paper [2] contains several explicit formulas for A^{-1} (Gohberg-Heinig formulas), here is one of them. For a given Toeplitz matrix $A = \left[a_{j-k} \right]$ its inverse is given by

$$
\begin{aligned}
A^{-1} = {} & \begin{bmatrix} x_0 & 0 & \cdots & 0 \\ x_1 & x_0 & \ddots & \vdots \\ \vdots & \vdots & \ddots & 0 \\ x_n & x_{n-1} & \cdots & x_0 \end{bmatrix} x_0^{-1} \begin{bmatrix} y_0 & y_{-1} & \cdots & y_{-n} \\ 0 & y_0 & \cdots & y_{1-n} \\ \vdots & \ddots & \ddots & \vdots \\ 0 & \cdots & 0 & y_0 \end{bmatrix} \\
& - \begin{bmatrix} 0 & 0 & \cdots & 0 \\ z_{-n} & 0 & \ddots & \vdots \\ \vdots & \ddots & \ddots & 0 \\ z_{-1} & \cdots & z_{-n} & 0 \end{bmatrix} z_0^{-1} \begin{bmatrix} 0 & w_n & \cdots & w_1 \\ 0 & 0 & \ddots & \vdots \\ \vdots & \ddots & \ddots & w_n \\ 0 & \cdots & 0 & 0 \end{bmatrix}
\end{aligned} \tag{0.1}
$$

where the parameters $\{x_i, y_i, z_i, w_i\}$ are obtained via solving four linear systems of equations

$$
A \begin{bmatrix} x_0 \\ x_1 \\ \vdots \\ x_n \end{bmatrix} = \begin{bmatrix} e \\ 0 \\ \vdots \\ 0 \end{bmatrix}, \qquad A \begin{bmatrix} z_{-n} \\ \vdots \\ z_{-1} \\ z_0 \end{bmatrix} = \begin{bmatrix} 0 \\ \vdots \\ 0 \\ e \end{bmatrix}, \tag{0.2}
$$

$$
\begin{aligned}
\left[\begin{matrix} w_0 & w_1 & \cdots & w_n \end{matrix} \right] A &= \left[\begin{matrix} e & 0 & \cdots & 0 \end{matrix} \right], \\
\left[\begin{matrix} y_{-n} & \cdots & y_{-1} & y_0 \end{matrix} \right] A &= \left[\begin{matrix} 0 & \cdots & 0 & e \end{matrix} \right].
\end{aligned} \tag{0.3}
$$

Due to its shift-invariant structure, a Toeplitz matrix A is defined by $2n+1$ entries $\{a_{-n}, \ldots, a_{-1}, a_0, a_1, \ldots, a_n\}$ appearing in its top row and first column. At the same time, the formula (0.1) describes the structure of A^{-1} using the redundant set of $4n+2$ parameters $\{x_i, y_i, w_i, z_i\}$. The second section of [2] deals with this discrepancy and proves that in fact, just $2n+1$ parameters $\{x_i, z_i\}$ are sufficient to completely describe the structure of A^{-1} (i.e., it is sufficient to solve two linear equations in (0.2) and to use the first and last columns of A^{-1} only). Alternatively, A^{-1} can also be described by $2n+1$ parameters $\{y_i, w_i\}$ only (i.e., it is sufficient to solve two linear equations in (0.3) and to use the first and last rows of A^{-1} only). In fact, the second section of [2] establishes a number of interconnections between the solution of the linear equations in (0.2) and (0.3).

The third section of [2] deals with a natural inverse problem, namely: given $4n+2$ parameters $\{x_i, y_i, w_i, z_i\}$, is there a Toeplitz matrix satisfying (0.2) and (0.3) The authors show that if the interconnections between the solution of linear equations in (0.2) and in (0.3) found in Section 2 are valid, then the desired Toeplitz matrix exists and it can be recovered via formula (0.1).

We would like to again emphasize that the above results of [2] fully cover a very important special case when A is a block Toeplitz matrix.

The above three sections fully generalize the inversion formulas of [GS72]. The next three sections of [2] generalize the inversion formulas of [GK72]. The difference is that [GS72] describes the structure of the entire complex matrix A^{-1} using its first and last columns only, while [GK72] describes the structure of A^{-1} using instead its first two columns. As was already mentioned, the first three sections of [2] generalize the results of [GS72] to matrices whose entries are taken from a noncommutative algebra with a unit. Correspondingly, the sections 4–6 of [2] follow the structure of its first three sections and contain full generalization of the results of [GK72].

We also refer the reader to alternative derivations of the Gohberg-Heinig formulas in [GKvS04] as well as to [BAS85] for a generalization of the results of [BAS86] to the block Toeplitz case.

Before turning to paper [3] let us again notice that the Gohberg-Semencul and the Gohberg-Heinig formulas have a very important computational consequence. Indeed, observe that the right-hand side of (0.1) involves only four triangular Toeplitz matrices. Hence A^{-1} can be *efficiently* applied to *any* vector using FFT (Fast Fourier Transform) in the case when the entries $\{a_k\}$ are either scalars or matrices of a reasonably small size. Furthermore, the linear equations in (0.2) and (0.3) can be *efficiently* solved using the Levinson algorithm [Wie49] in the scalar case or using block Levinson algorithm of [Mus88] in the block case. To sum up, the scheme just described means that the Gohberg-Semencul and the Gohberg-Heinig formulas imply the low cost $O(n^2)$ arithmetic operations for solving any block Toeplitz linear system when the blocks are reasonably small. It is worth mentioning that using Gohberg-Semencul-Heinig formulas in [BGY80, CK89, CK91] an algorithm requiring only $O(n \log^2 n)$ operations was developed.

Along with speed, one is also always concerned about numerical accuracy, and the Gohberg-Semencul formula was shown in [GH95] and [Hei01] to always provide numerically accurate results unless the condition number of A is too large.

The paper [GS72] contained not only Toeplitz inversion formulas but also their continual analogs. In fact, the authors designed a formula for the inverse of integral operators of the form

$$((I-K)\varphi)(t) = \varphi(t) - \int_0^\tau k(t-s)\varphi(s)ds \qquad (0 \le t \le \tau)$$

acting in the space $L_p(0,\tau)$, where the kernel $k(t)$ is a scalar function from $L_1(-\tau,\tau)$. Due to the difference nature of the kernel $k(t-s)$, $I-K$ is a continuous analog of a finite Toeplitz matrix $A = \left[\ a_{j-k}\ \right]$ (whose entries depend only on the difference $j-k$ of indices). The paper [3] generalizes these results of [GS72] to the case when the operator $I-K$ acts in the space $L_p^n(0,\tau)$, and $k(t)$ is an $n \times n$ matrix function from $L_1^n(-\tau,\tau)$. Thus, the paper [3] contains full continual analogs of the results in [2].

In particular, in the second section of [3] the authors present a continual analog of the formula (0.1). The four equations in (0.2) and (0.3) are replaced by the following four equations

$$x(t) - \int_0^\tau k(t-s)x(s)ds = k(t), \quad z(-t) - \int_0^\tau k(s-t)z(-s)ds = k(-t), \quad (0.4)$$

$$w(t) - \int_0^\tau w(s)k(t-s)ds = k(t), \quad y(-t) - \int_0^\tau y(-s)k(s-t)ds = k(-t), \quad (0.5)$$

In this case, the analog of the formula (0.1) is

$$((I-K)^{-1}f)(t) = f(t) + \int_0^\tau \gamma(t,s)f(s)ds,$$

where the kernel $\gamma(t,s)$ is determined from (0.4) and (0.5) via

$$\gamma(t,s) = x(t-s) + y(t-s) + \int_0^{\min(t,s)} [x(t-s)y(r-s) - z(t-r-\tau)w(r-s+\tau)]dr$$

The structure of the paper [3] mimics the structure of the first three sections of [2] (although the methods of [2] and [3] are absolutely different). Section 3 of [3] presents the results analogous to the ones of Section 2 of [2], and it describes the relations between the solutions $x(t), z(t)$ to (0.4) and the solutions $w(t), y(t)$ to (0.5). Finally, Section 4 of [3] is a counterpart of Section 3 of [2]. Specifically, it is devoted to the inverse problem of reconstructing the matrix function $k(t)$ from the matrix functions $x(t), y(-t), w(t), z(-t)$:

Given four matrix function $x, y, z, w \in L_1^{n\times n}[0,\tau]$, find $k \in L_1^{n\times n}[-\tau,\tau]$ such that the corresponding operator $I-K$ is invertible and the given functions are solutions of the equations (0.4)–(0.5).

Later, a certain refinement of these solutions turned out to be extremely useful in solving the inverse problems for Krein's orthogonal matrix polynomials and Krein's orthogonal matrix function, as well as in the study of Krein's canonical differential systems (see [GL88, GKL08, AGK$^+$09] and references therein). In fact, for the scalar case the connection of the inverse problems with Krein orthogonal polynomials has already been mentioned in the Russian edition of the monograph [GF71]. Finally, we mention that the solutions of the inverse problems in [2] inspired several other authors to deal with similar problems (see, e.g., [BAS86] and [KK86]).

One of the conditions in the solution of the inverse problem in [2] triggered the interest of Gohberg and Heinig in Sylvester resultant matrices. Paper [4] is devoted to the study of generalizations of the classical Sylvester resultant matrix to the case of polynomials with matrix coefficients.

For scalar polynomials $p(\lambda)$ and $q(\lambda)$ of degrees m and n, respectively, the resultant matrix is a square $(m+n) \times (m+n)$ matrix whose basic property is that its nullity is equal to the number of common zeros of the polynomials p and q (counting their multiplicities). This notion has been known for centuries (see [KN81] for history and details).

A simple example found in [4] shows that in the matrix polynomial case the (square) classical analog of the resultant matrix does not preserve its basic property concerning the common eigenvalues. Nevertheless, a certain non-square generalized resultant is introduced in [4] which does have the basic property of the resultant matrix, namely, the dimension of its kernel is equal to the number of common eigenvalues of the two given matrix polynomials counting their multiplicities (properly understood). In general, for two $r \times r$ matrix polynomials of degrees m and n this matrix has the size $(2\omega - m - n)r \times \omega r$ where $\omega \geq \min\{n+mr, m+nr\}$. In [4] the kernel of this generalized resultant is completely described in terms of common Jordan chains of the given polynomials. The proof of this result is rather involved. One of the important tools is the notion of multiple extension of a system of vectors which has been invented and studied by the authors.

Of course, in the case $r = 1$ the generalized resultant in [4] coincides with the classical one, and the paper [4] contains a refinement of the well known result on the Sylvester resultant matrix providing a complete description of its kernel in terms of common zeroes of the given (scalar) polynomials $p(\lambda)$ and $q(\lambda)$. [4] also includes applications to the solution of a system of two equations in two unknowns λ and μ in the case when each of these equations is a matrix polynomial in λ and a matrix polynomial in μ.

It is worth bearing in mind that the results of paper [4] had been obtained before the spectral theory of matrix polynomials was developed by I. Gohberg, P. Lancaster and L. Rodman (see [GLR83]) and from our point of view it is rather miraculous that the results of [4] were obtained without using this theory. Upon emergence of the latter theory, the notion of the generalized resultant was further analyzed in [GKLR81, GKLR83] in connection with some other classical notions (like Vandermonde, Bezoutian, etc.) Also, in [GKL08] and [KL09] necessary and sufficient conditions for the matrix polynomials p and q are found which ensure that the generalized resultant in [4] can be taken to be a square block matrix.

In paper [5], Gohberg and Heinig invented a continual analogue of the resultant. Namely, for two entire functions $\mathcal{A}(\lambda)$ and $\mathcal{B}(\lambda)$ $(\lambda \in \mathbb{C})$ of the form

$$\mathcal{A}(\lambda) = a_0 + \int_0^\tau a(t)e^{i\lambda t}dt, \ \mathcal{B}(\lambda) = b_0 + \int_{-\tau}^0 b(t)e^{i\lambda t}dt \tag{0.6}$$

where $a_0, b_0 \in \mathbb{C}^1$, $a \in L_1[0,\tau]$, $b \in L_1[-\tau, 0]$, and τ is some positive number, they define the operator $\mathcal{R}_0(\mathcal{A},\mathcal{B})$ acting on $L_1[-\tau,\tau]$ by the rule

$$(\mathcal{R}_0(\mathcal{A},\mathcal{B})f)(t) = \begin{cases} f(t) + \int\limits_{-\tau}^{\tau} a(t-s)f(s)ds \ (0 \le t \le \tau) \\ f(t) + \int\limits_{-\tau}^{\tau} b(t-s)f(s)ds \ (-\tau \le t < 0) \end{cases} \tag{0.7}$$

with the convention that $a(t) = 0$ for $t \notin [0,\tau]$ and $b(t) = 0$ for $t \notin [-\tau, 0]$. In [5] the operator $\mathcal{R}_0(\mathcal{A},\mathcal{B})$ is called the resultant operator of the functions $\mathcal{A}(\lambda)$ and $\mathcal{B}(\lambda)$. In the scalar case the kernel of the operator $\mathcal{R}_0(\mathcal{A},\mathcal{B})$ is completely described in [5] in terms of common zeroes of the functions $\mathcal{A}$ and $\mathcal{B}$. In particular, its dimension is precisely equal to the number of common zeroes of $\mathcal{A}$ and $\mathcal{B}$ (counting multiplicities). Thus in [5], an appropriate notion of a resultant for non-polynomial functions has been defined for the first time.

As in the discrete case simple examples show that this resultant is not working in the matrix valued case and hence a straightforward generalization of the above result to the case of matrix functions $\mathcal{A}(\lambda)$ and $\mathcal{B}(\lambda)$ turned out to be impossible. One can only state that $\dim \operatorname{Ker} \mathcal{R}_0(\mathcal{A},\mathcal{B}) \ge \#\{$ common eigenvalues of $\mathcal{A}$ and $\mathcal{B}$ (properly understood) $\}$. The reason for this phenomenon lies in the fact that in the matrix case (i.e., when $\mathcal{A}$ and $\mathcal{B}$ are $d \times d$ matrix functions) the kernel of $\mathcal{R}_0(\mathcal{A},\mathcal{B})$ may contain matrix functions that are not smooth enough (actually, they are not even absolutely continuous). This obstacle has been surmounted in [5] by a slight modification of the definition of $R_0(\mathcal{A},\mathcal{B})$ through introducing a generalized resultant $R_\varepsilon(\mathcal{A},\mathcal{B})$ acting from $L_1^d[-\tau,\tau+\varepsilon]$ to $L_1^d[-\tau-\varepsilon,\tau+\varepsilon]$ (see formula (2) in [5]). It is proved in [5] that for any $\varepsilon > 0$ the kernel for this operator is completely described by the common Jordan chains of the matrix functions $\mathcal{B}(\lambda)$ and $\mathcal{A}(\lambda)$, and in particular, its dimension equals the number of common eigenvalues of $\mathcal{B}(\lambda)$ and $\mathcal{A}(\lambda)$ (properly understood).

The proof of this result is far from being easy. The applications of the main result include a method of solving a system of two equation by two variables λ and μ in the case when the right-hand sides of the equations are of the form (0.6) in each variable. In another important application a continual analogue of the Bezoutian is introduced for the case $n = 1$, and its kernel is completely described in terms of common zeroes of the functions involved.

Paper [5] lent impetus to several other works in this area. Thus, in [GHKL05] the Bezoutian for matrix functions of form (0.6) has been studied. In [GKL07b, GKL07a] necessary and sufficient conditions for $\mathcal{A}$ and $\mathcal{B}$ have been found which ensure that the operator $\mathcal{R}_0(\mathcal{A}, \mathcal{B})$ describes completely the common spectral data of $\mathcal{A}$ and $\mathcal{B}$ in the matrix case.

The second part of the book deals with a very different topic. It consists of papers [6–13], written by I. Gohberg jointly with N. Krupnik[1]. In these papers, various aspects of the theory of one dimensional singular integral operators in the piecewise continuous setting are considered, but the general approach actually goes back to Gohberg's paper [Goh52]. There, Gelfand's theory of commutative Banach algebras was applied to such operators for the first time, though in the setting of continuous coefficients and closed curves. It was shown that the respective Calkin algebras (= the quotient algebras modulo the ideal of compact operators) are commutative, and therefore can be identified with their spaces of maximal ideals, which in turn are nothing but the curves on which the operators act. The Fredholmness criteria for these operators can be formulated in terms of their symbols, which in this setting are rather simple continuous functions on these curves, and do not depend on the choice of the space. Consequently, the essential spectra of the operators do not depend on the space as well. As was shown in joint work of I. Gohberg and M. Krein [GK58], this phenomenon persists for convolution type operators acting on vector functions on the half line. The case of singular integral operators with continuous (matrix) coefficients was disposed of, with the same outcome, by I. Simonenko [Sim61]. As became clear later (see, e.g., [BK97]), the smoothness of the curves (as long as they stay closed) and presence of weight (as long as the singular integral operator with Cauchy kernel stays bounded) are of no significance in the continuous coefficients setting.

The situation becomes much more complicated when one moves to a piecewise continuous setting: the symbols become matrix (as opposed to scalar) functions, with a domain of a rather complicated nature, in particular, depending on the Banach space in which the operators act. The (essential) spectra of the operators then also become functions of the space. More specifically, the content of [6–13] is as follows.

In [6], the spectra and essential spectra are described for Toeplitz operators generated by piecewise continuous functions on Hardy spaces H_p (Section 1) and, in parallel, of singular integral operators with piecewise continuous coefficients on Lebesgue spaces L_p (Section 2), $1 < p < \infty$. To this end, the notion of p-(non)singular functions and the p-index was introduced, and with their aid established the role of circular arcs filling the gaps of the spectrum originating from the discontinuities of the coefficients. In the subsequent paper [GK69b] the authors showed that the circular arcs persist when a power weight is introduced, though the curvature of the arcs now depends on the weight exponents. Further metamorphoses of the spectra were described in [Spi92] (for an arbitrary Muckenhoupt weight), and in a series of papers by A. Böttcher and Yu. Karlovich (where the transition to arbitrary Carleson curves was accomplished), summarized in their

[1]Current mailing address: 424 – 7805 Bayview Ave, Thornhill, Ontario L3T 7N1, Canada

monograph [BK97]. See also very nicely written survey articles [Böt95, BK01] for an entertaining, though still educational, guide through the subject.

As an application of these results, in Section 3 of [6] the authors give estimates from below for the essential norms of the singular integral operator S with the Cauchy kernel, and the related complementary projections P, Q. They show that the estimate for S is sharp when $p = 2^n$ or $p = 2^n/(2^n - 1)$, $n = 1, 2, \ldots$ Inspired by these results, Pichorides proved shortly afterwards [Pic72] that in fact this estimate coincides with the norm of S on the unit circle for all $p \in (1, \infty)$. The respective result for the projections P, Q was established much later [HV00]. See [HV10, Kru10] for the current state of the subject.

The singular integral operators in [6] are also considered on the so called symmetric spaces (see Section 4); we refer the reader to the translator's work [Kar98, Kar00].

Paper [7] contains a detailed description of the Banach algebra generated by individual Toeplitz operators considered in [6], Section 1. In particular, it is shown that its Calkin algebra is commutative, and its compact of maximal ideals is realized as a cylinder with an exotic topology. The symbol of this algebra is constructed, and the Fredholm criterion and the index formula for the operators from the algebra are stated in its terms. Note that the case $p = 2$ was considered by the authors earlier in [GK69a], and that a parallel theory for the algebra generated by Toeplitz matrices arising from the Fourier coefficients of piecewise continuous functions on the sequence space ℓ_p (which of course differs from the setting of [7] when $p \neq 2$) was developed by R. Duduchava in [Dud72].

In [8], it was observed for the first time that rather peculiar objects are contained in the (closed) algebra generated by singular integral operators with piecewise continuous coefficients; some operators with unbounded coefficients having power or logarithmic singularities among them. As a result, the Fredholmness criteria for such operators are obtained.

An auxiliary role in this paper is played by a (very useful on its own) observation that a sufficient Khvedelidze [Khve56] condition for the operator S to be bounded in L_p space with a power weight actually is also necessary. The resulting criterion was used repeatedly in numerous publications, though now it can of course be thought of as a direct consequence of the Hunt-Muckenhoupt-Weeden criterion.

In [9], the algebra generated by singular integral operators with piecewise continuous coefficients is considered in the case of a composed (that is, consisting of a finite number of closed or open simple Lyapunov curves) contour. The new difficulty arising in this setting is that S is not an involution any more; moreover, the operator $S^2 - I$ is not even compact. To overcome this difficulty, an approach is proposed in this paper (Lemma 1.1) which later received the name of the *linear dilation* procedure and was further developed in [GK70] (see also [Kru87], Theorem 1.7, Corollary 1.1 and Theorem 2.4). The symbol of the algebra is constructed, and in its terms the Fredholmness and the index formula are obtained. The spaces involved are L_p with power weights. The transition to arbitrary Muckenhoupt weights was accomplished in [GKS93], while the class of contours was extended to composed Carleson curves in [BBKS99].

Section 1 of paper [10] contains a new version of the *local principle.* The original version of the latter, invented by by I.B. Simonenko in [Sim65], is simplified[2] here via the introduction of *localizing classes* which at the same time makes the new version applicable in a wider setting. There also exists the Allan-Douglas version of the local principle; the relations between the three are discussed in [BKS88]. All three versions are currently in use, as a powerful tool in operator theory and numerical analysis; see, i.e., [BS93, DS08, FRS93, Kar98] for some further examples of the Gohberg-Krupnik local principle's applications. In the paper itself, the local principle is used to establish (i) the local closedness of the set of matrix functions generating Fredholm Toeplitz operators on the spaces L_2, (ii) the Fredholm theory of operators defined on ℓ_p via Toeplitz matrices generated by continuous matrix functions (and the algebras generated by them), and (iii) a parallel theory for the algebras generated by paired operators.

In [11], the symbol is constructed for the non-closed algebra generated by the singular integral operators with piecewise continuous coefficients on piecewise Lyapunov composed curves in the case of L_p spaces with power weights. This symbol is a matrix valued function (of variable size, depending on the geometry of the curve), the non-singularity of which is responsible for the Frednolmness of the operator. The index of the operator is calculated in terms of the symbol as well. These results were carried over to the closure of the algebra in a later authors' publication [GK93]. The transition to arbitrary (Muckenhoupt) weights and (Carleson) composed curves was accomplished in already mentioned papers [GKS93, BBKS99]; see also [BGK$^+$96] for an important intermediate step.

The paper [12] in its time was a breakthrough in the theory of singular integral operators with involutive shift. In clarification of the results accumulated earlier (see, i.e., [Lit67, ZL68, Ant70, KS72b, KS72a]), the authors came up with the relation which makes crystal clear the relation between the Fredholm properties and the index of the operator A (with shift) and the associated with it operator $\tilde{A}_W$ (without shift but acting on the space of vector functions with the size doubled). Moreover, for the operators with orientation preserving shift the relation actually holds for arbitrary measurable matrix coefficients, not just in the continuous case. On the other hand, for the orientation reversing shift the Fredholmness of A_W is still sufficient for the Fredholmness of A while the converse statement fails already in the case of piecewise continuous coefficients.

The paper [12] contains also the matrix symbol construction for the (non-closed) algebra generated by the singular integral operators with piecewise continuous coefficients and the shift $(W\phi)(t) = \phi(-t)$ acting on the unweighted space $L_2(-1,1)$. The generalization of this construction to the setting of arbitrary simple closed Lyapunov contours and changing orientation (sufficiently smooth) involutive shifts is carried out in [13].

[2]The presentation of the method per se in [10] takes only two pages.

References

[AGK+09] D. Alpay, I. Gohberg, M.A. Kaashoek, L. Lerer, and A. Sakhnovich, *Krein systems*, Modern analysis and applications, Oper. Theory Adv. Appl., vol. 191, Birkhäuser, Basel, 2009, pp. 19–36.

[Ant70] A.B. Antonevič, *The index of a pseudodifferential operator with a finite group of shifts*, Dokl. Akad. Nauk SSSR **190** (1970), 751–752 (in Russian), English translation: *Soviet Math. Dokl.* **11** (1970), 168–170.

[BAS85] A. Ben-Artzi and T. Shalom, *On inversion of block Toeplitz matrices*, Integral Equations Operator Theory **8** (1985), 751–779.

[BAS86] ———, *On inversion of Toeplitz and close to Toeplitz matrices*, Linear Algebra Appl. **75** (1986), 173–192.

[BBKS99] C.J. Bishop, A. Böttcher, Yu.I. Karlovich, and I.M. Spitkovsky, *Local spectra and index of singular integral operators with piecewise continuous coefficients on composed curves*, Math. Nachr. **206** (1999), 5–83.

[BGK+96] A. Böttcher, I. Gohberg, Yu. Karlovich, N. Krupnik, S. Roch, B. Silbermann, and I. Spitkovsky, *Banach algebras generated by N idempotents and applications*, Singular integral operators and related topics (Tel Aviv, 1995), Oper. Theory Adv. Appl., vol. 90, Birkhäuser, Basel, 1996, pp. 19–54.

[BGY80] R. Brent, F. Gustavson, and D. Yun, *Fast solutions of Toeplitz systems of equations and computation of Pade approximants*, Journal of Algorithms **1** (1980), 259–295.

[BK97] A. Böttcher and Yu.I. Karlovich, *Carleson curves, Muckenhoupt weights, and Toeplitz operators*, Birkhäuser Verlag, Basel and Boston, 1997.

[BK01] ———, *Cauchy's singular integral operator and its beautiful spectrum*, Systems, approximation, singular integral operators, and related topics (Bordeaux, 2000), Oper. Theory Adv. Appl., vol. 129, Birkhäuser, Basel, 2001, pp. 109–142.

[BKS88] A. Böttcher, N. Krupnik, and B. Silbermann, *A general look at local principles with special emphasis on the norm computation aspect*, Integral Equations Operator Theory **11** (1988), 455–479.

[Böt95] A. Böttcher, *Toeplitz operators with piecewise continuous symbols – a neverending story?*, Jber. d. Dt. Math.-Verein. **97** (1995), 115–129.

[BS93] A. Böttcher and I. Spitkovsky, *Wiener-Hopf integral operators with PC symbols on spaces with Muckenhoupt weight*, Revista Matématica Iberoamericana **9** (1993), 257–279.

[CK89] J. Chun and T. Kailath, *Divide-and-conquer solutions of least-squares problems for matrices with displacement structure*, Transactions of the Sixth Army Conference on Applied Mathematics and Computing (Boulder, CO, 1988), ARO Rep., vol. 89, U.S. Army Res. Office, Research Triangle Park, NC, 1989, pp. 1–21.

[CK91] ———, *Divide-and-conquer solutions of least-squares problems for matrices with displacement structure*, SIAM Journal of Matrix Analysis Appl. **12** (1991), 128–145.

[DS08] V.D. Didenko and B. Silbermann, *Approximation of additive convolution-like operators. Real C^*-algebra approach*, Frontiers in Mathematics, Birkhäuser Verlag, Basel, 2008.

[Dud72] R.V. Dudučava, *Discrete Wiener-Hopf equations in l_p spaces with weight*, Soobstich. Akad. Nauk Gruz. **67** (1972), 17–20 (in Russian).

[FRS93] T. Finck, S. Roch, and B. Silbermann, *Two projection theorems and symbol calculus for operators with massive local spectra*, Math. Nachr. **162** (1993), 167–185.

[GF71] I.C. Gohberg and I.A. Feldman, *Convolution equations and projection methods for their solution*, Nauka, Moscow, 1971 (in Russian), English translation *Amer. Math. Soc. Transl. of Math. Monographs* **41**, Providence, R.I. 1974.

[GH95] M. Gutknecht and M. Hochbruk, *The stability of inversion formulas for Toeplitz matrices*, Linear Algebra Appl. **223–224** (1995), 307–324.

[GHKL05] I. Gohberg, I. Haimovici, M.A. Kaashoek, and L. Lerer, *The Bezout integral operator: Main property and underlying abstract scheme*, The state space method. Generalizations and applications, Oper. Theory Adv. Appl., vol. 161, Birkhäuser, Basel, 2005, pp. 225–270.

[GK58] I. Gohberg and M.G. Krein, *Systems of integral equations on a half-line with kernel depending upon the difference of the arguments*, Uspekhi Mat. Nauk **13** (1958), no. 2, 3–72 (in Russian), English translation: *Amer. Math. Soc. Transl.* (2) **14** (1960), 217–287.

[GK69a] I. Gohberg and N. Krupnik, *On the algebra generated by the Toeplitz matrices*, Funkcional. Anal. i Priložen. **3** (1969), no. 2, 46–56 (in Russian), English translation: *Functional Anal. Appl.* **3** (1969), 119–127.

[GK69b] ______, *The spectrum of singular integral operators in L_p spaces with weight*, Dokl. Akad. Nauk SSSR **185** (1969), 745–748 (in Russian), English translation: *Soviet Math. Dokl.* **10** (1969), 406–410.

[GK70] ______, *Banach algebras generated by singular integral operators*, Colloquia Math. Soc. Janos Bolyai 5, Hilbert Space Operators (Tihany (Hungary)), 1970, pp. 239–267.

[GK72] ______, *A formula for inversion of Toeplitz matrices*, Matem. Issled. **7** (1972), no. 7(2), 272–283 (in Russian).

[GK93] ______, *Extension theorems for Fredholm and invertibility symbols*, Integral Equations Operator Theory **16** (1993), 514–529.

[GKL07a] I. Gohberg, M.A. Kaashoek, and L. Lerer, *The continuous analogue of the resultant and related convolution operators*, The extended field of operator theory, Oper. Theory Adv. Appl., vol. 171, Birkhäuser, Basel, 2007, pp. 107–127.

[GKL07b] ______, *Quasi-commutativity of entire matrix functions and the continuous analogue of the resultant*, Modern operator theory and applications, Oper. Theory Adv. Appl., vol. 170, Birkhäuser, Basel, 2007, pp. 101–106.

[GKL08] ______, *The resultant for regular matrix polynomials and quasi commutativity*, Indiana Univ. Math. J. **57** (2008), 2793–2813.

[GKLR81] I. Gohberg, M.A. Kaashoek, L. Lerer, and L. Rodman, *Common multiples and common divisors of matrix polynomials. I. Spectral method*, Indiana Univ. Math. J. **30** (1981), 321–356.

[GKLR83] ______, *Common multiples and common divisors of matrix polynomials. II. Vandermonde and resultant matrices*, Linear and Multilinear Algebra **12** (1982/83), 159–203.

[GKS93] I. Gohberg, N. Krupnik, and I. Spitkovsky, *Banach algebras of singular integral operators with piecewise continuous coefficients. General contour and weight*, Integral Equations Operator Theory **17** (1993), 322–337.

[GKvS04] I. Gohberg, M.A. Kaashoek, and F. van Schagen, *On inversion of finite Toeplitz matrices with elements in an algebraic ring*, Linear Algebra Appl. **385** (2004), 381–389.

[GL88] I. Gohberg and L. Lerer, *Matrix generalizations of M. G. Kreĭn theorems on orthogonal polynomials*, Orthogonal matrix-valued polynomials and applications (Tel Aviv, 1987–88), Oper. Theory Adv. Appl., vol. 34, Birkhäuser, Basel, 1988, pp. 137–202.

[GLR83] I. Gohberg, P. Lancaster, and L. Rodman, *Matrices and indefinite scalar products*, Oper. Theory Adv. Appl., vol. 8, Birkhäuser, Basel, 1983.

[GO92] I. Gohberg and V. Olshevsky, *Circulants, displacements and decompositions of matrices*, Integral Equations Operator Theory **15** (1992), 730–743.

[Goh52] I.C. Gohberg, *On an application of the theory of normed rings to singular integral equations*, Uspehi Matem. Nauk (N.S.) **7** (1952), no. 2(48), 149–156 (in Russian).

[GS72] I. Gohberg and L. Semencul, *The inversion of finite Toeplitz matrices and their continual analogues*, Matem. Issled. **7** (1972), no. 2(24), 201–223 (in Russian).

[Hei01] G. Heinig, *Stability of Toeplitz matrix inversion formulas*, Structured Matrices in Mathematics, Computer Science, and Engineering, vol. 2, American Mathematical Society Publications, Providence, RI. Contemporary Math **281** (2001), 101–116.

[HR84] G. Heinig and K. Rost, *Algebraic methods for Toeplitz-like matrices and operators*. Oper. Theory Adv. Appl., vol. 13, Birkhäuser, Basel, 1984.

[HV00] B. Hollenbeck and I.E. Verbitsky, *Best constants for the Riesz projection*, J. Functional Analysis **175** (2000), no. 2, 370–392.

[HV10] ______, *Best constants inequalities involving the analytic and co-analytic projection*, Topics in operator theory. Vol. 1. Operators, matrices and analytic functions, Oper. Theory Adv. Appl., vol. 202, Birkhäuser, Basel, 2010, 285–296.

[Kai86] T. Kailath, *A theorem of I. Schur and its impact on modern signal processing*, I. Schur methods in operator theory and signal processing, Oper. Theory Adv. Appl., vol. 18, Birkhäuser, Basel, 1986, pp. 9–30.

[Kar98] A.Yu. Karlovich, *Singular integral operators with piecewise continuous coefficients in reflexive rearrangement-invariant spaces*, Integral Equations Operator Theory **32** (1998), 436–481.

[Kar00] ———, *On the essential norm of the Cauchy singular integral operator in weighted rearrangement-invariant spaces*, Integral Equations Operator Theory **38** (2000), 28–50.

[KC89] T. Kailath and J. Chun, *Generalized Gohberg-Semencul formulas for matrix inversion*, The Gohberg anniversary collection, Oper. Theory Adv. Appl., vol. 40, Birkhäuser, Basel, 1989, pp. 231–245.

[Khve56] B.V. Khvedelidze, *Linear discontinuous boundary problems in the theory of functions, singular integral equations and some of their applications*, Akad. Nauk Gruzin. SSR. Trudy Tbiliss. Mat. Inst. Razmadze **23** (1956), 3–158 (in Russian).

[KK86] T. Kailath and I. Koltracht, *Matrices with block Toeplitz inverses*, Linear Algebra Appl. **75** (1986), 145–153.

[KKM79] T. Kailath, S. Kung, and M. Morf, *Displacement ranks of matrices and linear equations*, J. Math. Anal. Appl. **68** (1979), 395–407.

[KL09] M.A. Kaashoek and L. Lerer, *Quasi commutativity of regular matrix polynomials: Resultant and Bezoutian*, Modern analysis and applications, Oper. Theory Adv. Appl., vol. 191, Birkhäuser, Basel, 2009, pp. 19–36.

[KN81] M.G. Kreĭn and M.A. Naĭmark, *The method of symmetric and Hermitian forms in the theory of the separation of the roots of algebraic equations*, Linear and Multilinear Algebra **10** (1981), no. 4, 265–308, Translated from the Russian by O. Boshko and J.L. Howland.

[Kru87] N. Krupnik, *Banach algebras with symbol and singular integral operators*, Oper. Theory Adv. Appl., vol. 26, Birkhäuser, Basel, 1987.

[Kru10] ———, *Survey on the best constants in the theory of one dimensional singular integral operators*, Topics in operator theory. Vol. 1. Operators, matrices and analytic functions, Oper. Theory Adv. Appl., vol. 202, Birkhäuser, Basel, 2010, pp. 365–394.

[KS72a] N.K. Karapetjanc and S.G. Samko, *A certain new approach to the investigation of singular integral equations with a shift*, Dokl. Akad. Nauk SSSR **202** (1972), 273–276 (in Russian), English translation: *Soviet Math. Dokl.* **13** (1972), 79–83.

[KS72b] ———, *Singular integral operators with shift on an open contour*, Dokl. Akad. Nauk SSSR **204** (1972), 536–539 (in Russian), English translation: *Soviet Math. Dokl.* **13** (1972), 691–696.

[Lit67] G.S. Litvinčuk, *Noether theory of a system of singular integral equations with Carleman translation and complex conjugate unknowns*, Izv. Akad. Nauk SSSR Ser. Mat. **31** (1967), 563–586 (in Russian), English translation: *Math. USSR Izvestiya* **1** (1967), 545–567.

[LT86] L. Lerer and M. Tismenetsky, *Generalized Bezoutian and the inversion problem for block matrices*, I. General scheme, Integral Equations Operator Theory **9** (1986), 790–819.

[MG76] J.D. Markel and A.H. Gray, *Linear prediction of speech*, Springer Verlag, Berlin, 1976.

[Mus88] B.R. Musicus, *Levinson and fast Choleski algorithms for Toeplitz and almost Toeplitz matrices.* RLE Technical Report no. 538, December 1988, Research

Laboratory of Electronics, Massachusetts Institute of Technology, Cambridge, MA.

[Pic72] S.K. Pichorides, *On the best values of the constants in the theorems of M. Riesz, Zygmund and Kolmogorov*, Studia Math. **44** (1972), 165–179.

[Sim61] I.B. Simonenko, *The Riemann boundary value problem for n pairs of functions with continuous coefficients*, Izv. Vys. Uchebn. Zaved. Matematika (1961), no. 1 (20), 140–145 (in Russian).

[Sim65] ______, *A new general method of investigating linear operator equations of singular integral equation type. I*, Izv. Akad. Nauk SSSR Ser. Mat. **29** (1965), 567–586 (in Russian).

[Spi92] I.M. Spitkovsky, *Singular integral operators with PC symbols on the spaces with general weights*, J. Functional Analysis **105** (1992), 129–143.

[Tre64] W. Trench, *An algoithm for the inversion of finite Toeplitz matrices*, J. Soc. Indust. Appl. Math. **12** (1964), 515–522.

[Wie49] N. Wiener, *Extrapolation, interpolation and smoothing of stationary time series with engineering applications*, MIT Press, Boston, 1949.

[ZL68] È.I. Zverovič and G.S. Litvinčuk, *Boundary value problems with a shift for analytic functions, and singular functional equations*, Uspehi Mat. Nauk **23** (1968), no. 3 (141), 67–121 (in Russian), English translation: *Russ. Math. Surv.* **23** (1968), 67–124.

Leonid Lerer
Department of Mathematics
Technion – Israel Institute of Technology
Haifa, 32000, Israel
e-mail: `llerer@tx.technion.ac.il`

Vadim Olshevsky
Department of Mathematics
University of Connecticut
196 Auditorium Road, U-9
Storrs, CT 06269, USA
e-mail: `olshevsky@uconn.edu`

Ilya Spitkovsky
Department of Mathematics
College of William & Mary
Williamsburg, VA 23187-8795, USA
e-mail: `ilya@math.wm.edu`

Operator Theory:
Advances and Applications, Vol. 206, 1–6

Inversion of Finite Toeplitz Matrices

Israel Gohberg and Georg Heinig

In this communication Toeplitz matrices of the form $\|a_{j-k}\|_{j,k=0}^{n}$, where a_j ($j = 0, \pm 1, \dots, \pm n$) are elements of some noncommutative algebra, and their continual analogues are considered. The theorems presented here are generalizations of theorems from [1] to the noncommutative case.

Detailed proofs of the theorems stated below as well as generalizations of theorems from [2] to the noncommutative case and their continual analogues will be given elsewhere.

1. Inversion of Toeplitz matrices

Let $\mathcal{R}$ be a noncommutative ring with unit element e and a_j ($j = 0, \pm 1, \dots, \pm n$) be some elements of $\mathcal{R}$. For this collection of elements we consider in $\mathcal{R}$ the following systems of equations:

$$\sum_{k=0}^{n} a_{j-k} x_k = \delta_{0j} e \qquad (j = 0, 1, \dots, n), \tag{1}$$

$$\sum_{k=0}^{n} a_{k-j} z_{-k} = \delta_{0j} e \qquad (j = 0, 1, \dots, n), \tag{2}$$

$$\sum_{k=0}^{n} w_k a_{j-k} = \delta_{0j} e \qquad (j = 0, 1, \dots, n), \tag{3}$$

$$\sum_{k=0}^{n} y_{-k} a_{k-j} = \delta_{0j} e \qquad (j = 0, 1, \dots, n). \tag{4}$$

It is easy to see that if equations (1) and (4) are solvable then $x_0 = y_0$, and if equations (2) and (3) are solvable then $z_0 = w_0$.

The paper was originally published as И.Ц. Гохберг, Г. Хайниг, Об обращении конечных тёплицевых матриц, Матем. Исслед. 8 (1973), № 3(29), 151–156.
MR0341163 (49 #5913), Zbl 0337.15004.

Theorem 1. *Let $A = \|a_{j-k}\|_{j,k=0}^{n}$ be an invertible matrix with elements in $\mathcal{R}$ and x_j, z_{-j}, w_j, y_{-j} $(j = 0, 1, \dots, n)$ be solutions in $\mathcal{R}$ of equations (1)–(4). If at least one of the elements x_0 or z_0 is invertible, then the other is also invertible and the matrix inverse to A is constructed by the formula*

$$A^{-1} = \begin{Vmatrix} x_0 & 0 & \dots & 0 \\ x_1 & x_0 & \dots & 0 \\ \vdots & \vdots & \ddots & \vdots \\ x_n & x_{n-1} & \dots & x_0 \end{Vmatrix} x_0^{-1} \begin{Vmatrix} y_0 & y_{-1} & \dots & y_{-n} \\ 0 & y_0 & \dots & y_{1-n} \\ \vdots & \vdots & \ddots & \vdots \\ 0 & 0 & \dots & y_0 \end{Vmatrix}$$
$$- \begin{Vmatrix} 0 & \dots & 0 & 0 \\ z_{-n} & \dots & 0 & 0 \\ \vdots & \ddots & \vdots & \vdots \\ z_{-1} & \dots & z_{-n} & 0 \end{Vmatrix} z_0^{-1} \begin{Vmatrix} 0 & w_n & \dots & w_1 \\ \vdots & \vdots & \ddots & \vdots \\ 0 & 0 & \dots & w_n \\ 0 & 0 & \dots & 0 \end{Vmatrix}. \qquad (5)$$

In the case when $\mathcal{R}$ coincides with the algebra $L(m, \mathbb{C})$ of all complex matrices of order m and the matrix A is positive definite, this result was obtained, in fact, in [3].

Also, under the condition $\mathcal{R} = L(m, \mathbb{C})$ and other additional assumptions on the matrix A, results similar to Theorem 1 are contained in [4].

Notice that in the case when $\mathcal{R}$ is an algebra over $\mathbb{C}$ (or $\mathbb{R}$), from the solvability of equations (1)–(4) and the invertibility of at least one of the elements x_0, z_0 it follows that the matrix A is invertible[1]. In a number of cases this statement remains true if one requires only the solvability of two equations (1), (2) or (3), (4). For instance, this holds for $\mathcal{R} = L(m, \mathbb{C})$. The same takes place when the elements a_j have the form $\lambda_j I + T_j$, where λ_j are complex numbers and T_j are linear compact operators in a Banach space $\mathcal{L}$.

As an example consider the matrix

$$A = \begin{Vmatrix} e & a & \dots & a^n \\ b & e & \dots & a^{n-1} \\ \vdots & \vdots & \ddots & \vdots \\ b^n & b^{n-1} & \dots & e \end{Vmatrix}$$

where a and b are some elements of $\mathcal{R}$. The matrix A is invertible if and only if the element $e - ab$ is invertible. If the element $e - ab$ is invertible, then the element $e - ba$ is also invertible, and the solutions of equations (1)–(4) are given by the

[1]Probably this statement remains true also in the more general case of an arbitrary ring $\mathcal{R}$ with unit.

formulas

$$\begin{array}{llllllllllll} x_0 & = & c, & x_1 & = & -bc, & x_2 & = & x_3 & = \cdots = & x_n & = & 0, \\ y_0 & = & c, & y_{-1} & = & -ca, & y_{-2} & = & y_{-3} & = \cdots = & y_{-n} & = & 0, \\ w_0 & = & d, & w_1 & = & -db, & w_2 & = & w_3 & = \cdots = & w_n & = & 0, \\ z_0 & = & d, & z_{-1} & = & -ad, & z_{-2} & = & z_{-3} & = \cdots = & z_{-n} & = & 0, \end{array}$$

where $c = (e - ab)^{-1}$ and $d = (e - ba)^{-1}$. In view of Theorem 1, we obtain that in the case under consideration

$$A^{-1} = \left\| \begin{array}{ccccccc} c & -ca & 0 & \dots & 0 & 0 & 0 \\ -bc & bca + c & -ca & \dots & 0 & 0 & 0 \\ 0 & -bc & bca + c & \dots & 0 & 0 & 0 \\ \dots & \dots & \dots & \dots & \dots & \dots & \dots \\ 0 & 0 & 0 & \dots & -bc & bca + c & -ca \\ 0 & 0 & 0 & \dots & 0 & -bc & d \end{array} \right\|.$$

In the case $a = b \in \mathbb{C}$ a similar example was considered in [5].

Theorem 1 implies the following.

Corollary 1. *Let the conditions of Theorem 1 be fulfilled. Then the matrix* $A_{n-1} = \|a_{j-k}\|_{j,k=0}^{n-1}$ *is invertible and*

$$A_{n-1}^{-1} = \left\| \begin{array}{cccc} x_0 & 0 & \dots & 0 \\ x_1 & x_0 & \dots & 0 \\ \vdots & \vdots & \ddots & \vdots \\ x_{n-1} & x_{n-2} & \dots & x_0 \end{array} \right\| x_0^{-1} \left\| \begin{array}{cccc} y_0 & y_{-1} & \dots & y_{1-n} \\ 0 & y_0 & \dots & y_{2-n} \\ \vdots & \vdots & \ddots & \vdots \\ 0 & 0 & \dots & y_0 \end{array} \right\|$$
$$- \left\| \begin{array}{cccc} z_{-n} & 0 & \dots & 0 \\ z_{1-n} & z_{-n} & \dots & 0 \\ \vdots & \vdots & \ddots & \vdots \\ z_{-1} & z_{-2} & \dots & z_{-n} \end{array} \right\| z_0^{-1} \left\| \begin{array}{cccc} w_n & w_{n-1} & \dots & w_1 \\ 0 & w_n & \dots & w_2 \\ \vdots & \vdots & \ddots & \vdots \\ 0 & 0 & \dots & w_n \end{array} \right\|. \qquad (6)$$

2. Inverse problem

Consider the problem of reconstruction of the matrix $\|a_{j-k}\|_{j,k=0}^{n}$ from the solutions of equations (1)–(4).

Theorem 2. *Let* w_j, x_j, y_{-j}, z_{-j} $(j = 0, 1, \dots, n)$ *be given systems of elements in* $\mathcal{R}$ *and the elements* w_0, x_0, y_0, z_0 *be invertible. For the existence of an invertible Toeplitz matrix* $A = \|a_{j-k}\|_{j,k=0}^{n}$ *with elements* $a_j \in \mathcal{R}$ $(j = 0, \pm 1, \dots, \pm n)$ *for which* w_j, x_j, y_{-j}, z_{-j} *are solutions of equations* (1)–(4) *it is necessary and sufficient that the following three conditions be fulfilled:*

1) $x_0 = y_0$ *and* $z_0 = w_0$*;*

2)
$$\left\| \begin{matrix} x_0 & 0 & \dots & 0 \\ x_1 & x_0 & \dots & 0 \\ \vdots & \vdots & \ddots & \vdots \\ x_n & x_{n-1} & \dots & x_0 \end{matrix} \right\| x_0^{-1} \left\| \begin{matrix} y_{-n} & 0 & \dots & 0 \\ y_{1-n} & y_{-n} & \dots & 0 \\ \vdots & \vdots & \ddots & \vdots \\ y_0 & y_{-1} & \dots & y_{-n} \end{matrix} \right\|$$
$$= \left\| \begin{matrix} z_{-n} & 0 & \dots & 0 \\ z_{1-n} & z_{-n} & \dots & 0 \\ \vdots & \vdots & \ddots & \vdots \\ z_{-1} & z_{-2} & \dots & z_{-n} \end{matrix} \right\| z_0^{-1} \left\| \begin{matrix} w_0 & 0 & \dots & 0 \\ w_1 & w_0 & \dots & 0 \\ \vdots & \vdots & \ddots & \vdots \\ w_n & w_{n-1} & \dots & w_0 \end{matrix} \right\|$$

and

$$\left\| \begin{matrix} x_n & 0 & \dots & 0 \\ x_{n-1} & x_n & \dots & 0 \\ \vdots & \vdots & \ddots & \vdots \\ x_0 & x_1 & \dots & x_n \end{matrix} \right\| x_0^{-1} \left\| \begin{matrix} y_0 & 0 & \dots & 0 \\ y_{-1} & y_0 & \dots & 0 \\ \vdots & \vdots & \ddots & \vdots \\ y_{-n} & y_{1-n} & \dots & y_0 \end{matrix} \right\|$$
$$= \left\| \begin{matrix} z_0 & 0 & \dots & 0 \\ z_{-1} & z_0 & \dots & 0 \\ \vdots & \vdots & \ddots & \vdots \\ z_{-n} & z_{1-n} & \dots & z_0 \end{matrix} \right\| z_0^{-1} \left\| \begin{matrix} w_n & 0 & \dots & 0 \\ w_{n-1} & w_n & \dots & 0 \\ \vdots & \vdots & \ddots & \vdots \\ w_0 & w_1 & \dots & w_n \end{matrix} \right\|;$$

3) *at least one of the matrices*

$$M = \left\| \begin{matrix} x_0 & \dots & 0 & z_{-n} & \dots & 0 \\ \vdots & \ddots & \vdots & \vdots & \ddots & \vdots \\ x_{n-1} & \ddots & x_0 & z_{-1} & \ddots & z_{-n} \\ x_n & \ddots & x_1 & z_0 & \ddots & z_{1-n} \\ \vdots & \ddots & \vdots & \vdots & \ddots & \vdots \\ 0 & \dots & x_n & 0 & \dots & z_0 \end{matrix} \right\|,$$

$$N = \left\| \begin{matrix} w_0 & \dots & w_{n-1} & w_n & \dots & 0 \\ \vdots & \ddots & \ddots & \ddots & \ddots & \vdots \\ 0 & \dots & w_0 & w_1 & \dots & w_n \\ y_{-n} & \dots & y_{-1} & y_0 & \dots & 0 \\ \vdots & \ddots & \ddots & \ddots & \ddots & \vdots \\ 0 & \dots & y_{-n} & y_{1-n} & \dots & y_0 \end{matrix} \right\|$$

is invertible.

If conditions 1)–3) hold, then the matrix A is uniquely reconstructed by formula (5). The matrix A can be reconstructed also by one pair of solutions of equations (1)–(4).

Theorem 3. *Let x_j, z_{-j} $(j = 0, 1, \ldots, n)$ be given systems of elements in $\mathcal{R}$ and the elements x_0 and z_0 be invertible. For the existence of an invertible Toeplitz matrix $A = \|a_{j-k}\|_{j,k=0}^{n}$ with elements $a_j \in \mathcal{R}$ $(j = 0, \pm 1, \ldots, \pm n)$ for which x_j and z_{-j} are solutions of equations (1) and (2) it is necessary and sufficient that the matrix M be invertible and for the vector*

$$\begin{Vmatrix} \chi_{-n} \\ \vdots \\ \chi_0 \\ \vdots \\ \chi_{n-1} \end{Vmatrix} = M^{-1} \begin{Vmatrix} 0 \\ \vdots \\ 0 \\ z_{-n} z_0^{-1} x_n - x_0 \\ \vdots \\ z_{-1} z_0^{-1} x_n - x_{n-1} \end{Vmatrix}$$

the condition $\chi_0 = -e$ be fulfilled.

An analogous theorem holds for the pair of solutions w_j and y_{-j}.

3. Continual analogue

Let $\mathcal{A}$ be a Banach algebra and $k(t)$ $(-\tau \le t \le \tau,\ \tau > 0)$ be a measurable function with values in $\mathcal{A}$ such that

$$\int_{-\tau}^{\tau} \|k(t)\|\, dt < \infty,$$

that is, $k(t) \in L_1([-\tau, \tau], \mathcal{A})$.

The role of equations (1)–(4) in the continual case is played by the following equations:

$$g_+(t) - \int_0^{\tau} k(t-s) g_+(s)\, ds = k(t) \quad (0 \le t \le \tau), \tag{7}$$

$$h_+(t) - \int_0^{\tau} h_+(s) k(t-s)\, ds = k(t) \quad (0 \le t \le \tau), \tag{8}$$

$$g_-(-t) - \int_0^{\tau} k(s-t) g_-(-s)\, ds = k(-t) \quad (0 \le t \le \tau), \tag{9}$$

$$h_-(-t) - \int_0^{\tau} h_-(s) k(s-t)\, ds = k(-t) \quad (0 \le t \le \tau). \tag{10}$$

Theorem 4. *Let the operator $I - K$, where*

$$(K\varphi)(t) = \int_0^{\tau} k(t-s)\varphi(s)\, ds \quad (0 \le t \le \tau),$$

be invertible in the space $L_1([0, \tau], \mathcal{A})$. Then

$$((I-K)^{-1}\varphi)(t) = \varphi(t) + \int_0^{\tau} \gamma(t,s)\varphi(s)\, ds,$$

where

$$\gamma(t,s) = g_+(t-s) + h_-(t-s) + \int_0^{\min(t,s)} g_+(t-u)h_-(u-s)\,du - \int_\tau^{\tau+\min(t,s)} g_-(t-u)h_+(u-s)\,du,$$

$$g_+(-t) = h_-(t) = 0 \quad \textit{for} \quad t > 0,$$

and $g_+(t), h_+(t), g_-(t), h_-(t)$ *be solutions of equations* (7)–(10) *such that*

$$g_+(t), h_+(t) \in L_1([0,\tau], \mathcal{A}), \quad g_-(t), h_-(t) \in L_1([-\tau,0], \mathcal{A}).$$

Note that if $\mathcal{A} = L(m, \mathbb{C})$ and the matrix function $k(t)$ is selfadjoint or differentiable and $dk(t)/dt \in L_1([-\tau,\tau], \mathcal{A})$, then the solvability of equations (7)–(10) implies the invertibility of the operator $I - K$ [2].

We will not formulate here continual analogs of inverse theorems.

References

[1] I.C. Gohberg and A.A. Semencul, *The inversion of finite Toeplitz matrices and their continual analogues.* Matem. Issled. **7** (1972), no. 2(24), 201–223 (in Russian). MR0353038 (50 #5524), Zbl 0288.15004.

[2] I.C. Gohberg and N.Ya. Krupnik, *A formula for the inversion of finite Toeplitz matrices.* Matem. Issled. **7** (1972), no. 2(24), 272–283 (in Russian). MR0353039 (50 #5525), Zbl 0288.15005.

[3] L.M. Kutikov, *The structure of matrices which are the inverse of the correlation matrices of random vector processes.* Zh. Vychisl. Matem. Matem. Fiz. **7** (1967), 764–773 (in Russian). English translation: U.S.S.R. Comput. Math. Math. Phys. **7** (1967), no. 4, 58–71. MR0217863 (36 #952), Zbl 0251.15023.

[4] I.I. Hirschman, Jr., *Matrix-valued Toeplitz operators.* Duke Math. J. **34** (1967), 403–415. MR0220002 (36 #3071), Zbl 0182.46203.

[5] L.M. Kutikov, *Inversion of correlation matrices.* Izv. Akad. Nauk SSSR Tehn. Kibernet. (1965), no. 5, 42–47 (in Russian). English translation: Engineering Cybernetics (1965), no. 5, 35–39. MR0203871 (34 #3718).

[2] Probably this statement remains true without additional conditions on the matrix $k(t)$.

Operator Theory:
Advances and Applications, Vol. 206, 7–46

Inversion of Finite Toeplitz Matrices Consisting of Elements of a Noncommutative Algebra

Israel Gohberg and Georg Heinig

Abstract. Theorems on the inversion of Toeplitz matrices $\|a_{j-k}\|_{j,k=0}^n$ consisting of complex numbers are obtained in [1, 2]. In this paper those results are generalized to the case where a_j $(j = 0, \pm 1, \dots, \pm n)$ are elements of some noncommutative algebra with unit. The paper consists of six sections. The results of [1] are generalized in the first three sections, the results of [2] are extended in the last three sections. Continual analogues of results of this paper will be presented elsewhere.

1. Theorems on inversion of Toeplitz matrices

Everywhere in what follows $\mathfrak{A}$ denotes some (in general, noncommutative) algebra[1] with unit e.

In this section the inverse matrix to a matrix $\|a_{j-k}\|_{j,k=0}^n$ with elements $a_j \in \mathfrak{A}$ $(j = 0, \pm 1, \dots, \pm n)$ is constructed from solutions of the following equations

$$\sum_{k=0}^{n} a_{j-k} x_k = \delta_{0j} e \quad (j = 0, 1, \dots, n), \tag{1.1}$$

$$\sum_{k=0}^{n} a_{k-j} z_{-k} = \delta_{0j} e \quad (j = 0, 1, \dots, n), \tag{1.2}$$

$$\sum_{k=0}^{n} w_k a_{j-k} = \delta_{0j} e \quad (j = 0, 1, \dots, n), \tag{1.3}$$

$$\sum_{k=0}^{n} y_{-k} a_{k-j} = \delta_{0j} e \quad (j = 0, 1, \dots, n). \tag{1.4}$$

The paper was originally published as И.Ц. Гохберг, Г. Хайниг, Обращение конечных тёплицевых матриц, составленных из элементов некоммутативной алгебры, Rev. Roumaine Math. Pures Appl. **19** (1974), 623–663. MR0353040 (50 #5526), Zbl 0337.15005.

[1]A major part of the results of the paper remains true also in the case when $\mathfrak{A}$ is a ring with a unit.

First of all, note that if equations (1.1) and (1.4) are solvable, then $x_0 = y_0$. Indeed,

$$y_0 = \sum_{j=0}^{k} y_{-j} \left(\sum_{k=0}^{n} a_{j-k} x_k \right) = \sum_{k=0}^{n} \left(\sum_{j=0}^{n} y_{-j} a_{j-k} \right) x_k = x_0.$$

From the equalities

$$z_0 = \sum_{j=0}^{n} w_j \left(\sum_{k=0}^{n} a_{k-j} z_{-k} \right), \quad w_0 = \sum_{k=0}^{n} \left(\sum_{j=0}^{n} w_j a_{k-j} \right) z_{-k}$$

it follows that if equations (1.2) and (1.3) are solvable, then $z_0 = w_0$.

Theorem 1.1. *Let elements a_j ($j = 0, \pm 1, \ldots, \pm n$) belong to the algebra $\mathfrak{A}$. If systems (1.1)–(1.4) are solvable and one of the elements z_0, x_0 is invertible, then the matrix $A = \|a_{j-k}\|_{j,k=0}^{n}$ and the second element are invertible and the matrix inverse to A is constructed by the formula*

$$A^{-1} = \left\| \begin{array}{cccc} x_0 & 0 & \ldots & 0 \\ x_1 & x_0 & \ldots & 0 \\ \vdots & \vdots & \ddots & \vdots \\ x_n & x_{n-1} & \ldots & x_0 \end{array} \right\| x_0^{-1} \left\| \begin{array}{cccc} y_0 & y_{-1} & \ldots & y_{-n} \\ 0 & y_0 & \ldots & y_{1-n} \\ \vdots & \vdots & \ddots & \vdots \\ 0 & 0 & \ldots & y_0 \end{array} \right\|$$
$$- \left\| \begin{array}{cccc} 0 & \ldots & 0 & 0 \\ z_{-n} & \ldots & 0 & 0 \\ \vdots & \ddots & \vdots & \vdots \\ z_{-1} & \ldots & z_{-n} & 0 \end{array} \right\| z_0^{-1} \left\| \begin{array}{cccc} 0 & w_n & \ldots & w_1 \\ \vdots & \vdots & \ddots & \vdots \\ 0 & 0 & \ldots & w_n \\ 0 & 0 & \ldots & 0 \end{array} \right\|. \tag{1.5}$$

This theorem was obtained in [3] in the case when $\mathfrak{A}$ coincides with the algebra $\mathbb{C}_{m \times m}$ of all complex matrices of order m and the matrix A is positive definite.

Under the condition $\mathfrak{A} = \mathbb{C}_{m \times m}$ and other additional assumptions on the matrix A, results similar to Theorem 1.1 are contained in [4].

Note that in the proof of Theorem 1.1 given below some ideas from [3] are used.

Preceding the proof of Theorem 1.1 we state the following lemma.

Lemma 1.1. *Let $\mathbb{V}$ be a linear space, $L(\mathbb{V})$ be the algebra of all linear operators acting in $\mathbb{V}$, and $A = \|A_{j-k}\|_{j,k=0}^{n}$ be a Toeplitz matrix with elements A_j ($j = 0, \pm 1, \ldots, \pm n$) in $L(\mathbb{V})$. If the systems of equations*

$$\sum_{k=0}^{n} A_{j-k} X_k = \delta_{0j} I \quad (j = 0, 1, \ldots, n), \tag{1.6}$$

$$\sum_{k=0}^{n} A_{k-j} Z_{-k} = \delta_{0j} I \quad (j = 0, 1, \ldots, n), \tag{1.7}$$

$$\sum_{k=0}^{n} W_k A_{j-k} = \delta_{0j} I \quad (j = 0, 1, \ldots, n), \tag{1.8}$$

$$\sum_{k=0}^{n} Y_{-k} A_{k-j} = \delta_{0j} I \quad (j = 0, 1, \ldots, n) \tag{1.9}$$

are solvable in $L(\mathbb{V})$ *and at least one of the operators* W_0, X_0, Y_0, Z_0 *is invertible, then the matrix* A *is also invertible.*

Proof. Suppose all the hypotheses of the lemma are fulfilled. By $\mathbb{V}_n$ denote the linear space of one-row matrices $\varphi = \{\varphi_j\}_{j=0}^n$ with entries $\varphi_j \in \mathbb{V}$.

Consider the homogeneous equation $A\varphi = 0$ $(\varphi \in \mathbb{V}_n)$. If $\varphi = \{\varphi_j\}_{j=0}^n$ is a solution of this equation, then in view of (1.8) we obtain

$$0 = \sum_{j=0}^{n} W_{n-j} \sum_{k=0}^{n} A_{j-k}\varphi_k = \sum_{k=0}^{n} \left(\sum_{j=0}^{n} W_{n-j} A_{j-k} \right) \varphi_k = \varphi_n.$$

Analogously, in view of (1.9),

$$0 = \sum_{j=0}^{n} Y_{-j} \sum_{k=0}^{n} A_{j-k}\varphi_k = \sum_{k=0}^{n} \left(\sum_{j=0}^{n} Y_{-j} A_{j-k} \right) \varphi_k = \varphi_0.$$

By $\operatorname{Ker} A$ denote the subspace of $\mathbb{V}_n$ that consists of all solutions of the equation $A\varphi = 0$. To each nonzero vector $\varphi = \{\varphi_j\}_{j=0}^n \in \operatorname{Ker} A$ we assign the number $p(\varphi) = p$ such that $\varphi_0 = \varphi_1 = \cdots = \varphi_{p-1} = 0$ and $\varphi_p \neq 0$. By what has just been proved, $0 < p(\varphi) < n$. Let p^* be the maximal value of the function $p(\varphi)$ $(\varphi \in \operatorname{Ker} A,\ \varphi \neq 0)$ and let $h \in \operatorname{Ker} A$ be a vector such that $p^* = p(h)$. Consider the vector $h^- = \{h_j^-\}_{j=0}^n \in V_n$ defined by the equality

$$h_j^- = \begin{cases} h_{j-1} & \text{if } \ j = 1, 2, \ldots, n, \\ 0 & \text{if } \ j = 0. \end{cases}$$

Obviously,

$$\sum_{k=0}^{n} A_{j-k} h_k^- = \sum_{k=0}^{n} A_{j-k-1} h_k = 0 \quad (j = 1, 2, \ldots, n).$$

From here and (1.9) it follows that

$$\begin{aligned} Y_0 \sum_{k=0}^{n} A_{-k} h_k^- &= Y_0 \sum_{k=0}^{n} A_{-k} h_k^- + \sum_{j=1}^{n} Y_{-j} \sum_{k=0}^{n} A_{j-k} h_k^- \\ &= \sum_{k=0}^{n} \left(\sum_{j=0}^{n} Y_{-j} A_{j-k} \right) h_k^- = h_0^- = 0. \end{aligned}$$

Therefore, if the element Y_0 (or the element X_0 coinciding with it) is invertible, then $h^- \in \operatorname{Ker} A$. Because $p(h^-) > p(h)$, we conclude that in this case the subspace $\operatorname{Ker} A$ consists of zero only.

Analogously, we define the function $q = q(\varphi)$ ($\varphi \in \operatorname{Ker} A$, $\varphi \neq 0$, $\varphi = \{\varphi_j\}_{j=0}^n$) such that $\varphi_q \neq 0$ and $\varphi_{q+1} = \cdots = \varphi_n = 0$. By what has just been proved, $0 < q(\varphi) < n$. Put $q^* = \min q(\varphi)$ ($\varphi \in \operatorname{Ker} A$, $\varphi \neq 0$). Let g be a vector in $\operatorname{Ker} A$ such that $q(g) = q^*$. Then for the vector $g^+ = \{g_j^+\}_{j=0}^n$ defined by the equality

$$g_j^+ = \begin{cases} g_{j+1} & \text{if} \quad j = 0, 1, \ldots, n-1, \\ 0 & \text{if} \quad j = n, \end{cases}$$

we have

$$\sum_{k=0}^n A_{j-k} g_k^+ = \sum_{k=0}^n A_{j-k+1} g_k = 0 \quad (j = 0, 1, \ldots, n-1).$$

From here and (1.8) it follows that

$$\begin{aligned} W_0 \sum_{k=0}^n A_{n-k} g_k^+ &= W_0 \sum_{k=0}^n A_{n-k} g_k^+ + \sum_{j=0}^{n-1} W_{n-j} \sum_{k=0}^r A_{j-k} g_k^+ \\ &= \sum_{k=0}^n \left(\sum_{j=0}^n W_{n-j} A_{j-k} \right) g_k^+ = g_n^+ = 0. \end{aligned}$$

Hence, if the element W_0 (or the element Z_0 coinciding with it) is invertible, then $g \in \operatorname{Ker} A$ and $q(g^+) > q(g)$. It follows that $\operatorname{Ker} A = \{0\}$ in this case, too.

Thus, we have proved that in all cases $\operatorname{Ker} A = \{0\}$.

Passing to the dual space and the adjoint operator equations, it is easy to see that the the adjoint matrix $A^* = \|A_{k-j}^*\|_{j,k=0}^n$ also satisfies the hypotheses of the lemma. Hence, by what has been proved, we obtain that $\operatorname{Ker} A^* = \{0\}$. Thus, the operator generated by the matrix A in the space $\mathbb{V}_n$ is invertible. Therefore, the matrix A is invertible in $L(\mathbb{V})$.

The lemma is proved. □

Note that in some cases Lemma 1.1 remains true if in its formulation one considers only the two equations (1.8), (1.9) or (1.6), (1.7) instead of all four equations (1.6)–(1.9). For instance, this happens if $\mathbb{V} = \mathbb{C}^m$, because in this case $\dim \operatorname{Ker} A = \dim \operatorname{Ker} A^*$. The same happens if $\mathbb{V}$ is a linear topological space and the operators A_j have the form $\varepsilon_j I + T_j$, where ε_j are complex numbers and T_j are compact operators.

Proof of Theorem 1.1. First, let us show that the matrix A is invertible. Assume that the linear space $\mathbb{V}$ coincides with $\mathfrak{A}$. Obviously, one can consider $\mathfrak{A}$ as a part of the algebra $L(\mathfrak{A})$ of all linear operators acting in $\mathfrak{A}$. The hypotheses of Lemma 1.1 are satisfied for the matrix $A = \|a_{j-k}\|_{j,k=0}^n$. Hence the matrix A is invertible. The entries of the inverse matrix belong to the algebra $\mathfrak{A}$. Indeed, the equations

$$\sum_{k=0}^n a_{j-k} t_{kr} = \delta_{rj} e \quad (j = 0, 1, \ldots, n)$$

have solutions t_{kr} in $\mathfrak{A}$ for every $r = 0, 1, \dots, n$. From here it follows that the matrix A^{-1} coincides with the matrix $\|t_{kr}\|_{k,r=0}^{n}$ whose entries belong to the algebra $\mathfrak{A}$.

We construct the matrices

$$W = \left\| \begin{array}{cccc} e & \dots & 0 & 0 \\ \vdots & \ddots & \vdots & \vdots \\ 0 & \dots & e & 0 \\ w_n & \dots & w_1 & w_0 \end{array} \right\|, \quad X = \left\| \begin{array}{cccc} x_0 & 0 & \dots & 0 \\ x_1 & e & \dots & 0 \\ \vdots & \vdots & \ddots & \vdots \\ x_n & 0 & \dots & e \end{array} \right\|,$$

$$Y = \left\| \begin{array}{cccc} y_0 & y_{-1} & \dots & y_{-n} \\ 0 & e & \dots & 0 \\ \vdots & \vdots & \ddots & \vdots \\ 0 & 0 & \dots & e \end{array} \right\|, \quad Z = \left\| \begin{array}{cccc} e & \dots & 0 & z_{-n} \\ \vdots & \ddots & \vdots & \vdots \\ 0 & \dots & e & z_{-1} \\ 0 & \dots & 0 & z_0 \end{array} \right\|.$$

Obviously, the equalities

$$AZ = \left\| \begin{array}{ccc|c} & & & 0 \\ & A_{n-1} & & \vdots \\ & & & 0 \\ \hline a_n & \dots & a_1 & e \end{array} \right\|, \quad AX = \left\| \begin{array}{c|ccc} e & a_{-1} & \dots & a_{-n} \\ \hline 0 & & & \\ \vdots & & A_{n-1} & \\ 0 & & & \end{array} \right\|$$

hold. Here and in what follows, by A_r $(r = 0, 1, \dots, n)$ we denote the matrix $\|a_{j-k}\|_{j,k=0}^{r}$. From these equalities it follows that the element x_0 or z_0 is invertible if and only if the matrix A_{n-1} is invertible. Hence the invertibility of one of the elements x_0 or z_0 implies the invertibility of the other one.

It is easily seen that

$$WAZ = \left\| \begin{array}{cc} A_{n-1} & 0 \\ 0 & w_0 \end{array} \right\|, \quad YAX = \left\| \begin{array}{cc} x_0 & 0 \\ 0 & A_{n-1} \end{array} \right\|, \tag{1.10}$$

whence

$$A^{-1} = Z \left\| \begin{array}{cc} A_{n-1}^{-1} & 0 \\ 0 & w_0^{-1} \end{array} \right\| W = X \left\| \begin{array}{cc} x_0^{-1} & 0 \\ 0 & A_{n-1}^{-1} \end{array} \right\| Y. \tag{1.11}$$

Setting $A_r^{-1} = \|c_{jk}^{r}\|_{j,k=0}^{n}$, we obtain from (1.11) the following recurrent formulas:

$$\begin{aligned} c_{jk}^{n} &= c_{jk}^{n-1} + z_{j-n} w_0^{-1} w_{n-k} \qquad && (j, k = 0, 1, \dots, n-1), \\ c_{jk}^{n} &= c_{j-1,k-1}^{n-1} + x_j x_0^{-1} y_{-k} \qquad && (j, k = 1, 2, \dots, n), \end{aligned} \tag{1.12}$$

and the equalities

$$c_{nk}^{n} = w_{n-k}, \quad c_{jn}^{n} = z_{j-n}, \quad c_{0k}^{n} = y_{-k}, \quad c_{j0}^{n} = x_j \tag{1.13}$$

for $j, k = 0, 1, \dots, n$.

From equalities (1.12) and (1.13) it follows that

$$c_{jk}^{n} = c_{j-1,k-1}^{n} + x_j x_0^{-1} y_{-k} + z_{j-n-1} z_0^{-1} w_{n-k+1} \tag{1.14}$$

for $j,k = 1,2,\dots,n$. If one takes $c_{jk} = 0$, whenever one of the indices j,k is negative or is greater than n, and $z_{-n-1} = w_{n+1}$, then it is easy to see that equality (1.14) remains true also in the cases $j,k = 0, n+1$. Thus, the equality

$$c_{jk}^n = \sum_{r=0}^{\min(j,k)} (x_{j-r}x_0^{-1}y_{r-k} - z_{j-n-1-r}w_0^{-1}w_{n-k+1+r}) \tag{1.15}$$

holds. This formula coincides with (1.5).

The theorem is proved. □

Corollary 1.1. *Let the hypotheses of Theorem* 1.1 *be fulfilled. Then for the matrix* A^{-1} *the equality*

$$\begin{aligned} A^{-1} = &\left\| \begin{matrix} z_0 & z_{-1} & \dots & z_{-n} \\ 0 & z_0 & \dots & z_{1-n} \\ \vdots & \vdots & \ddots & \vdots \\ 0 & 0 & \dots & z_0 \end{matrix} \right\| z_0^{-1} \left\| \begin{matrix} w_0 & 0 & \dots & 0 \\ w_1 & w_0 & \dots & 0 \\ \vdots & \vdots & \ddots & \vdots \\ w_n & w_{n-1} & \dots & w_0 \end{matrix} \right\| \\ &- \left\| \begin{matrix} 0 & x_n & \dots & x_1 \\ \vdots & \vdots & \ddots & \vdots \\ 0 & 0 & \dots & x_n \\ 0 & 0 & \dots & 0 \end{matrix} \right\| x_0^{-1} \left\| \begin{matrix} 0 & \dots & 0 & 0 \\ y_{-n} & \dots & 0 & 0 \\ \vdots & \ddots & \vdots & \vdots \\ y_{-1} & \dots & y_{-n} & 0 \end{matrix} \right\| \end{aligned} \tag{1.16}$$

also holds.

Proof. Indeed, from equality (1.14) it follows that

$$c_{jk}^n = c_{j+1,k+1}^n + z_{j-n}z_0^{-1}w_{n-k} - x_{j+1}x_0^{-1}y_{1-k} \tag{1.17}$$

for $j,k = 0,1,\dots,n-1$. From here and (1.13) it follows that[2]

$$c_{jk}^n = \sum_{r=0}^{\min(n-j,n-k)} (z_{j-n+r}z_0^{-1}w_{n-k-r} - x_{j+1+r}x_0^{-1}y_{-k-1-r}).$$

This formula coincides with (1.16). □

As an example consider the matrix

$$A = \left\| \begin{matrix} e & a & \dots & a^n \\ b & e & \dots & a^{n-1} \\ \vdots & \vdots & \ddots & \vdots \\ b^n & b^{n-1} & \dots & e \end{matrix} \right\|,$$

where a and b are some elements of the algebra $\mathfrak{A}$. Assume that the element $e-ab$ is invertible. Then the element $e-ba$ is also invertible (see [5, Chap. II, Section 7.5],

[2]Here it is assumed that $x_{n+1} = y_{-n-1} = 0$.

p. 197 of the 2nd Russian edition). It is easy to see that solutions of equations (1.1)–(1.4) are determined by the formulas

$$\begin{aligned}
&x_0 = (e-ab)^{-1}, \quad && x_1 = -b(e-ab)^{-1}, \quad && x_2 = x_3 = \cdots = x_n = 0;\\
&y_0 = (e-ab)^{-1}, && y_{-1} = -(e-ab)^{-1}a, && y_{-2} = y_{-3} = \cdots = y_{-n} = 0;\\
&w_0 = (e-ba)^{-1}, && w_1 = -(e-ba)^{-1}b, && w_2 = w_3 = \cdots = w_n = 0;\\
&z_0 = (e-ba)^{-1}, && z_{-1} = -a(e-ba)^{-1}, && z_{-2} = z_{-3} = \cdots = z_{-n} = 0.
\end{aligned}$$

In view of Theorem 1.1, we obtain that in the considered case the matrix A is invertible and its inverse is given by the formula

$$A^{-1} = \left\| \begin{array}{ccccccc}
c & -ca & 0 & \dots & 0 & 0 & 0\\
-bc & bca+c & -ca & \dots & 0 & 0 & 0\\
0 & -bc & bca+c & \dots & 0 & 0 & 0\\
\dots & \dots & \dots & \dots & \dots & \dots & \dots\\
0 & 0 & 0 & \dots & -bc & bca+c & -ca\\
0 & 0 & 0 & \dots & 0 & -bc & d
\end{array} \right\|,$$

where $c = (e-ab)^{-1}$ and $d = bca + c - a(e-ba)^{-1}b = (e-ba)^{-1}$.

Note that the invertibility of the matrix A implies the invertibility of the element $e - ab$ because

$$A \left\| \begin{array}{ccccc}
e & 0 & 0 & \dots & 0\\
-b & e & 0 & \dots & 0\\
0 & 0 & e & \dots & 0\\
\vdots & \vdots & \vdots & \ddots & \vdots\\
0 & 0 & 0 & \dots & e
\end{array} \right\| = \left\| \begin{array}{ccccc}
e-ab & 0 & 0 & \dots & 0\\
0 & e & 0 & \dots & 0\\
0 & 0 & e & \dots & 0\\
\vdots & \vdots & \vdots & \ddots & \vdots\\
0 & 0 & 0 & \dots & e
\end{array} \right\|.$$

This example was considered in [6] in the case $b = a \in \mathbb{C}$.

Theorem 1.2. *Let the hypotheses of Theorem* 1.1 *be fulfilled. Then the matrix* A_{n-1} *is invertible and its inverse is constructed by the formula*

$$\begin{aligned}
A_{n-1}^{-1} = {} & \left\| \begin{array}{cccc}
x_0 & 0 & \dots & 0\\
x_1 & x_0 & \dots & 0\\
\vdots & \vdots & \ddots & \vdots\\
x_{n-1} & x_{n-2} & \dots & x_0
\end{array} \right\| x_0^{-1} \left\| \begin{array}{cccc}
y_0 & y_{-1} & \dots & y_{1-n}\\
0 & y_0 & \dots & y_{2-n}\\
\vdots & \vdots & \ddots & \vdots\\
0 & 0 & \dots & y_0
\end{array} \right\| \\
& - \left\| \begin{array}{cccc}
z_{-n} & 0 & \dots & 0\\
z_{1-n} & z_{-n} & \dots & 0\\
\vdots & \vdots & \ddots & \vdots\\
z_{-1} & z_{-2} & \dots & z_{-n}
\end{array} \right\| z_0^{-1} \left\| \begin{array}{cccc}
w_n & w_{n-1} & \dots & w_1\\
0 & w_n & \dots & w_2\\
\vdots & \vdots & \ddots & \vdots\\
0 & 0 & \dots & w_n
\end{array} \right\|.
\end{aligned} \tag{1.18}$$

Proof. The invertibility of the matrix A_{n-1} follows from Theorem 1.1 and equalities (1.10). From equalities (1.12) and (1.5) it follows that

$$\begin{aligned}
c_{jk}^{n-1} &= c_{jk}^{n} - z_{j-n}w_0^{-1}w_{n-k} \\
&= \sum_{r=0}^{\min(j,k)} (x_{j-r}x_0^{-1}y_{r-k} - z_{j-n-1-r}w_0^{-1}w_{n+1-k+r}) - z_{j-n}w_0^{-1}w_{n-k} \\
&= \sum_{r=0}^{\min(j,k)} (x_{j-r}x_0^{-1}y_{r-k} - z_{j-n-r}w_0^{-1}w_{n-k+r})
\end{aligned}$$

for $j, k = 0, 1, \ldots, n-1$. The last formula coincides with (1.18).

The theorem is proved. □

Corollary 1.2. *Let the hypotheses of Theorem* 1.1 *be fulfilled. Then the matrix* A_{n-1} *is invertible and the equality*

$$A_{n-1}^{-1} = \left\| \begin{matrix} z_0 & z_{-1} & \ldots & z_{1-n} \\ 0 & z_0 & \ldots & z_{2-n} \\ \vdots & \vdots & \ddots & \vdots \\ 0 & 0 & \ldots & z_0 \end{matrix} \right\| z_0^{-1} \left\| \begin{matrix} w_0 & 0 & \ldots & 0 \\ w_1 & w_0 & \ldots & 0 \\ \vdots & \vdots & \ddots & \vdots \\ w_{n-1} & w_{n-2} & \ldots & w_0 \end{matrix} \right\|$$
$$- \left\| \begin{matrix} x_n & x_{n-1} & \ldots & x_1 \\ 0 & x_n & \ldots & x_2 \\ \vdots & \vdots & \ddots & \vdots \\ 0 & 0 & \ldots & x_n \end{matrix} \right\| x_0^{-1} \left\| \begin{matrix} y_{-n} & 0 & \ldots & 0 \\ y_{1-n} & y_{-n} & \ldots & 0 \\ \vdots & \vdots & \ddots & \vdots \\ y_{-1} & y_{-2} & \ldots & y_0 \end{matrix} \right\| \tag{1.19}$$

holds.

This corollary is derived from Theorem 1.2 in the same way as Corollary 1.1 is deduced from Theorem 1.1.

Formulas (1.15) and (1.18) also admit another representation, namely, the next statement holds.

Corollary 1.3. *Let the hypotheses of Theorem* 1.1 *be fulfilled and* $A_r^{-1} = \|c_{jk}^r\|_{j,k=0}^r$ $(r = n-1, n)$. *Put*

$$c^r(\zeta, \theta) = \sum_{j,k=0}^{r} c_{jk}^r \zeta^j \theta^{-k} \quad (r = n-1, n),$$

where ζ *and* θ *are variables in* $\mathbb{C}$. *Then the equalities*

$$c^n(\zeta,\theta) = (1-\zeta\theta^{-1})^{-1}(x(\zeta)x_0^{-1}y(\theta) - (\zeta\theta^{-1})^{n+1}z(\zeta)z_0^{-1}w(\theta)) \tag{1.20}$$

and

$$c^{n-1}(\zeta,\theta) = (1-\zeta\theta^{-1})^{-1}(x(\zeta)x_0^{-1}y(\theta) - (\zeta\theta^{-1})^{n}z(\zeta)z_0^{-1}w(\theta)) \tag{1.21}$$

hold, where

$$w(\theta) = \sum_{k=0}^{n} w_k\theta^k, \quad x(\theta) = \sum_{k=0}^{n} x_k\theta^k, \quad y(\theta) = \sum_{k=-n}^{0} y_k\theta^k, \quad z(\theta) = \sum_{k=-n}^{0} z_k\theta^k.$$

Proof. Indeed, we have

$$(1-\zeta\theta^{-1})c(\zeta,\theta) = \sum_{j,k=0}^{n} c_{jk}(\zeta^j\theta^{-k} - \zeta^{j+1}\theta^{-k-1}) = \sum_{j,k=0}^{n+1} (c_{jk} - c_{j-1.k-1})\zeta^j\theta^{-k},$$

where it is assumed that $c_{jk} = 0$ for $j > n$, $j < 0$, $j < 0$, $k > n$, or $k < 0$. Using formulas (1.13) and (1.14), we get

$$\begin{aligned}(1-\zeta\theta^{-1})c(\zeta,\theta) &= \sum_{j,k=0}^{n} x_j x_0^{-1} y_{-k}\zeta^j\theta^{-k} - \sum_{j,k=0}^{n+1} z_{j-n-1} z_0^{-1} w_{n-k+1}\zeta^j\theta^{-k} \\ &= \sum_{j=0}^{n} x_j\zeta^j x_0 \sum_{k=0}^{n} y_{-k}\theta^{-k} - \sum_{j=1}^{n+1} z_{j-n-1}\zeta^j z_0^{-1} \sum_{k=1}^{n+1} w_{n-k+1}\theta^{-k}.\end{aligned}$$

This immediately implies formula (1.20).

Formula (1.21) is proved analogously. □

2. Properties of solutions of equations (1.1)–(1.4)

Under the hypotheses of Theorem 1.1, the solutions of equations (1.1)–(1.4) are connected by a series of relations. In particular, the solutions of equations (1.3) and (1.4) are uniquely determined by the solutions of equations (1.1) and (1.2), and vice versa. The main properties of the solutions of equations (1.1)–(1.4) are presented below. Let us fix the following notation. For elements $t_1, \dots, t_m \in \mathfrak{A}$, by T_{km} and T^{km} denote the following square Toeplitz matrices:

$$T_{km} = \left\| \begin{matrix} t_k & & 0 \\ \vdots & \ddots & \\ t_m & \dots & t_k \end{matrix} \right\|, \quad T^{km} = \left\| \begin{matrix} t_k & \dots & t_m \\ & \ddots & \vdots \\ 0 & & t_k \end{matrix} \right\|.$$

Through this section we will suppose that the matrix $A = \|a_{j-k}\|_{j,k=0}^{n}$ satisfies the hypotheses of Theorem 1.1.

Proposition 2.1. *For the solutions of equations* (1.1)–(1.4) *the following relations*

$$X_{0n} x_0^{-1} Y_{-n,0} = Z_{-n,0} z_0^{-1} W_{0n}, \tag{2.1}$$

$$X_{n0} x_0^{-1} Y_{0,-n} = Z_{0,-n} z_0^{-1} W_{n0} \tag{2.2}$$

hold.

Proof. Indeed, from Theorem 1.1 it follows that

$$w_{n-k} = \sum_{r=0}^{n} (x_{n-r} x_0^{-1} y_{r-k} - z_{-1-r} z_0^{-1} w_{n+1+r-k})$$

and

$$z_{j-n} = \sum_{r=0}^{n} (x_{j-r} x_0^{-1} y_{r-n} - z_{-n-1+j-r} z_0^{-1} w_{1+r})$$

for $k = 0, 1, \dots, n$ and $j = 0, 1, \dots, n$.

Obviously, from these equalities it follows that

$$X_{0n} x_0^{-1} \left\| \begin{matrix} y_{-n} \\ \vdots \\ y_0 \end{matrix} \right\| = Z_{-n,0} z_0^{-1} \left\| \begin{matrix} w_0 \\ \vdots \\ w_n \end{matrix} \right\| \tag{2.3}$$

and

$$\| \, x_0 \quad \dots \quad x_n \, \| x_0^{-1} Y_{0,-n} = \| \, z_{-n} \quad \dots \quad z_0 \, \| z_0^{-1} W_{n0}. \tag{2.4}$$

Since the product of lower triangular Toeplitz matrices is again a lower triangular Toeplitz matrix and two such matrices coincide if their first columns coincide, we see that (2.3) and (2.4) imply equalities (2.1) and (2.2). □

By J_n denote the matrix $\|\delta_{j+k,n} e\|_{j,k=0}^{n}$. For every Toeplitz matrix A the equality

$$J_n A J_n = A' \tag{2.5}$$

holds, where A' is the matrix transpose of A. Multiplying equalities (2.3) and (2.4) from the left and the right by J_n and taking into account the identity $J_n^2 = I$, we arrive at the following statement.

Proposition 2.2. *For the solutions of equations* (1.1)–(1.4) *the equalities*

$$X^{0n} x_0^{-1} Y^{-n,0} = Z^{-n,0} z_0^{-1} W^{0n} \tag{2.6}$$

and

$$X^{n0} x_0^{-1} Y^{-n,0} = Z^{0,-n} z_0^{-1} W^{n0} \tag{2.7}$$

hold.

Let us introduce the matrices P_k by

$$P_k = \left\| \begin{matrix} \overbrace{e \quad & & }^{k} & & & 0 \\ & e & & & & \\ & & \ddots & & & \\ & & & e & & \\ & & & & 0 & \\ & & & & & \ddots & \\ 0 & & & & & & 0 \end{matrix} \right\| .$$

Since

$$P_k X_{0n} = P_k X_{0n} P_k, \quad X^{0n} P_k = P_k X^{0n} P_k,$$

it is easy to see that from equalities (2.1) and (2.2) it follows that

$$X_{0k} x_0^{-1} Y_{-n,k-n} = Z_{-n,k-n} z_0^{-1} W_{0k} \tag{2.8}$$

and

$$X_{n,n-k} x_0^{-1} Y_{0,-k} = Z_{0,-k} z_0^{-1} W_{n,n-k} \tag{2.9}$$

for $k = 0, 1, \ldots, n$.

These equalities and (2.5) also imply the equalities

$$X^{0k} x_0^{-1} Y^{-n,k-n} = Z^{-n,k-n} z_0^{-1} W^{0k} \tag{2.10}$$

and

$$X^{n,n-k} x_0^{-1} Y^{0,-k} = Z^{0,-k} z_0^{-1} W^{n,n-k} \tag{2.11}$$

for $k = 0, 1, \ldots, n$.

Proposition 2.3. *The block matrices*

$$M = \left\| \begin{matrix} X_{0,n-1} & Z_{-n,-1} \\ X^{n1} & Z^{0,1-n} \end{matrix} \right\|, \quad N = \left\| \begin{matrix} W^{0,n-1} & W_{n1} \\ Y^{-n,-1} & Y_{0,1-n} \end{matrix} \right\|$$

are invertible.

Proof. Let us prove the invertibility of the matrix M. By direct verification we check that

$$M = \left\| \begin{matrix} E & R_1 \\ 0 & E \end{matrix} \right\| \left\| \begin{matrix} R_2 & 0 \\ 0 & R_3 \end{matrix} \right\| \left\| \begin{matrix} E & 0 \\ R_4 & E \end{matrix} \right\|$$

where $E = \|\delta_{jk} e\|_{j,k=0}^{n}$ and

$$R_1 = Z_{-n,1}(Z^{0,1-n})^{-1}, \quad R_2 = X_{0,n-1} - R_1 X^{n1},$$
$$R_3 = Z^{0,1-n}, \qquad R_4 = (Z^{0,1-n})^{-1} X^{n1}.$$

From this equality it follows that the matrix M is invertible if and only if the matrix R_2 is invertible. In view of (2.11), we obtain

$$X^{n1} x_0^{-1} Y^{0,1-n} = Z^{0,1-n} z_0^{-1} W^{n1}.$$

Therefore,

$$R_2 = X_{0,n-1} - Z_{-n,-1} z_0^{-1} W^{n1} (Y^{0,1-n})^{-1} x_0$$
$$= (X_{0,n-1} x_0^{-1} Y^{0,1-n} - Z_{-n,-1} z_0^{-1} W^{n,1})(Y^{0,1-n})^{-1} x_0.$$

Thus, according to Theorem 1.2, $R_2 = A_{n-1}^{-1}(Y^{0,1-n})^{-1} x_0$. From here it follows that the matrix R_2 is invertible, and whence the matrix M is invertible, too. Analogously it is proved that the matrix N is invertible as well. □

Proposition 2.4. *For the matrix*

$$\widetilde{M} = \begin{Vmatrix} x_0 & & 0 & z_{-n} & & 0 \\ \vdots & \ddots & & \vdots & \ddots & \\ x_n & \dots & x_0 & z_0 & \dots & z_{-n} \\ & \ddots & \vdots & & \ddots & \vdots \\ 0 & & x_n & 0 & & z_0 \end{Vmatrix},$$

the homogeneous equation

$$\widetilde{M}\chi = 0 \tag{2.12}$$

has a unique solution

$$\chi = \{\chi_{-n}, \dots, \chi_{-0}, \chi_{+0}, \dots, , \chi_n\} \tag{2.13}$$

with the property

$$\chi_{-0} = e, \quad \chi_{+0} = -e. \tag{2.14}$$

For the solutions w_j and y_{-j} $(j = 0, 1, \dots, n)$ of equations (1.3) *and* (1.4) *the equalities*

$$w_j = -z_0\chi_j, \quad y_{-j} = x_0\chi_{-j} \quad (j = 1, 2, \dots, n) \tag{2.15}$$

hold.

Proof. Indeed, from (2.3) and the equality

$$X^{n1}x_0^{-1} \begin{Vmatrix} y_{-n} \\ \vdots \\ y_{-1} \end{Vmatrix} = Z^{0,1-n}z_0^{-1} \begin{Vmatrix} w_1 \\ \vdots \\ w_n \end{Vmatrix},$$

which results from (2.11), it follows that the vector χ defined by equalities (2.13)–(2.15) is a solution of equation (2.12).

Let a vector χ with the property (2.14) satisfy condition (2.12). Obviously, then the equality

$$\begin{Vmatrix} x_0 & & 0 & z_{-n} & & 0 \\ \vdots & \ddots & & \vdots & \ddots & \\ x_{n-1} & \dots & x_0 & z_{-1} & \dots & z_{-n} \\ x_n & \dots & x_1 & z_0 & \dots & z_{1-n} \\ & \ddots & \vdots & & \ddots & \vdots \\ 0 & & x_n & 0 & & z_0 \end{Vmatrix} \begin{Vmatrix} \chi_{-n} \\ \vdots \\ \chi_{-1} \\ \chi_{+0} \\ \vdots \\ \chi_{n-1} \end{Vmatrix} = \begin{Vmatrix} 0 \\ \vdots \\ 0 \\ z_{-n}z_0^{-1}x_n - x_0 \\ z_{1-n}z_0^{-1}x_n - x_1 \\ \vdots \\ z_{-1}z_0^{-1}x_n - x_{n-1} \end{Vmatrix} \tag{2.16}$$

holds. Since the vector χ satisfies this equality, its components satisfy equality (2.15) and the matrix on the left-hand side of equality (2.16) coincides with the invertible matrix M, we conclude that the vector χ defined by equalities (2.13)–(2.15) is a unique solution of equation (2.12). □

Note that the following statement is obtained as a byproduct.

Proposition 2.5. *Equation* (2.16) *has a unique solution. For its components, equalities* (2.15) *hold and* $\chi_{+0} = -e$.

The following statements are proved analogously.

Proposition 2.6. *For the matrix*

$$\widetilde{N} = \left\| \begin{array}{ccccc} w_0 & \dots & w_n & & 0 \\ & \ddots & \vdots & \ddots & \\ 0 & & w_0 & \dots & w_n \\ y_{-n} & \dots & y_0 & & 0 \\ & \ddots & \vdots & \ddots & \\ 0 & & y_{-n} & \dots & y_0 \end{array} \right\|,$$

the homogeneous equation

$$\omega \widetilde{N} = 0$$

has a unique solution

$$\omega = \{\omega_{-n}, \dots, \omega_{-0}, \omega_{+0}, \dots, \omega_n\}$$

with the property

$$\omega_{-0} = e, \quad \omega_{+0} = -e.$$

For the solutions x_j *and* z_{-j} $(j = 0, 1, \dots, n)$ *of equations* (1.1) *and* (1.2), *the equalities*

$$x_j = -\omega_j y_0, \quad z_{-j} = \omega_{-j} w_0 \quad (j = 1, 2, \dots, n) \tag{2.17}$$

hold.

Relations between the solutions of equations (1.1)–(1.4) for the matrix $A = \|a_{j-k}\|_{j,k=0}^{n}$ and the matrix $A_{n-1} = \|a_{j-k}\|_{j,k=0}^{n-1}$ are obtained in the next statements.

Proposition 2.7. *The elements* x_j^{n-1}, z_{-j}^{n-1}, w_j^{n-1}, y_{-j}^{n-1} $(j = 0, 1, \dots, n-1)$ *defined by the equalities*

$$\begin{aligned} x_j^{n-1} &= x_j - z_{j-n} z_0^{-1} x_n & z_{-j}^{n-1} &= z_{-j} - x_{n-j} x_0^{-1} z_{-n}, \\ w_j^{n-1} &= w_j - w_n y_0^{-1} y_{j-n}, & y_{-j}^{n-1} &= y_{-j} - y_{-n} w_0^{-1} w_{n-j} \end{aligned} \tag{2.18}$$

are solutions of the following equations

$$\sum_{k=0}^{n-1} a_{j-k} x_k^{n-1} = \delta_{0j} e \quad (j = 0, 1, \dots, n-1), \tag{2.19}$$

$$\sum_{k=0}^{n-1} a_{k-j} z_{-k}^{n-1} = \delta_{0j} e \quad (j = 0, 1, \dots, n-1), \tag{2.20}$$

$$\sum_{k=0}^{n-1} w_k^{n-1} a_{j-k} = \delta_{0j} e \quad (j = 0, 1, \ldots, n-1), \tag{2.21}$$

$$\sum_{k=0}^{n-1} y_{-k}^{n-1} a_{k-j} = \delta_{0j} e \quad (j = 0, 1, \ldots, n-1). \tag{2.22}$$

It is straightforward to verify this statement.

Proposition 2.8. *Let the solutions* x_j^{n-1} *and* z_{-j}^{n-1} $(j = 0, 1, \ldots, n-1)$ *of equations* (2.19) *and* (2.20) *exist and the elements* x_0^{n-1} *and* z_0^{n-1} *be invertible. Then the elements*

$$\begin{aligned} \xi_n &= (x_0^{n-1})^{-1} + \sum_{k=0}^{n-1} a_{-k-1} z_{k+1-n}^{n-1} (z_0^{n-1})^{-1} \alpha_n, \\ \zeta_n &= (z_0^{n-1})^{-1} + \sum_{k=0}^{n-1} a_{k+1} x_{n-k-1}^{n-1} (x_0^{n-1})^{-1} \beta_n \end{aligned} \tag{2.23}$$

are also invertible, where

$$\begin{aligned} \alpha_n &= -z_0^{n-1} \sum_{k=0}^{n-1} a_{n-k} x_k^{n-1} (x_0^{n-1})^{-1}, \\ \beta_n &= -x_0^{n-1} \sum_{k=0}^{n-1} a_{k-n} z_{-k}^{n-1} (z_0^{n-1})^{-1}, \end{aligned} \tag{2.24}$$

and the equalities

$$\begin{aligned} &x_0 = \xi_n^{-1}, \quad z_0 = \zeta_n^{-1}, \quad x_n = \alpha_n \xi_n^{-1}, \quad z_{-n} = \beta_n \zeta_n^{-1}, \\ &x_j = \left(x_j^{n-1} (x_0^{n-1})^{-1} + z_{j-n}^{n-1} (z_0^{n-1})^{-1} \alpha_n\right) \xi_n^{-1} \quad (j = 1, 2, \ldots, n-1), \\ &z_{-j} = \left(z_{-j}^{n-1} \alpha_0 (z_0^{n-1})^{-1} + x_{n-j}^{n-1} (x_0^{n-1})^{-1} \beta_n\right) \zeta_n^{-1} \quad (j = 1, 2, \ldots, n-1) \end{aligned} \tag{2.25}$$

hold, where as usual solutions of equations (1.1), (1.2) *are denoted by* x_j, z_{-j}.

Proof. Indeed, putting

$$\alpha_j = x_j^{n-1} (x_0^{n-1})^{-1} + z_{j-n}^{n-1} (z_0^{n-1})^{-1} \alpha_n \quad (j = 1, 2, \ldots, n-1)$$

and $\alpha_0 = e$, we get

$$\sum_{k=0}^{n} a_{j-k} \alpha_k = \sum_{k=1}^{n-1} a_{j-k} x_k^{n-1} (x_0^{n-1})^{-1} + \sum_{k=1}^{n-1} a_{j-k} z_{k-n}^{n-1} (z_0^{n-1})^{-1} \alpha_n + a_j + a_{j-n} \alpha_n.$$

From here it follows that

$$\sum_{k=0}^{n} a_{j-k} \alpha_k = \sum_{k=0}^{n-1} a_{j-k} x_k^{n-1} (x_0^{n-1})^{-1} + \sum_{k=1}^{n} a_{j-k} z_{k-n}^{n-1} (z_0^{n-1})^{-1} \alpha_n. \tag{2.26}$$

According to the definition of the elements x_j^{n-1} and z_{-j}^{n-1}, from here we get

$$\sum_{k=0}^{n} a_{j-k}\alpha_k = 0 \quad (j = 1, 2, \ldots, n-1). \tag{2.27}$$

Additionally, from (2.24) and (2.26) it follows that

$$\begin{aligned}\sum_{k=0}^{n} a_{n-k}\alpha_k &= \sum_{k=0}^{n-1} a_{n-k}x_0^{n-1}(x_0^{n-1})^{-1} \\ &\quad - \sum_{k=1}^{n} a_{n-k}z_{k-n}^{n-1}(z_0^{n-1})^{-1}z_0^{n-1}\sum_{k=0}^{n-1} a_{n-k}x_k^{n-1}(x_0^{n-1})^{-1} \\ &= \left(e - \sum_{k=1}^{n} a_{n-k}z_{k-n}^{n-1}\right)\sum_{k=0}^{n-1} a_{n-k}x_k^{n-1}(x_0^{n-1})^{-1} = 0.\end{aligned} \tag{2.28}$$

From equalities (2.26) and (2.13) it follows that

$$\sum_{k=0}^{n} a_{-k}\alpha_k = x_0^{n-1} + \sum_{k=0}^{n-1} a_{-k-1}z_{k+1-r}^{r-1}(z_0^{r-1})^{-1}\alpha_n = \xi_n. \tag{2.29}$$

For solutions x_j of equation (1.1) the equality

$$\sum_{k=0}^{n} a_{j-k}x_k\xi_n = \delta_{j0}\xi_n \quad (j = 0, 1, \ldots, n)$$

holds. From the uniqueness of the solution of system (1.1) it follows that

$$\alpha_j = x_j\xi_n \quad (j = 0, 1, \ldots, n). \tag{2.30}$$

In particular, $e = \alpha_0 = x_0\xi_n$. Hence the element ξ_n is invertible and is the inverse of the element x_0. This fact and equalities (2.27)–(2.30) immediately imply relations (2.25) for ξ_n, α_n, and x_j. Relations (2.25) for ξ_n, β_n, and z_{-j} are proved analogously.

The statement is proved. □

The following statement is proved analogously.

Proposition 2.9. *Let the solutions w_j^{n-1} and y_{-j}^{n-1} $(j = 0, 1, \ldots, n-1)$ of equations (2.21) and (2.22) exist and the elements w_0^{n-1} and y_0^{n-1} be invertible. Then the elements*

$$\theta_n = (w_0^{n-1})^{-1} - \gamma_n \sum_{k=0}^{n-1}(z_0^{n-1})^{-1}z_{k+1-n}^{n-1}a_{-k-1},$$

$$\eta_n = (y_0^{n-1})^{-1} + \delta_n \sum_{k=0}^{n-1}(y_0^{n-1})^{-1}y_{n-k-1}^{n-1}a_{k+1}$$

are also invertible, where

$$\gamma_n = -\sum_{k=0}^{n-1}(w_0^{n-1})^{-1}w_k^{n-1}a_{n-k}y_0^{n-1},$$

$$\delta_n = -\sum_{k=0}^{n-1}(y_0^{n-1})^{-1}y_{-k}^{n-1}a_{k-n}w_0^{n-1},$$

and the equalities

$$w_0 = \theta_n^{-1}, \quad y_0 = \eta_n^{-1}, \quad w_n = \theta_n^{-1}\gamma_n, \quad y_{-n} = \eta_n^{-1}\delta_n,$$

$$w_j = \theta_n^{-1}\big((w_0^{n-1})^{-1}w_j^{n-1} + \gamma_n(y_0^{n-1})^{-1}y_{j-n}^{n-1}\big) \quad (j = 1, 2, \dots, n-1), \tag{2.31}$$

$$y_{-j} = \eta_n^{-1}\big((y_0^{n-1})^{-1}y_{-j}^{n-1} + \delta_n(w_0^{n-1})^{-1}w_{n-j}^{n-1}\big) \quad (j = 1, 2, \dots, n-1)$$

hold.

The successive application of formulas (2.25) and (2.31) to all minors A_r $(r = 0, 1, \dots, n)$ allows us to give a rule for the effective calculation of the solutions of equations (1.1)–(1.4) in the case when all minors A_r are invertible.

Formulas (2.25) and (2.31) can also be applied without the assumption that all the minors A_r are invertible. Consider the case when $\mathfrak{A}$ is a Banach algebra and not all minors A_r are invertible. Suppose

$$A(\lambda) = \|a_{jk}(\lambda)\|_{j,k=0}^{n}$$

is a matrix function that is holomorphic and invertible in some (connected) domain $G \subset \mathbb{C}$ and whose entries belong to $\mathfrak{A}$. Assume that $0 \in G$, $A(0) = A$, and there exists some point $\lambda_0 \in G$ such that all minors

$$A_r(\lambda_0) = \|a_{j-k}(\lambda_0)\|_{j,k=0}^{r}$$

are invertible. Then for all λ in some neighborhood of the point λ_0 the solutions $x_j^r(\lambda)$, $z_{-j}^r(\lambda)$, $w_j^r(\lambda)$, and $y_{-j}^r(\lambda)$ of equations (1.1)–(1.4) for the matrix $A_r(\lambda)$ exist. These solutions are holomorphic functions. Moreover, the functions $x_0^r(\lambda)$ $(= y_0^r(\lambda))$ and $z_0^r(\lambda)$ $(= w_0^r(\lambda))$ are invertible in a neighborhood of the point λ_0. Thus the functions $x_j^n(\lambda)$, $z_{-j}^n(\lambda)$, $w_j^n(\lambda)$, and $y_{-j}^n(\lambda)$ can be calculated with the aid of the recurrent formulas (2.25) and (2.31). Since the matrix $A(\lambda)$ is invertible for all $\lambda \in G$, we see that for all $\lambda \in G$, there exist solutions $x_j(\lambda)$, $z_{-j}(\lambda)$, $w_j(\lambda)$, and $y_{-j}(\lambda)$ of equations (1.1)–(1.4) for the matrix $A(\lambda)$. These solutions depend on λ holomorphically in G. It follows that the functions $x_j^n(\lambda)$, $z_{-j}^n(\lambda)$, $w_j^n(\lambda)$, and $y_{-j}^n(\lambda)$ admit continuations holomorphic in G, which coincide with the functions $x_j(\lambda)$, $z_{-j}(\lambda)$, $w_j(\lambda)$, and $y_{-j}(\lambda)$. Obviously, $x_j = x_j(0)$, $z_{-j} = z_{-j}(0)$, $w_j = w_j(0)$, and $y_{-j} = y_{-j}(0)$.

Note that in the case when $\mathfrak{A}$ is the algebra of quadratic matrices $\mathbb{C}_{m\times m}$, a function $A(\lambda)$ satisfying the above assumptions always exists. For instance, one can take $A(\lambda) = A - \lambda I$ or $A(\lambda) = A - \lambda(A - I)$.

3. Inverse problem

Theorem 3.1. *Let w_j, x_j, y_{-j}, z_{-j} $(j = 0, 1, \dots, n)$ be given systems of elements in $\mathfrak{A}$ and the elements w_0, x_0, y_0, z_0 be invertible. For the existence of an invertible Toeplitz matrix $A = \|a_{j-k}\|_{j,k=0}^{n}$ with elements $a_j \in \mathfrak{A}$ $(j = 0, \pm 1, \dots, \pm n)$ such that w_j, x_j, y_{-j}, z_{-j} are solutions of equations (1.1)–(1.4) it is necessary and sufficient that the following three conditions be fulfilled:*

1) $x_0 = y_0$ *and* $z_0 = w_0$*;*
2)

$$X_{0n} x_0^{-1} Y_{-n,0} = Z_{-n,0} z_0^{-1} W_{0n}, \tag{3.1}$$

$$X_{n0} x_0^{-1} Y_{0,-n} = Z_{0,-n} z_0^{-1} W_{n0}; \tag{3.2}$$

3) *at least one of the matrices*

$$M = \left\| \begin{matrix} X_{0,n-1} & Z_{-n,-1} \\ X^{n1} & Z^{0,n-1} \end{matrix} \right\|, \quad N = \left\| \begin{matrix} W^{0,n-1} & W_{n1} \\ Y^{-n,-1} & Y_{0,1-n} \end{matrix} \right\|$$

is invertible.

If conditions 1)–3) are fulfilled, then both matrices M and N are invertible and the matrix $A = \|a_{j-k}\|_{j,k=0}^{n}$ is uniquely determined by formula (1.5).

Proof. The necessity of condition 1) is obtained in the proof of Theorem 1.1 and the necessity of conditions 2) and 3) is proved in the previous section. We shall prove the sufficiency of the hypotheses of the theorem. For definiteness, suppose that the matrix M is invertible. Consider the matrix $B = \|b_{jk}\|_{j,k=0}^{n}$ defined by the equality

$$B = X_{0n} x_0^{-1} Y^{0,-n} - \widetilde{Z}_{-n,-1} z_0^{-1} \widetilde{W}^{n1},$$

where

$$\widetilde{Z}_{-n,-1} = \left\| \begin{matrix} 0 & \dots & 0 & 0 \\ z_{-n} & \ddots & \ddots & 0 \\ \vdots & \ddots & \ddots & \vdots \\ z_{-1} & \dots & z_{-n} & 0 \end{matrix} \right\|, \quad \widetilde{W}^{n1} = \left\| \begin{matrix} 0 & w_n & \dots & w_1 \\ \vdots & \ddots & \ddots & \vdots \\ 0 & \ddots & \ddots & w_n \\ 0 & 0 & \dots & 0 \end{matrix} \right\|.$$

Let us show that the matrix B is invertible.

It is easy to see that in view of the invertibility of the elements x_0 and z_0, the invertibility of the matrix M implies the invertibility of the matrix

$$\widetilde{M} = \left\| \begin{matrix} X_{0n} & \widetilde{Z}_{-n,-1} \\ \widetilde{X}^{n1} & Z^{0,-n} \end{matrix} \right\|.$$

By a straightforward verification it is obtained that

$$\widetilde{M} = \left\| \begin{matrix} E & \widetilde{R}_1 \\ 0 & E \end{matrix} \right\| \left\| \begin{matrix} \widetilde{R}_2 & 0 \\ 0 & \widetilde{R}_3 \end{matrix} \right\| \left\| \begin{matrix} E & 0 \\ \widetilde{R}_4 & E \end{matrix} \right\|, \tag{3.3}$$

where

$$\widetilde{R}_1 = \widetilde{Z}_{-n,-1}(Z^{0,-n})^{-1}, \quad \widetilde{R}_2 = X_{0n} - \widetilde{R}_1\widetilde{X}^{n1},$$

$$\widetilde{R}_3 = Z^{0,-n}, \qquad \widetilde{R}_4 = (Z^{0,-n})^{-1}\widetilde{X}^{n1}.$$

Hence the matrix $\widetilde{R}_2$ is invertible. From equality (3.2) it follows immediately that

$$\widetilde{X}^{n1}x_0^{-1}Y^{0,-n} = Z^{0,-n}z_0^{-1}\widetilde{W}^{n1},$$

whence

$$\widetilde{R}_2 = (X_{0n}x_0^{-1}Y^{0,-n} - \widetilde{Z}_{-n,-1}z_0^{-1}\widetilde{W}^{n1}(Y^{0,-n})^{-1}x_0.$$

Thus

$$B = \widetilde{R}_2 x_0^{-1}Y^{0,-n}$$

and this implies that B is an invertible matrix.

Put $A = \|a_{jk}\|_{j,k=0}^n$. Obviously, in view of condition 1),

$$b_{0j} = y_{-j}, \quad b_{j0} = x_j \quad (j = 0, 1, \dots, n). \tag{3.4}$$

Moreover, equality (3.2) implies, in particular, that the equality

$$\sum_{r=0}^{k} x_{n-r}x_0^{-1}y_{r-k} = \sum_{r=0}^{k} z_{-r}z_0^{-1}w_{n-k+r} \quad (k = 0, 1, \dots, n)$$

holds and equality (3.1) implies that

$$\sum_{r=0}^{j} x_{j-r}x_0^{-1}y_{r-n} = \sum_{r=0}^{j} z_{j-n-r}z_0^{-1}w_r \quad (j = 0, 1, \dots, n).$$

From here it follows that

$$w_{n-k} = \sum_{r=0}^{k}(x_{n-r}x_0^{-1}y_{r-k} - z_{-1-r}z_0^{-1}w_{n+1+r-k})$$

and

$$z_{j-n} = \sum_{r=0}^{j}(x_{j-r}x_0^{-1}y_{r-n} - z_{-n-1+j-r}z_0^{-1}w_{1+r}).$$

By the definition of the matrix B, the expressions on the right-hand sides of the last equalities coincide with the elements b_{nk} and b_{jn}, respectively. Hence

$$b_{nk} = w_{n-k}, \quad b_{jn} = z_{j-n}. \tag{3.5}$$

From equalities (3.4) and (3.5) it follows immediately that the elements x_j, z_{-j}, w_j, and y_{-j} are solutions of equations (1.1)–(1.4), respectively, for the matrix A.

It remains to prove that the matrix A is a Toeplitz matrix. We put

$$A^{(1)} = \|a_{jk}\|_{j,k=0}^{n-1}, \quad A^{(2)} = \|a_{jk}\|_{j,k=1}^{n}.$$

Clearly, it is sufficient to prove that $A^{(1)} = A^{(2)}$.

Obviously, the equalities

$$\left\| \begin{matrix} e & \dots & 0 & 0 \\ \vdots & \ddots & \vdots & \vdots \\ 0 & \dots & e & 0 \\ w_n & \dots & w_1 & w_0 \end{matrix} \right\| A \left\| \begin{matrix} e & \dots & 0 & z_{-n} \\ \vdots & \ddots & \vdots & \vdots \\ 0 & \dots & e & z_{-1} \\ 0 & \dots & 0 & z_0 \end{matrix} \right\| = \left\| \begin{matrix} A^{(1)} & 0 \\ 0 & z_0 \end{matrix} \right\|$$

and

$$\left\| \begin{matrix} y_0 & y_{-1} & \dots & y_{-n} \\ 0 & e & \dots & 0 \\ \vdots & \vdots & \ddots & \vdots \\ 0 & 0 & \dots & e \end{matrix} \right\| A \left\| \begin{matrix} x_0 & 0 & \dots & 0 \\ x_1 & e & \dots & 0 \\ \vdots & \vdots & \ddots & \vdots \\ x_n & 0 & \dots & e \end{matrix} \right\| = \left\| \begin{matrix} x_0 & 0 \\ 0 & A^{(2)} \end{matrix} \right\|$$

hold. From here it follows that the matrices $A^{(1)}$ and $A^{(2)}$ are invertible and

$$B = A^{-1} = \left\| \begin{matrix} e & \dots & 0 & z_{-n} \\ \vdots & \ddots & \vdots & \vdots \\ 0 & \dots & e & z_{-1} \\ 0 & \dots & 0 & z_0 \end{matrix} \right\| \left\| \begin{matrix} (A^{(1)})^{-1} & 0 \\ 0 & z_0^{-1} \end{matrix} \right\| \left\| \begin{matrix} e & \dots & 0 & 0 \\ \vdots & \ddots & \vdots & \vdots \\ 0 & \dots & e & 0 \\ w_n & \dots & w_1 & w_0 \end{matrix} \right\|$$

$$= \left\| \begin{matrix} x_0 & 0 & \dots & 0 \\ x_1 & e & \dots & 0 \\ \vdots & \vdots & \ddots & \vdots \\ x_n & 0 & \dots & e \end{matrix} \right\| \left\| \begin{matrix} x_0^{-1} & 0 \\ 0 & (A^{(2)})^{-1} \end{matrix} \right\| \left\| \begin{matrix} y_0 & y_{-1} & \dots & y_{-n} \\ 0 & e & \dots & 0 \\ \vdots & \vdots & \ddots & \vdots \\ 0 & 0 & \dots & e \end{matrix} \right\| .$$

Putting $\|b^r_{jk}\|_{jk=0}^{n-1} = (A^{(r)})^{-1}$, from the last equality we deduce that

$$b_{jk} = b^1_{jk} + z_{j-n} w_0^{-1} w_{n-k}, \quad b_{jk} = b^2_{j-1,k-1} + x_j x_0^{-1} y_{-k}.$$

Therefore,

$$\begin{aligned} b^1_{jk} &= \sum_{r=0}^{n} (x_{j-r} x_0^{-1} y_{r-k} - z_{-n-1+j-r} z_0^{-1} w_{n+1+r-k}) - z_{j-n} w_0^{-1} w_{n-k} \\ &= \sum_{r=0}^{n} (x_{j-r} x_0^{-1} y_{r-k} - z_{-n+j-r} z_0^{-1} w_{n+r-k}) \end{aligned}$$

and

$$\begin{aligned} b^2_{jk} &= \sum_{r=0}^{n} (x_{j+1-r} x_0^{-1} y_{r-k-1} - z_{-n+j-r} z_0^{-1} w_{n+r-k}) - x_{j+1} x_0^{-1} y_{-k} \\ &= \sum_{r=0}^{n} (x_{j-r} x_0^{-1} y_{r-k} - z_{-n+j-r} z_0^{-1} w_{n+r-k}). \end{aligned}$$

Hence $A^{(1)} = A^{(2)}$. This means that A is a Toeplitz matrix.

The uniqueness of the matrix A follows from Theorem 1.1. According to it, equality (1.5) holds for the matrix A.

The theorem is proved under the assumption that the matrix M is invertible. Analogously, it is also proved in the case when the matrix N is invertible.

The theorem is proved. □

Theorem 3.2. *Let x_j and z_{-j} $(j = 0, 1, \ldots, n)$ be given systems of elements in $\mathfrak{A}$ and the elements x_0 and z_0 be invertible.*

For the existence of a Toeplitz matrix $A = \|a_{j-k}\|_{j,k=0}^{n}$ with elements $a_j \in \mathfrak{A}$ $(j = 0, \pm 1, \ldots, \pm n)$ such that x_j and z_{-j} are solutions of equations (1.1) *and* (1.2), *respectively, it is necessary and sufficient that the matrix M be invertible and that for the vector*

$$\left\| \begin{matrix} \chi_{-n} \\ \vdots \\ \chi_0 \\ \vdots \\ \chi_{n-1} \end{matrix} \right\| = M^{-1} \left\| \begin{matrix} 0 \\ \vdots \\ 0 \\ z_{-n} z_0^{-1} x_n - x_0 \\ \vdots \\ z_{-1} z_0^{-1} x_n - x_{n-1} \end{matrix} \right\| \tag{3.6}$$

the condition $\chi_0 = -e$ be fulfilled.

Under these conditions, solutions of equations (1.3) *and* (1.4) *are given by the equalities*

$$y_{-j} = x_0 \chi_{-j} \quad (j = 1, 2, \ldots, n), \qquad y_0 = x_0, \tag{3.7}$$

$$w_j = -z_0 \chi_j \quad (j = 0, 1, \ldots, n-1), \qquad w_n = x_n, \tag{3.8}$$

and the matrix A is uniquely determined by equality (1.5).

Proof. Assume that the hypotheses of the theorem are fulfilled. With the aid of equalities (3.7) and (3.8) we introduce the systems y_{-j} and w_j $(j = 0, 1, \ldots, n)$. Equality (3.6) can be rewritten in the form

$$\left\| \begin{matrix} x_0 & & 0 & z_{-n} & & 0 \\ \vdots & \ddots & & \vdots & \ddots & \\ x_n & \cdots & x_0 & z_0 & \cdots & z_{-n} \\ & \ddots & \vdots & & \ddots & \vdots \\ 0 & & x_n & 0 & & z_0 \end{matrix} \right\| \left\| \begin{matrix} x_0^{-1} & y_{-n} \\ \vdots & \vdots \\ x_0^{-1} & y_0 \\ -z_0^{-1} & w_0 \\ \vdots & \vdots \\ -z_0^{-1} & w_n \end{matrix} \right\| = 0. \tag{3.9}$$

It is easy to check that the last equality implies relations (3.1) and (3.2) and vice versa. Since additionally $x_0 = y_0$ and $z_0 = w_0$, we see that the systems x_j, z_{-j}, w_j, and and y_{-j} satisfy all the hypotheses of Theorem 3.1.

Let us prove that the converse statement also holds. If the matrix A exists, then in view of Theorem 3.1 equality (3.9) holds and the matrix M is invertible. From here it follows that equalities (3.7) and (3.8) hold. In view of condition 1) of Theorem 3.1 from (3.8) it follows that $\chi_0 = -e$.

The theorem is proved. □

The following theorem is proved analogously.

Theorem 3.3. *Let w_j and y_{-j} $(j = 0, 1, \dots, n)$ be given systems of elements in $\mathfrak{A}$ and the elements w_0 and y_0 be invertible.*

For the existence of a Toeplitz matrix $A = \|a_{j-k}\|_{j,k=0}^{n}$ with elements $a_j \in \mathfrak{A}$ $(j = 0, \pm 1, \dots, \pm n)$ such that w_j and y_{-j} are solutions of equation (1.3) and (1.4), respectively, it is necessary and sufficient that the matrix N be invertible and that the vector

$$\|\omega_{-n}, \dots, \omega_0, \dots, \omega_{n-1}\| = \|0, \dots, 0, w_n y_0^{-1} y_{-n} - w_0, \dots, w_n y_0^{-1} y_{-1} - w_{n-1}\| N^{-1}$$

have the property $\omega_0 = -e$.

Under these conditions, solutions of equations (1.1) and (1.2) are given by the equalities

$$\begin{aligned} x_j &= -\omega_j y_0 \quad (j = 0, 1, \dots, n-1), & x_n &= w_n, \\ z_{-j} &= \omega_{-j} w_0 \quad (j = 1, 2, \dots, n), & z_0 &= w_0, \end{aligned}$$

and the matrix A is uniquely determined by equality (1.5).

The next theorem gives a rule for calculating the elements a_j by solutions of equations (1.1), (1.2) or (1.3), (1.4).

Theorem 3.4. *Suppose x_j, z_{-j} $(j = 0, 1, \dots, n)$ are given systems of elements in $\mathfrak{A}$ that are solutions of equations (1.1) and (1.2) for some matrix $A = \|a_{j-k}\|_{j,k=0}^{n}$ with the invertible minors $A_r = \|a_{j-k}\|_{j,k=0}^{r}$ $(r = 0, 1, \dots, n)$. Then the elements a_j $(j = 0, \pm 1, \dots, \pm n)$ are determined by the recurrent equalities*

$$\begin{aligned} a_r &= -\sum_{k=1}^{r} a_{r-k} x_k^r (x_0^r)^{-1}, \\ a_{-r} &= -\sum_{k=1}^{r} a_{k-r} z_{-k}^r (z_0^r)^{-1}, \quad (r = 1, 2, \dots, n), \\ a_0 &= (x_0^0)^{-1} = (z_0^0)^{-1}, \end{aligned} \tag{3.10}$$

where the elements x_k^r and z_{-k}^r are given by the recurrent formulas

$$\begin{aligned} x_k^n &= x_k, & z_{-k}^n &= z_{-k}, \\ z_{-k}^{r-1} &= z_{-k}^r - z_{r-k}^r (x_0^r)^{-1} z_{-r}^r, & x_k^{r-1} &= x_k^r - z_{k-r}^r (z_0^r)^{-1} x_r^r, \end{aligned}$$

and the elements x_0^r and z_0^r $(r = 0, 1, \dots, r)$ are invertible.

Proof. Successively applying Proposition 2.8, we see that for the elements x_j^r and z_j^r the equalities

$$\sum_{k=0}^{r} a_{j-k} x_k^r = \delta_{j0} e, \quad \sum_{k=0}^{r} a_{k-j} z_{-k}^r = \delta_{j0} e \quad (j = 0, 1, \dots, r) \tag{3.11}$$

hold. The invertibility of the elements x_0^r and z_0^r follows from the invertibility of the minors A_r and the equalities

$$A_r \left\| \begin{array}{cccc} x_0^r & 0 & \dots & 0 \\ x_1^r & e & \dots & 0 \\ \vdots & \vdots & \ddots & \vdots \\ x_r^r & 0 & \dots & e \end{array} \right\| = \left\| \begin{array}{c|ccc} e & a_{-1} & \dots & a_{-r} \\ \hline 0 & & & \\ \vdots & & A_{r-1} & \\ 0 & & & \end{array} \right\|$$

and

$$A_r \left\| \begin{array}{cccc} e & \dots & 0 & z_{-r}^r \\ \vdots & \ddots & \vdots & \vdots \\ 0 & \dots & e & z_{-1}^r \\ 0 & \dots & 0 & z_0^r \end{array} \right\| = \left\| \begin{array}{ccc|c} & & & 0 \\ & A_{r-1} & & \vdots \\ & & & 0 \\ \hline a_r & \dots & a_1 & e \end{array} \right\| .$$

Equality (3.11) with $j = r$ implies formulas (3.10).

The theorem is proved. □

4. Other theorems on inversion of Toeplitz matrices

In this section the inverse of the matrix $A = \|a_{j-k}\|_{j,k=0}^n$, where $a_j \in \mathfrak{A}$ and $j = 0, \pm 1, \dots, \pm n$, is constructed with the aid of solutions of the following equations

$$\sum_{k=0}^n a_{j-k} x_k = \delta_{0j} e \quad (j = 0, 1, \dots, n), \tag{4.1}$$

$$\sum_{k=0}^n w_k a_{j-k} = \delta_{0j} e \quad (j = 0, 1, \dots, n), \tag{4.2}$$

$$\sum_{k=0}^n a_{j-k} s_k = \delta_{1j} e \quad (j = 1, 2, \dots, n+1), \tag{4.3}$$

$$\sum_{k=0}^n t_k a_{j-k} = \delta_{1j} e \quad (j = 1, 2, \dots, n+1), \tag{4.4}$$

where a_{n+1} is an arbitrary element in $\mathfrak{A}$, as well as with the aid of solutions of equations (4.1), (4.2) and the systems

$$\sum_{k=0}^n a_{j-k} u_k = \delta_{j1} e \quad (j = 0, 1, \dots, n), \tag{4.5}$$

$$\sum_{k=0}^n v_k a_{j-k} = \delta_{j1} e \quad (j = 0, 1, \dots, n). \tag{4.6}$$

It is easy to see that if the systems x_j, w_j, s_j, and t_j are solutions of systems (4.1)–(4.4), then the systems x_j, w_j, u_j, and v_j, where

$$u_j = s_j - x_j \sum_{r=0}^{n} a_{-r} s_r$$

and

$$v_j = t_j - \sum_{r=0}^{n} t_r a_{-r} w_j,$$

satisfy equations (4.1), (4.2), (4.5), and (4.6).

Note also that if equations (4.1) and (4.2) are solvable, then $x_n = w_n$. Indeed,

$$w_n = \sum_{k=0}^{n} w_{n-k} \sum_{j=0}^{n} a_{k-j} x_j = \sum_{j=0}^{n} \left(\sum_{k=0}^{n} w_{n-k} a_{k-j} \right) x_j = x_n.$$

The equalities $s_n = w_n$, $u_{n-1} = v_{n-1}$, $u_n = w_{n-1}$, and $v_n = x_{n-1}$ are proved analogously.

Theorem 4.1. *Let $A = \|a_{j-k}\|_{j,k=0}^{n}$ be a Toeplitz matrix with elements $a_j \in \mathfrak{A}$ $(j = 0, \pm 1, \dots, \pm n)$ and let for some $a_{n+1} \in \mathfrak{A}$ equations (4.1)–(4.4) be solvable and the element $x_n (= w_n)$ be invertible.*

Then the matrix A is invertible and its inverse is given by the formula

$$A^{-1} = \|x_j x_n^{-1} w_{n-k}\|_{j,k=0}^{n} + \left\| \begin{matrix} s_0 & 0 & \dots & 0 \\ s_1 & s_0 & \dots & 0 \\ \vdots & \vdots & \ddots & \vdots \\ s_n & s_{n-1} & \dots & s_0 \end{matrix} \right\| x_n^{-1} \left\| \begin{matrix} 0 & w_n & \dots & w_1 \\ \vdots & \vdots & \ddots & \vdots \\ 0 & 0 & \dots & w_n \\ 0 & 0 & \dots & 0 \end{matrix} \right\| - \left\| \begin{matrix} x_0 & 0 & \dots & 0 \\ x_1 & x_0 & \dots & 0 \\ \vdots & \vdots & \ddots & \vdots \\ x_n & x_{n-1} & \dots & x_0 \end{matrix} \right\| x_n^{-1} \left\| \begin{matrix} 0 & t_n & \dots & t_1 \\ \vdots & \vdots & \ddots & \vdots \\ 0 & 0 & \dots & t_n \\ 0 & 0 & \dots & 0 \end{matrix} \right\|. \tag{4.7}$$

The proof of this theorem is based on the next lemma similar to Lemma 1.1.

Lemma 4.1. *Let $\mathbb{V}$ be a linear space, $L(\mathbb{V})$ be the algebra of all linear operators acting in $\mathbb{V}$, $A = \|A_{j-k}\|_{j,k=0}^{n}$ be a Toeplitz matrix with elements A_j $(j = 0, \pm 1, \dots, \pm n)$ in $L(\mathbb{V})$, and $A_{n+1} \in L(\mathbb{V})$.*

If the systems of equations

$$\sum_{k=0}^{n} A_{j-k} X_k = \delta_{0j} I \quad (j = 0, 1, \dots, n), \tag{4.8}$$

$$\sum_{k=0}^{n} W_k A_{j-k} = \delta_{0j} I \quad (j = 0, 1, \dots, n), \tag{4.9}$$

$$\sum_{k=0}^{n} A_{j-k}S_k = \delta_{1j}I \quad (j = 1, 2, \dots, n+1), \tag{4.10}$$

$$\sum_{k=0}^{n} T_k A_{j-k} = \delta_{1j}I \quad (j = 1, 2, \dots, n+1) \tag{4.11}$$

are solvable in $L(\mathbb{V})$, and the operator X_n $(= W_n)$ is invertible, then the matrix A is also invertible.

Proof. Let the hypotheses of the lemma be fulfilled. As above, by $\mathbb{V}_n$ denote the linear space of one-row matrices $\varphi = \{\varphi_j\}_{j=0}^n$ with entries $\varphi_j \in \mathbb{V}$.

If a vector $\varphi = \{\varphi_j\}_{j=0}^n$ $(\neq 0)$ belongs to $\operatorname{Ker} A$, then

$$\varphi_n = \sum_{k=0}^{n} \left(\sum_{j=0}^{n} W_{n-j} A_{j-k} \right) \varphi_k = \sum_{j=0}^{n} W_{n-j} \left(\sum_{k=0}^{n} A_{j-k}\varphi_k \right) = 0,$$

$$\varphi_{n-1} = \sum_{k=0}^{n} \left(\sum_{j=0}^{n} V_{n-j} A_{j-k} \right) \varphi_k = \sum_{j=0}^{n} V_{n-j} \left(\sum_{k=0}^{n} A_{j-k}\varphi_k \right) = 0.$$

Here we take

$$V_k = T_k - \sum_{r=0}^{n} T_r A_{-r} W_k \quad (k = 0, 1, \dots, n).$$

These operators satisfy the equation

$$\sum_{j=0}^{n} V_j A_{k-j} = \delta_{k1} I \quad (k = 0, 1, \dots, n).$$

For an arbitrary vector $\varphi = \{\varphi_j\}_{j=0}^n \in \operatorname{Ker} A$ $(\neq 0)$, by $p = p(\varphi)$ denote the number such that $\varphi_1 = \varphi_2 = \cdots = \varphi_p = 0$ and $\varphi_{p+1} \neq 0$. It is obvious that $p(\varphi) \leq n - 2$ for every $\varphi \in \operatorname{Ker} A$ $(\neq 0)$.

Let $p^* = \max p(\varphi)$ $(\varphi \in \operatorname{Ker} A)$ and let $h (\neq 0)$ be a vector in $\operatorname{Ker} A$ for which $p(h) = p^*$. Introduce the vector $h^+ = \{h_j^+\}_{j=0}^n$ by setting

$$h_j^+ = \begin{cases} h_{j-1} & \text{if } \ j = 1, 2, \dots, n, \\ 0 & \text{if } \ j = 0. \end{cases}$$

From the equalities

$$\sum_{k=0}^{n} A_{j-k} h_k^+ = \sum_{k=1}^{n} A_{j-k} h_{k-1} = \sum_{k=0}^{n} A_{j-1-k} h_k \quad (j = 1, 2, \dots, n),$$

$$\sum_{j=0}^{n} W_{n-j} \sum_{k=0}^{n} A_{j-k} h_k^+ = W_n \sum_{k=0}^{n} A_{-k} h_k^+ + \sum_{j=1}^{n} W_{n-j} \sum_{k=0}^{n} A_{j-k} h_k^+$$

and the invertibility of the operator W_n it follows that $h^+ \in \operatorname{Ker} A$. Since $h^+ \neq 0$ and $p(h^+) > p^*$, this leads to the conclusion that $\operatorname{Ker} A = \{0\}$.

To finish the proof, consider the operator B defined in the space $\mathbb{V}_n^*$ by the matrix $\|A_{j-k}^*\|_{j,k=0}^n$ consisting of the adjoint operators to the operators A_{j-k}. Passing to the adjoint operators in equalities (4.1)–(4.4), we obtain that the operator B satisfies all the hypotheses of Lemma 4.1. Hence, in view of what has been proved above, $\operatorname{Ker} B = \{0\}$. Obviously, $B' = A^*$, where B' is the matrix transposed to B. Moreover, $J_n B J_n = B'$, where $J_n = \|\delta_{j+k,n} I\|_{j,k=0}^n$. Thus $\dim \operatorname{Ker} A^* = \dim \operatorname{Ker} B' = 0$. Since $\dim \operatorname{Ker} A^* = \dim \mathbb{V}_n / A\mathbb{V}_n$, we see that the operator A is invertible.

The lemma is proved. □

Note that, in fact, we have proved the following statement simultaneously with Lemma 4.1.

Lemma 4.2. *Let $A = \|A_{j-k}\|_{j,k=0}^n$ be a Toeplitz matrix with elements A_j $(j = 0, \pm 1, \dots, \pm n)$ in $L(\mathbb{V})$.*

If for some operator $A_{n+1} \in L(\mathbb{V})$ the systems of equations (4.8), (4.9), *and the systems*

$$\sum_{k=0}^{n} A_{j-k} U_k = \delta_{j1} I \quad (j = 0, 1, \dots, n),$$

$$\sum_{k=0}^{n} V_k A_{j-k} = \delta_{j1} I \quad (j = 0, 1, \dots, n)$$

are solvable in $L(\mathbb{V})$, and the operator X_n $(= W_n)$ is invertible, then the matrix A is invertible.

One can make a remark about Lemmas 4.1 and 4.2 similar to the remark following Lemma 1.1.

Proof of Theorem 4.1. From Lemma 4.1 it follows that the matrix A is invertible. This statement is proved in the same way as in the proof of Theorem 1.1. Let us prove formula (4.7). In view of (4.1) and (4.2) the equality

$$\left\| \begin{matrix} w_n & w_{n-1} & \dots & w_0 \\ 0 & e & \dots & 0 \\ \vdots & \vdots & \ddots & \vdots \\ 0 & 0 & \dots & e \end{matrix} \right\| A \left\| \begin{matrix} e & \dots & 0 & x_0 \\ \vdots & \ddots & \vdots & \vdots \\ 0 & \dots & e & x_{n-1} \\ 0 & \dots & 0 & x_n \end{matrix} \right\| = \left\| \begin{matrix} 0 & x_n \\ \widehat{A} & 0 \end{matrix} \right\| \tag{4.12}$$

holds, where $\widehat{A} = \|a_{j-k+1}\|_{j,k=0}^{n-1}$. From here it follows that the matrix $\widehat{A}$ is invertible and

$$A^{-1} = \left\| \begin{matrix} e & \dots & 0 & x_0 \\ \vdots & \ddots & \vdots & \vdots \\ 0 & \dots & e & x_{n-1} \\ 0 & \dots & 0 & x_n \end{matrix} \right\| \left\| \begin{matrix} 0 & \widehat{A}^{-1} \\ x_n^{-1} & 0 \end{matrix} \right\| \left\| \begin{matrix} w_n & w_{n-1} & \dots & w_0 \\ 0 & e & \dots & 0 \\ \vdots & \vdots & \ddots & \vdots \\ 0 & 0 & \dots & e \end{matrix} \right\|.$$

Putting $A^{-1} = \|c_{jk}\|_{j,k=0}^{n}$ and $\widehat{A}^{-1} = \|\widehat{c}_{jk}\|_{j,k=0}^{n-1}$, we get

$$\begin{aligned} c_{jk} &= \widehat{c}_{j,k-1} + x_j x_n^{-1} w_{n-k} \quad (j = 0,1,\dots,n-1,\ k = 1,\dots,n), \\ c_{j0} &= x_j, \quad c_{nk} = w_{n-k} \quad (j,k = 0,1,\dots,n). \end{aligned} \tag{4.13}$$

Relations (4.1)–(4.4) imply the equalities

$$\begin{aligned} &\left\| \begin{matrix} w_n & w_{n-1} & \dots & w_0 \\ 0 & e & \dots & 0 \\ \vdots & \vdots & \ddots & \vdots \\ 0 & 0 & \dots & e \end{matrix} \right\| \widetilde{A} \left\| \begin{matrix} s_0 & 0 & \dots & 0 \\ s_1 & e & \dots & 0 \\ \vdots & \vdots & \ddots & \vdots \\ s_n & 0 & \dots & e \end{matrix} \right\| = \left\| \begin{matrix} x_n & 0 \\ 0 & \widehat{A} \end{matrix} \right\|, \\ &\left\| \begin{matrix} e & \dots & 0 & 0 \\ \vdots & \ddots & \vdots & \vdots \\ 0 & \dots & e & 0 \\ t_n & \dots & t_1 & t_0 \end{matrix} \right\| \widetilde{A} \left\| \begin{matrix} e & \dots & 0 & x_0 \\ \vdots & \ddots & \vdots & \vdots \\ 0 & \dots & e & x_{n-1} \\ 0 & \dots & 0 & x_n \end{matrix} \right\| = \left\| \begin{matrix} \widehat{A} & 0 \\ 0 & x_n \end{matrix} \right\|, \end{aligned} \tag{4.14}$$

where $\widetilde{A} = \|a_{j-k+1}\|_{j,k=0}^{n}$. Since the elements x_n and w_n are invertible, from the first equality it follows that the matrix $\widetilde{A}$ is right-invertible. The second equality yields the left-invertibility of the matrix $\widetilde{A}$. Thus, the matrix $\widetilde{A}$ is invertible. Also, (4.14) implies the invertibility of the elements s_0, t_0 and the equality

$$\begin{aligned} A^{-1} &= \left\| \begin{matrix} s_0 & 0 & \dots & 0 \\ s_1 & e & \dots & 0 \\ \vdots & \vdots & \ddots & \vdots \\ s_n & 0 & \dots & e \end{matrix} \right\| \left\| \begin{matrix} x_n^{-1} & 0 \\ 0 & \widehat{A}^{-1} \end{matrix} \right\| \left\| \begin{matrix} w_n & w_{n-1} & \dots & w_0 \\ 0 & e & \dots & 0 \\ \vdots & \vdots & \ddots & \vdots \\ 0 & 0 & \dots & e \end{matrix} \right\| \\ &= \left\| \begin{matrix} e & \dots & 0 & x_0 \\ \vdots & \ddots & \vdots & \vdots \\ 0 & \dots & e & x_{n-1} \\ 0 & \dots & 0 & x_n \end{matrix} \right\| \left\| \begin{matrix} \widehat{A}^{-1} & 0 \\ 0 & x_n^{-1} \end{matrix} \right\| \left\| \begin{matrix} e & \dots & 0 & 0 \\ \vdots & \ddots & \vdots & \vdots \\ 0 & \dots & e & 0 \\ t_n & \dots & t_1 & t_0 \end{matrix} \right\|. \end{aligned} \tag{4.15}$$

Thus, for the entries $\widetilde{c}_{jk}$ $(j,k = 0,1,\dots,n)$ of the matrix $\widetilde{A}^{-1}$ the equalities

$$\begin{aligned} \widetilde{c}_{jk} &= \widehat{c}_{j-1,k-1} + s_j x_n^{-1} w_{n-k} = \widehat{c}_{jk} + x_j x_n^{-1} t_{n-k} && (j,k = 1,2,\dots,n), \\ \widetilde{c}_{0k} &= s_0 x_n^{-1} w_{n-k} = \widehat{c}_{0k} + x_0 x_n^{-1} t_{n-k} && (k = 0,1,\dots,n), \\ \widetilde{c}_{j0} &= s_j x_n^{-1} w_n = \widehat{c}_{j0} + x_j x_n^{-1} t_n && (j = 0,1,\dots,n) \end{aligned}$$

hold. Therefore,

$$\widehat{c}_{jk} = \widehat{c}_{j-1,k-1} + s_j x_n^{-1} w_{n-k} - x_j x_n^{-1} t_{n-k} \quad (j,k = 1,2,\dots,n), \tag{4.16}$$

$$\widehat{c}_{0k} = s_0 x_n^{-1} w_{n-k} - x_0 x_n^{-1} t_{n-k} \quad (k = 0,1,\dots,n), \tag{4.17}$$

$$\widehat{c}_{j0} = s_j x_n^{-1} w_n - x_j x_n^{-1} t_n \quad (j = 0,1,\dots,n). \tag{4.18}$$

Equalities (4.16)–(4.18) immediately imply the formula

$$\widehat{c}_{jk} = \sum_{r=0}^{n-1} (s_{j-r} x_n^{-1} w_{n-k+r} - x_{j-r} x_n^{-1} t_{n-k+r}). \tag{4.19}$$

Now taking into account (4.13), we obtain

$$c_{jk} = \sum_{r=0}^{n} (s_{j-k} x_n^{-1} w_{n+1-k+r} - x_{j-r} x_n^{-1} t_{n+1-k+r}) + x_j x_n^{-1} w_{n-k} \ (j,k = 0,1,\dots,n).$$

It is easy to see that this formula coincides with formula (4.7).

The theorem is proved. □

The following theorem is analogous to Theorem 1.2.

Theorem 4.2. *Let the hypotheses of Theorem* 4.1 *be fulfilled. Then the matrix* $\widehat{A} = \|a_{j-k+1}\|_{j,k=0}^{n-1}$ *is invertible and its inverse is given by the formula*

$$\begin{aligned}\widehat{A}^{-1} &= \left\| \begin{matrix} s_0 & 0 & \dots & 0 \\ s_1 & s_0 & \dots & 0 \\ \vdots & \vdots & \ddots & \vdots \\ s_{n-1} & s_{n-2} & \dots & s_0 \end{matrix} \right\| x_n^{-1} \left\| \begin{matrix} w_n & w_{n-1} & \dots & w_1 \\ 0 & w_n & \dots & w_2 \\ \vdots & \vdots & \ddots & \vdots \\ 0 & 0 & \dots & w_n \end{matrix} \right\| \\ &= \left\| \begin{matrix} x_0 & 0 & \dots & 0 \\ x_1 & x_0 & \dots & 0 \\ \vdots & \vdots & \ddots & \vdots \\ x_{n-1} & x_{n-2} & \dots & x_0 \end{matrix} \right\| x_n^{-1} \left\| \begin{matrix} t_n & t_{n-1} & \dots & t_1 \\ 0 & t_n & \dots & t_2 \\ \vdots & \vdots & \ddots & \vdots \\ 0 & 0 & \dots & t_n \end{matrix} \right\|. \end{aligned} \tag{4.20}$$

Proof. The invertibility of the matrix $\widehat{A}$ follows from equality (4.12) and the invertibility of the element x_n. Formula (4.20) follows immediately from formula (4.19). The theorem is proved. □

Corollary 4.1. *Let the hypotheses of Theorem* 4.1 *be fulfilled. Then the inverse of* $\widehat{A}$ *is calculated by the formula*

$$\begin{aligned}\widehat{A}^{-1} = &\left\| \begin{matrix} x_n & x_{n-1} & \dots & x_1 \\ 0 & x_n & \dots & x_2 \\ \vdots & \vdots & \ddots & \vdots \\ 0 & 0 & \dots & x_n \end{matrix} \right\| x_n^{-1} \left\| \begin{matrix} t_0 & 0 & \dots & 0 \\ t_1 & t_0 & \dots & 0 \\ \vdots & \vdots & \ddots & \vdots \\ t_{n-1} & t_{n-2} & \dots & t_0 \end{matrix} \right\| \\ &- \left\| \begin{matrix} s_n & s_{n-1} & \dots & s_1 \\ 0 & s_n & \dots & s_2 \\ \vdots & \vdots & \ddots & \vdots \\ 0 & 0 & \dots & s_n \end{matrix} \right\| x_n^{-1} \left\| \begin{matrix} w_0 & 0 & \dots & 0 \\ w_1 & w_0 & \dots & 0 \\ \vdots & \vdots & \ddots & \vdots \\ w_{n-1} & w_{n-2} & \dots & w_0 \end{matrix} \right\|. \end{aligned} \tag{4.21}$$

Proof. Indeed, analogously to eqns. (4.17) and (4.18) from (4.15) it follows that

$$\widehat{c}_{n-1,k-1} = t_{n-k} - s_n x_n^{-1} w_{n-k}, \quad \widehat{c}_{j-1,n-1} = x_j x_n^{-1} t_n - s_j x_n^{-1} w_0.$$

These equalities and equality (4.16) immediately imply (4.21). □

Corollary 4.1 and equality (4.13) imply the following.

Corollary 4.2. *Let the hypotheses of Theorem* 4.1 *be fulfilled. Then the inverse of* A *is given by the formula*

$$A^{-1} = \|x_j x_n^{-1} w_{n-k}\|_{j,k=0}^n + \left\| \begin{matrix} 0 & x_n & \dots & x_1 \\ \vdots & \vdots & \ddots & \vdots \\ 0 & 0 & \dots & x_n \\ 0 & 0 & \dots & 0 \end{matrix} \right\| x_n^{-1} \left\| \begin{matrix} t_0 & 0 & \dots & 0 \\ t_1 & t_0 & \dots & 0 \\ \vdots & \vdots & \ddots & \vdots \\ t_n & t_{n-1} & \dots & t_0 \end{matrix} \right\| - \left\| \begin{matrix} 0 & s_n & \dots & s_1 \\ \vdots & \vdots & \ddots & \vdots \\ 0 & 0 & \dots & s_n \\ 0 & 0 & \dots & 0 \end{matrix} \right\| x_n^{-1} \left\| \begin{matrix} w_0 & 0 & \dots & 0 \\ w_1 & w_0 & \dots & 0 \\ \vdots & \vdots & \ddots & \vdots \\ w_n & w_{n-1} & \dots & w_0 \end{matrix} \right\| .$$

Theorem 4.3. *If for a matrix* $A = \|a_{j-k}\|_{j,k=0}^n$ $(a_j \in \mathfrak{A}, j = 0, \pm 1, \dots, \pm n)$ *equations* (4.1), (4.2), (4.5), *and* (4.6) *are solvable and the element* x_n *is invertible, then the matrix* A *is invertible and the equality*

$$\begin{aligned} A^{-1} &= \|x_j x_n^{-1} w_{n-k}\|_{j,k=0}^n \\ &+ \left\| \begin{matrix} u_0 & 0 & \dots & 0 \\ u_1 & u_0 & \dots & 0 \\ \vdots & \vdots & \ddots & \vdots \\ u_n & u_{n-1} & \dots & u_0 \end{matrix} \right\| x_n^{-1} \left\| \begin{matrix} 0 & w_n & \dots & w_1 \\ \vdots & \vdots & \ddots & \vdots \\ 0 & 0 & \dots & w_n \\ 0 & 0 & \dots & 0 \end{matrix} \right\| \\ &- \left\| \begin{matrix} x_0 & 0 & \dots & 0 \\ x_1 & x_0 & \dots & 0 \\ \vdots & \vdots & \ddots & \vdots \\ x_n & x_{n-1} & \dots & x_0 \end{matrix} \right\| x_n^{-1} \left\| \begin{matrix} 0 & v_n & \dots & v_1 \\ \vdots & \vdots & \ddots & \vdots \\ 0 & 0 & \dots & v_n \\ 0 & 0 & \dots & 0 \end{matrix} \right\| \\ &+ \left\| \begin{matrix} x_0 & 0 & \dots & 0 \\ x_1 & x_0 & \dots & 0 \\ \vdots & \vdots & \ddots & \vdots \\ x_n & x_{n-1} & \dots & x_0 \end{matrix} \right\| x_n^{-1}(v_n - u_n)x_n^{-1} \left\| \begin{matrix} 0 & w_n & \dots & w_1 \\ \vdots & \vdots & \ddots & \vdots \\ 0 & 0 & \dots & w_n \\ 0 & 0 & \dots & 0 \end{matrix} \right\| \end{aligned} \tag{4.22}$$

holds.

This theorem generalizes one statement from [2] proved in the case $\mathfrak{A} = \mathbb{C}$. In that case the last summand on the right-hand side of equality (4.22) is equal to zero. Note that, probably, Theorems 4.1 and 4.2 are new even in the case $\mathfrak{A} = \mathbb{C}$.

Proof. The invertibility of the operator A follows from Lemma 4.2.

It is easy to see that

$$\widehat{x}_j = s_j - x_j x_n^{-1} s_n = u_j - x_j x_n^{-1} u_n \tag{4.23}$$

and

$$\widehat{w}_j = t_j - s_n x_n^{-1} w_j = v_j - v_n x_n^{-1} w_j \quad (j = 0, 1, \ldots, n),$$

where $\widehat{x}_j$ and $\widehat{w}_j$ are solutions of the equations

$$\sum_{k=0}^{n-1} a_{j-k}\widehat{x}_k = \delta_{j0}e \quad (j = 1, 2, \ldots, n), \tag{4.24}$$

$$\sum_{k=0}^{n-1} \widehat{w}_k a_{j-k} = \delta_{j0}e \quad (j = 1, 2, \ldots, n). \tag{4.25}$$

From Theorem 4.1 and formulas (4.23) it follows that

$$\begin{aligned} c_{jk} &= \sum_{r=0}^{n} (s_{j-r}x_n^{-1}w_{n+1-k+r} - x_{j-r}x_n^{-1}t_{n+1-k+r}) + x_j x_n^{-1} w_{n-k} \\ &= \sum_{r=0}^{n} \Big[\big(u_{j-r} - x_{j-r}x_n^{-1}u_n + x_{j-r}x_n^{-1}s_n\big)x_n^{-1}w_{n+1-k+r} \\ &\qquad - x_{j-r}x_n^{-1}\big(v_{n+1-k+r} - s_n x_n^{-1} w_{n+1-k+r} + v_n x_n^{-1} w_{n+1-k+r}\big)\Big] \\ &\quad + x_j x_n^{-1} w_{n-k}. \end{aligned}$$

Hence,

$$\begin{aligned} c_{jk} = \sum_{r=0}^{n} \big(&u_{j-r}x_n^{-1}w_{n+1-k+r} - x_{j-r}x_n^{-1}v_{n+1-k+r} \\ &+ x_{j-r}x_n^{-1}(v_n - u_n)x_n^{-1}w_{n+1-k+r}\big) + x_j x_n^{-1} w_{n-k}. \end{aligned}$$

This formula coincides with (4.22).

The theorem is proved. □

The following statement is analogous to Theorem 4.2, and its proof is similar to the proof of the above theorem.

Corollary 4.3. *Let the hypotheses of Theorem 4.1 be fulfilled. Then the matrix* $\widehat{A} = \|a_{j-k+1}\|_{j,k=0}^{n-1}$ *is invertible and its inverse is constructed by the formula*

$$\begin{aligned} \widehat{A}^{-1} = &\left\| \begin{matrix} u_0 & 0 & \ldots & 0 \\ u_1 & u_0 & \ldots & 0 \\ \vdots & \vdots & \ddots & \vdots \\ u_{n-1} & u_{n-2} & \ldots & u_0 \end{matrix} \right\| x_n^{-1} \left\| \begin{matrix} w_n & w_{n-1} & \ldots & w_1 \\ 0 & w_n & \ldots & w_2 \\ \vdots & \vdots & \ddots & \vdots \\ 0 & 0 & \ldots & w_n \end{matrix} \right\| \\ &- \left\| \begin{matrix} x_0 & 0 & \ldots & 0 \\ x_1 & x_0 & \ldots & 0 \\ \vdots & \vdots & \ddots & \vdots \\ x_{n-1} & x_{n-2} & \ldots & x_0 \end{matrix} \right\| x_n^{-1} \left\| \begin{matrix} v_n & v_{n-1} & \ldots & v_1 \\ 0 & v_n & \ldots & v_2 \\ \vdots & \vdots & \ddots & \vdots \\ 0 & 0 & \ldots & v_n \end{matrix} \right\| \end{aligned}$$

$$+\left\|\begin{array}{cccc} x_0 & 0 & \dots & 0 \\ x_1 & x_0 & \dots & 0 \\ \vdots & \vdots & \ddots & \vdots \\ x_{n-1} & x_{n-2} & \dots & x_0 \end{array}\right\| x_n^{-1}(v_n - u_n)x_n^{-1} \left\|\begin{array}{cccc} w_n & w_{n-1} & \dots & w_1 \\ 0 & w_n & \dots & w_2 \\ \vdots & \vdots & \ddots & \vdots \\ 0 & 0 & \dots & w_n \end{array}\right\|.$$

We mention also the following statement.

Corollary 4.4. *Let the hypotheses of Theorem* 4.1 *be fulfilled. Then the matrix* A *is invertible and its inverse is constructed by the formula*

$$\begin{aligned} A^{-1} = {} & \|x_j x_n^{-1} w_{n-k}\|_{j,k=0}^n \\ & + \left\|\begin{array}{cccccc} \widehat{x}_0 & 0 & \dots & 0 & 0 \\ \widehat{x}_1 & \widehat{x}_0 & \dots & 0 & 0 \\ \vdots & \vdots & \ddots & \vdots & \vdots \\ \widehat{x}_{n-1} & \widehat{x}_{n-2} & \dots & \widehat{x}_0 & 0 \\ 0 & \widehat{x}_{n-1} & \dots & \widehat{x}_1 & \widehat{x}_0 \end{array}\right\| x_n^{-1} \left\|\begin{array}{cccccc} 0 & w_n & \dots & w_2 & w_1 \\ 0 & 0 & \dots & w_3 & w_2 \\ \vdots & \vdots & \ddots & \vdots & \vdots \\ 0 & 0 & \dots & 0 & w_n \\ 0 & 0 & \dots & 0 & 0 \end{array}\right\| \\ & - \left\|\begin{array}{cccccc} x_0 & 0 & \dots & 0 & 0 \\ x_1 & x_0 & \dots & 0 & 0 \\ \vdots & \vdots & \ddots & \vdots & \vdots \\ x_{n-1} & x_{n-2} & \dots & x_0 & 0 \\ x_n & x_{n-1} & \dots & x_1 & x_0 \end{array}\right\| x_n^{-1} \left\|\begin{array}{cccccc} 0 & 0 & \widehat{w}_{n-1} & \dots & \widehat{w}_1 \\ \vdots & \vdots & \vdots & \ddots & \vdots \\ 0 & 0 & 0 & \dots & \widehat{w}_{n-1} \\ 0 & 0 & 0 & \dots & 0 \\ 0 & 0 & 0 & \dots & 0 \end{array}\right\|, \end{aligned}$$

where $\widehat{x}_j$ *and* $\widehat{w}_j$ $(j = 0, 1, \dots, n-1)$ *are solutions of equations* (4.24) *and* (4.25), *respectively.*

This statement is proved in the same way as Theorem 4.3 with the aid of equalities (4.23).

Note that for all presented propositions one can formulate dual statements. They can be obtained by passing to the transposed matrices with the aid of the transformation $J_n A J_n$.

Corollary 4.5. *Let the hypotheses of Theorem* 4.1 *be fulfilled,* $A^{-1} = \|c_{jk}\|_{j,k=0}^n$ *and* $\widehat{A}^{-1} = \|\widehat{c}_{jk}\|_{j,k=0}^{n-1}$. *Put*

$$c(\zeta, \theta) = \sum_{j,k=0}^{n} c_{jk} \zeta^j \theta^{-k}, \quad \widehat{c}(\zeta, \theta) = \sum_{j,k=0}^{n-1} \widehat{c}_{jk} \zeta^j \theta^{-k},$$

where ζ *and* θ *are complex variables. Then*

$$\begin{aligned} c(\zeta,\theta) &= (\theta - \zeta)^{-1}\big(s(\zeta)x_n^{-1}w(\theta) - x(\zeta)x_n^{-1}t(\theta)\big)\theta^{-n} \\ &\quad + x(\zeta)x_n^{-1}w(\theta)\theta^{-n}, && (4.26) \\ \widehat{c}(\zeta,\theta) &= (1 - \zeta\theta^{-1})\big(s(\zeta)x_n^{-1}w(\theta) - x(\zeta)x_n^{-1}t(\theta)\big)\theta^{-n}, && (4.27) \end{aligned}$$

where

$$x(\zeta)=\sum_{j=0}^{n}x_j\zeta^j,\quad s(\zeta)=\sum_{j=0}^{n}s_j\zeta^j,\quad w(\theta)=\sum_{k=0}^{n}w_k\theta^k,\quad t(\theta)=\sum_{k=0}^{n}t_k\theta^k.$$

Proof. Indeed, the equality

$$(1-\zeta\theta^{-1})c(\zeta,\theta)=\sum_{j,k=0}^{n+1}(c_{jk}-c_{j-1,k-1})\zeta^j\theta^{-k}$$

holds, where we set $c_{jk}=0$ if one of the numbers j,k is negative or is greater than n. In view of formulas (4.13), (4.16)–(4.18), we obtain

$$\begin{aligned}(1-\zeta\theta^{-1})c(\zeta,\theta)=\sum_{j,k=0}^{n+1}\big(&x_jx_nw_{n-k}-x_{j-1}x_n^{-1}w_{n-k+1}+s_jx_n^{-1}w_{n-k+1}\\&+s_jx_n^{-1}w_{n-k+1}-x_jx_n^{-1}t_{n-k+1}\big)\zeta^j\theta^{-k}.\end{aligned}$$

This implies formula (4.26). Formula (4.27) is proved analogously. □

5. Properties of solutions of equations (4.1)–(4.4)

This section is similar to Section 2. In this section main properties of the solutions of equations (4.1)–(4.4) are obtained. Here the notations of Section 2 are used. Suppose that $A=\|a_{j-k}\|_{j,k=0}^{n}$ ($a_j\in\mathfrak{A}$, $j=0,\pm1,\dots,\pm n$) is a Toeplitz matrix and a_{n+1} is an element such that all equations (4.1)–(4.1) have solutions and, moreover, the element x_n ($=w_n$) is invertible.

Proposition 5.1. *For solutions x_j, w_j, s_j, t_j ($j=0,1,\dots,n$) of equations* (4.1)–(4.4), *the relations*

$$S^{n1}x_n^{-1}W^{n1}=X^{n1}x_n^{-1}T^{n1},\tag{5.1}$$

$$S_{0n}x_n^{-1}W_{0n}=X_{0n}x_n^{-1}T_{0n}\tag{5.2}$$

hold.

Proof. Indeed, since $c_{nk}=w_{n-k}$, where $A^{-1}=\|c_{jk}\|_{j,k=0}^{n}$, according to Theorem 4.1 we have

$$w_{n-k}=w_{n-k}+\sum_{r=0}^{k-1}\big(s_{n-r}x_n^{-1}w_{n+1-k+r}-x_{n-r}x_n^{-1}t_{n+1-k+r}\big).$$

From here it follows that

$$\|\ s_n\ \dots\ s_1\ \|x_n^{-1}W^{n1}=\|\ x_n\ \dots\ x_1\ \|x_n^{-1}T^{n1}.$$

Taking into account that a product of two upper triangular Toeplitz matrices is again an upper triangular Toeplitz matrix and that two such matrices coincide if their first columns coincide, we see that the last equality implies equality (5.1).

Let us prove equality (5.2). From equality (4.15) it is easy to derive the relation

$$s_0 x_n^{-1} w_0 = x_0 x_n^{-1} t_0. \tag{5.3}$$

From Theorem 4.1 it follows that

$$c_{jn} = \sum_{r=0}^{j} (s_{j-r} x_n^{-1} w_{1+r} - x_{j-r} x_n^{-1} t_{1+r}) + x_j x_n^{-1} w_0.$$

According to Corollary 4.2, we have

$$\begin{aligned} c_{jn} &= x_{j+1} x_n^{-1} t_0 - s_{j+1} x_n^{-1} w_0 + x_j x_n^{-1} w_0 \quad (j = 0, 1, \dots, n-1), \\ c_{nn} &= w_0. \end{aligned}$$

Comparing the last two equalities, we get

$$\left\| \begin{matrix} s_1 & s_0 & \dots & 0 \\ \vdots & \ddots & \ddots & \vdots \\ s_n & \dots & s_1 & s_0 \end{matrix} \right\| x_n^{-1} \left\| \begin{matrix} w_0 \\ \vdots \\ w_n \end{matrix} \right\| = \left\| \begin{matrix} x_1 & x_0 & \dots & 0 \\ \vdots & \ddots & \ddots & \vdots \\ x_n & \dots & x_1 & x_0 \end{matrix} \right\| x_n^{-1} \left\| \begin{matrix} t_0 \\ \vdots \\ t_n \end{matrix} \right\|.$$

From here and (5.3) it follows that

$$S_{0n} x_n^{-1} \left\| \begin{matrix} w_0 \\ \vdots \\ w_n \end{matrix} \right\| = X_{0n} x_n^{-1} \left\| \begin{matrix} t_0 \\ \vdots \\ t_n \end{matrix} \right\|.$$

The last equality immediately implies (5.2).

The proposition is proved. □

Multiplying equalities (5.1) and (5.2) from the left and from the right by a matrix J_n, in view of (2.5) we obtain the following equalities:

$$S_{n1} x_n^{-1} W_{n1} = X_{n1} x_n^{-1} T_{n1}, \tag{5.4}$$

$$S^{0n} x_n^{-1} W^{0n} = X^{0n} x_n^{-1} T^{0n}. \tag{5.5}$$

Notice also that the equalities

$$S^{nk} x_n^{-1} W^{nk} = X^{nk} x_n^{-1} T^{nk}, \quad S_{nk} x_n^{-1} W_{nk} = X_{nk} x_n^{-1} T_{nk}, \tag{5.6}$$

where $k = 1, 2, \dots, n$, and

$$S_{0k} x_n^{-1} W_{0k} = X_{0k} x_n^{-1} T_{0k}, \quad S^{0k} x_n^{-1} W^{0k} = X^{0k} x_n^{-1} T^{0k} \tag{5.7}$$

where $k = 0, 1, \dots, n$, are easily derived from equalities (5.1), (5.2), (5.4), and (5.5).

Proposition 5.2. *The block matrices*

$$K = \left\| \begin{matrix} S_{0,n-1} & X_{0,n-1} \\ S^{n1} & X^{n1} \end{matrix} \right\|, \quad L = \left\| \begin{matrix} T^{0,n-1} & T_{n1} \\ W^{0,n-1} & W_{n1} \end{matrix} \right\|$$

are invertible.

Proof. Indeed, from the obvious equality

$$K = \begin{Vmatrix} E & R_1 \\ 0 & E \end{Vmatrix} \begin{Vmatrix} R_2 & 0 \\ 0 & R_3 \end{Vmatrix} \cdot \begin{Vmatrix} E & 0 \\ R_4 & E \end{Vmatrix},$$

where $E = \|\delta_{jk} e\|_{j,k=0}^{n}$ and

$$R_1 = X_{0,n-1}(X^{n1})^{-1}, \quad R_2 = S_{0,n-1} - R_1 S^{n1},$$
$$R_3 = X^{n1}, \qquad\qquad R_4 = (X^{n1})^{-1} S^{n1},$$

it follows that the matrix K is invertible if and only if the matrix R_2 is invertible. By Proposition 5.1, the equalities

$$R_2 = S_{0,n-1} - X_{0,n-1}(X^{n1})^{-1} S^{n1} = S_{0,n-1} - X_{0,n-1} x_n^{-1} T^{n1} (W^{n1})^{-1} x_n$$

hold. Hence

$$R_2 = (S_{0,n-1} x_n^{-1} W^{n1} - X_{0,n-1} x_n^{-1} T^{n1})(W^{n1})^{-1} x_n.$$

Thus, according to Theorem 4.2,

$$R_2 = \widehat{A}^{-1} (W^{n1})^{-1} x_n. \tag{5.8}$$

This implies the invertibility of the matrix R_2, and of the matrix K as well.

The invertibility of the operator L is proved analogously. The statement is proved. □

Proposition 5.3. *Let*

$$\widetilde{K} = \begin{Vmatrix} s_0 & 0 & \dots & 0 & x_0 & 0 & \dots & 0 \\ s_1 & s_0 & \dots & 0 & x_1 & x_0 & \dots & 0 \\ \vdots & \vdots & \ddots & \vdots & \vdots & \vdots & \ddots & \vdots \\ s_n & s_{n-1} & \dots & s_0 & x_n & x_{n-1} & \dots & x_0 \\ 0 & s_n & \dots & s_1 & 0 & x_n & \dots & x_1 \\ \vdots & \vdots & \ddots & \vdots & \vdots & \vdots & \ddots & \vdots \\ 0 & 0 & \dots & s_n & 0 & 0 & \dots & x_n \end{Vmatrix}.$$

Then the homogeneous equation

$$\widetilde{K}\chi = 0 \tag{5.9}$$

has a unique solution $\chi = \{\chi_0, \dots, \chi_n, \chi'_0, \dots, \chi'_n\}$ *with the property* $\chi_n = e$. *Moreover, for this solution the equalities*

$$w_k = x_n \chi_k, \quad t_k = -x_n \chi'_k \quad (k = 0, 1, \dots, n) \tag{5.10}$$

hold.

Proof. Indeed, equalities (5.1) and (5.2) immediately imply equality (5.9) if the elements χ_k are given by formula (5.10). Let us show the uniqueness of this solution. Let χ be a solution of equation (5.9) with the property $\chi_n = e$. Then from (5.9) it follows that

$$K \left\| \begin{matrix} \chi_0 \\ \vdots \\ \chi_{n-1} \\ \chi'_0 \\ \vdots \\ \chi'_{n-1} \end{matrix} \right\| = \left\| \begin{matrix} 0 \\ \vdots \\ 0 \\ x_0 x_n^{-1} s_n - s_0 \\ \vdots \\ x_{n-1} x_n^{-1} s_n - s_{n-1} \end{matrix} \right\| \tag{5.11}$$

and $\chi'_n = -x_n^{-1} s_n$.

In view of Proposition 5.2, this implies the uniqueness of the solution with the given property.

The proposition is proved. □

The following statement is proved analogously.

Proposition 5.4. *For the matrix*

$$\widetilde{L} = \left\| \begin{matrix} t_0 & t_1 & \dots & t_n & 0 & \dots & 0 \\ 0 & t_0 & \dots & t_{n-1} & t_n & \dots & 0 \\ \vdots & \vdots & \ddots & \vdots & \vdots & \ddots & \vdots \\ 0 & 0 & \dots & t_0 & t_1 & \dots & t_n \\ w_0 & w_1 & \dots & w_n & 0 & \dots & 0 \\ 0 & w_0 & \dots & w_{n-1} & w_n & \dots & 0 \\ \vdots & \vdots & \ddots & \vdots & \vdots & \ddots & \vdots \\ 0 & 0 & \dots & w_0 & w_1 & \dots & w_n \end{matrix} \right\| ,$$

the homogeneous equation

$$\omega \widetilde{L} = 0 \tag{5.12}$$

has a unique solution $\omega = \{\omega_0, \dots, \omega_n, \omega'_0, \dots, \omega'_n\}$ *with the property* $\omega_n = e$. *Moreover, for this solution the equalities*

$$x_k = \omega_k w_n, \quad s_k = -\omega'_k w_n \quad (k = 0, 1, \dots, n) \tag{5.13}$$

hold.

The following statement is obtained by a straightforward verification.

Proposition 5.5. *Suppose equations (4.1)–(4.4) have solutions* x_j, w_j, s_j, *and* t_j $(j = 0, 1, \dots, n)$ *and the elements* x_0, w_0, *and* x_n *are invertible. Then the elements* x_j^{n-1}, w_j^{n-1}, s_j^{n-1}, *and* t_j^{n-1} $(j = 0, 1, \dots, n-1)$ *defined by the equalities*

$$\begin{aligned} x_j^{n-1} &= s_{j+1} - x_{j+1} x_0^{-1} s_0, & w_j^{n-1} &= t_{j+1} - t_0 w_0^{-1} w_{j+1}, \\ s_j^{n-1} &= s_j - x_j x_n^{-1} s_n, & t_j^{n-1} &= t_j - s_n x_n^{-1} w_j \end{aligned} \tag{5.14}$$

are solutions of equations (4.1)–(4.4) for the matrix A_{n-1}, *respectively.*

6. Inverse problem for equations (4.1)–(4.4)

In this section the problem of reconstruction of the matrix A from solutions of equations (4.1)–(4.4) is solved.

Theorem 6.1. *Let x_j, w_j, s_j, and t_j $(j = 0, 1, \dots, n)$ be given systems of elements in $\mathfrak{A}$ and the elements x_n and w_n be invertible. For the existence of an element $a_{n+1} \in \mathfrak{A}$ and a Toeplitz matrix $A = \|a_{j-k}\|_{j,k=0}^{n}$ with elements $a_j \in \mathfrak{A}$ $(j = 0, \pm 1, \dots, \pm n)$ such that x_j, w_j, s_j, and t_j are solutions of equations (4.1)–(4.4), it is necessary and sufficient that the following conditions be fulfilled:*

1) $x_n = w_n$;
2) $S^{n1} x_n^{-1} W^{n1} = X^{n1} x_n^{-1} T^{n1}$ *and* $S_{0n} x_n^{-1} W_{0n} = X_{0n} x_n^{-1} T_{0n}$;
3) *one of the matrices*

$$K = \begin{Vmatrix} S_{0,n-1} & X_{0,n-1} \\ S^{n1} & X^{n1} \end{Vmatrix}, \quad L = \begin{Vmatrix} T^{0,n-1} & T_{n1} \\ W^{0,n-1} & W_{n1} \end{Vmatrix}$$

is invertible;

4) *one of the elements s_0, t_0 is invertible.*

If conditions 1)–4) are fulfilled, then both matrices K and L and both elements s_0 and t_0 are invertible. Moreover, the matrix A and the element a_{n+1} are uniquely determined by the systems x_j, w_j, s_j, and t_j.

Proof. The necessity of the first condition was obtained in Section 3, the necessity of conditions 2) and 3) was proved in Section 4. The invertibility of the elements s_0 and t_0 was established in the proof of Theorem 4.1.

Let us show the sufficiency of the conditions 1)–4). Assume for definiteness that the matrix K is invertible. Put

$$\widehat{C} = \|\widehat{c}_{jk}\|_{j,k=0}^{n-1} = S_{0,n-1} x_n^{-1} W^{n1} - X_{0,n-1} x_n^{-1} T^{n1}.$$

The matrix $\widehat{C}$ is invertible. Indeed, it is easy to show that the invertibility of the matrix K implies the invertibility of the matrix $D = S_{0,n-1} - X_{0,n-1}(X^{n1})^{-1} S^{n1}$. From condition 2) it follows that

$$D = S_{0,n-1} - X_{0,n-1} x_n^{-1} T^{n1} (W^{n1})^{-1} x_n,$$

whence,

$$D = \widehat{C} (W^{n1})^{-1} x_n.$$

Thus the matrix $\widehat{C}$ is invertible. Put $\widehat{A} = \|a_{jk}\|_{j,k=0}^{n-1}$.

Consider the matrix $\widetilde{C}$ defined by the equality

$$\widetilde{C} = \|\widetilde{c}_{jk}\|_{j,k=0}^{n} = \begin{Vmatrix} s_0 & 0 & \dots & 0 \\ s_1 & e & \dots & 0 \\ \vdots & \vdots & \ddots & \vdots \\ s_n & 0 & \dots & e \end{Vmatrix} \begin{Vmatrix} x_n^{-1} & 0 \\ 0 & \widehat{C} \end{Vmatrix} \begin{Vmatrix} w_n & w_{n-1} & \dots & w_0 \\ 0 & e & \dots & 0 \\ \vdots & \vdots & \ddots & \vdots \\ 0 & 0 & \dots & e \end{Vmatrix}.$$

Obviously,

$$\begin{aligned} \widetilde{c}_{jk} &= c_{j-1,k-1} + s_j x_n^{-1} w_{n-k} \quad (j,k = 1,2,\dots,n), \\ \widetilde{c}_{j0} &= s_j, \quad \widetilde{c}_{0k} = s_0 x_n^{-1} w_{n-k} \quad (j,k = 0,1,\dots,n). \end{aligned} \tag{6.1}$$

According to the definition of the elements we have

$$\widehat{c}_{jk} - \widehat{c}_{j-1,k-1} = s_j x_n^{-1} w_{n-k} - x_j x_n^{-1} t_{n-k} \quad (j,k = 1,2,\dots,n). \tag{6.2}$$

This equality together with (6.1) implies the equality

$$\widetilde{c}_{jk} = \widehat{c}_{jk} + x_j x_n^{-1} t_{n-k} \quad (j,k = 0,1,\dots,n-1). \tag{6.3}$$

From condition 2) it follows that

$$\sum_{r=0}^{j} s_{j-r} x_n^{-1} w_r = \sum_{r=0}^{j} x_{j-r} x_n^{-1} t_r \quad (j = 0,1,\dots,n).$$

This implies that

$$\sum_{r=0}^{j-1} (s_{j-1-r} x_n^{-1} w_{1+r} - x_{j-1-r} x_n^{-1} t_{1+r}) = x_j x_n^{-1} t_0 - s_j x_n^{-1} w_0.$$

Obviously, the left-hand side of this equality coincides with the element $\widehat{c}_{j-1,n-1}$. Therefore, in view of (6.1), the equality

$$\widetilde{c}_{jn} = x_j x_n^{-1} t_0 \tag{6.4}$$

holds. The equality

$$\widetilde{c}_{nk} = t_{n-k} \tag{6.5}$$

is proved analogously. From equalities (6.3)–(6.5) it follows that

$$\widetilde{C} = \left\| \begin{array}{cccc} e & \dots & 0 & x_0 \\ \vdots & \ddots & \vdots & \vdots \\ 0 & \dots & e & x_{n-1} \\ 0 & \dots & 0 & x_n \end{array} \right\| \left\| \begin{array}{cc} \widehat{C} & 0 \\ 0 & x_n^{-1} \end{array} \right\| \left\| \begin{array}{cccc} e & \dots & 0 & 0 \\ \vdots & \ddots & \vdots & \vdots \\ 0 & \dots & e & 0 \\ t_n & \dots & t_1 & t_0 \end{array} \right\|. \tag{6.6}$$

In view of condition 4), the matrix $\widetilde{C}$ is invertible. Put $\widetilde{A} = \|\widetilde{a}_{jk}\|_{j,k=0}^n = \widetilde{C}^{-1}$. From the equalities $\widetilde{c}_{j0} = s_j$ and $\widetilde{c}_{nk} = t_{n-k}$ it follows that the elements s_j and t_{n-k} are solutions of equations (4.1) and (4.2), respectively, for the matrix $\widetilde{A}$.

From (6.1) and (6.6) it follows that

$$\left\| \begin{array}{cccc} w_n & w_{n-1} & \dots & w_0 \\ 0 & e & \dots & 0 \\ \vdots & \vdots & \ddots & \vdots \\ 0 & 0 & \dots & e \end{array} \right\| \widetilde{A} \left\| \begin{array}{cccc} s_0 & 0 & \dots & 0 \\ s_1 & e & \dots & 0 \\ \vdots & \vdots & \ddots & \vdots \\ s_n & 0 & \dots & e \end{array} \right\| = \left\| \begin{array}{cc} x_n & 0 \\ 0 & \widehat{A} \end{array} \right\| \tag{6.7}$$

and

$$\begin{Vmatrix} e & \dots & 0 & 0 \\ \vdots & \ddots & \vdots & \vdots \\ 0 & \dots & e & 0 \\ t_n & \dots & t_1 & t_0 \end{Vmatrix} \widetilde{A} \begin{Vmatrix} e & \dots & 0 & x_0 \\ \vdots & \ddots & \vdots & \vdots \\ 0 & \dots & e & x_{n-1} \\ 0 & \dots & 0 & x_n \end{Vmatrix} = \begin{Vmatrix} \widehat{A} & 0 \\ 0 & x_n \end{Vmatrix}. \tag{6.8}$$

From the first equality it follows that $\widetilde{a}_{jk} = \widehat{a}_{j-1,k-1}$ for $j,k = 1,2,\dots,n$, and from the second equality it follows that $\widetilde{a}_{jk} = \widehat{a}_{j-1,k-1}$ for $j,k = 0,1,\dots,n-1$. Therefore the matrix $\widetilde{A}$ is a Toeplitz matrix. Put $\widetilde{a}_j = \widetilde{a}_{j0}$ and $\widetilde{a}_{-j} = \widetilde{a}_{0j}$ ($j = 0,1,\dots,n$). Consider the matrix

$$C = \begin{Vmatrix} e & \dots & 0 & x_0 \\ \vdots & \ddots & \vdots & \vdots \\ 0 & \dots & e & x_{n-1} \\ 0 & \dots & 0 & x_n \end{Vmatrix} \begin{Vmatrix} 0 & \widehat{C} \\ x_n^{-1} & 0 \end{Vmatrix} \begin{Vmatrix} w_n & w_{n-1} & \dots & w_0 \\ 0 & e & \dots & 0 \\ \vdots & \vdots & \ddots & \vdots \\ 0 & 0 & \dots & e \end{Vmatrix} \tag{6.9}$$

and set $A = \|a_{jk}\|_{j,k=0}^n = C^{-1}$. From (6.7) it follows that

$$\begin{Vmatrix} w_n & w_{n-1} & \dots & w_0 \\ 0 & e & \dots & 0 \\ \vdots & \vdots & \ddots & \vdots \\ 0 & 0 & \dots & e \end{Vmatrix} A \begin{Vmatrix} e & \dots & 0 & x_0 \\ \vdots & \ddots & \vdots & \vdots \\ 0 & \dots & e & x_{n-1} \\ 0 & \dots & 0 & x_n \end{Vmatrix} = \begin{Vmatrix} 0 & x_n \\ \widehat{A} & 0 \end{Vmatrix}. \tag{6.10}$$

From this equality and equalities (6.7)–(6.8) the following relations can be easily derived:

$$A \begin{Vmatrix} e & \dots & 0 & x_0 \\ \vdots & \ddots & \vdots & \vdots \\ 0 & \dots & e & x_{n-1} \\ 0 & \dots & 0 & x_n \end{Vmatrix} = \widetilde{A} \begin{Vmatrix} 0 & \dots & 0 & s_0 \\ e & \dots & 0 & s_1 \\ \vdots & \ddots & \vdots & \vdots \\ 0 & \dots & e & s_n \end{Vmatrix} \tag{6.11}$$

and

$$\begin{Vmatrix} w_n & w_{n-1} & \dots & w_0 \\ 0 & e & \dots & 0 \\ \vdots & \vdots & \ddots & \vdots \\ 0 & 0 & \dots & e \end{Vmatrix} A = \begin{Vmatrix} t_n & \dots & t_1 & t_0 \\ e & \dots & 0 & 0 \\ \vdots & \ddots & \vdots & \vdots \\ 0 & \dots & e & 0 \end{Vmatrix} \widetilde{A}. \tag{6.12}$$

From equality (6.11) it follows that

$$a_{jk} = \widetilde{a}_{j,1+k} = \widetilde{a}_{j-1-k} \quad (j = 0,1,\dots,n;\ k = 0,1,\dots,n-1)$$

and from (6.12) in turn it follows that

$$a_{jk} = \widetilde{a}_{j-1,k} = \widetilde{a}_{j-1-k} \quad (j = 1,2,\dots,n;\ k = 0,1,\dots,n).$$

Therefore the matrix A is a Toeplitz matrix. From equality (6.10) it follows immediately that the elements x_j and w_j are solutions of equations (4.1) and (4.2) for the matrix A. Since the elements s_j and t_j are solutions of equations (4.1) and (4.2) for the matrix $\widetilde{A}$, they are solutions of equations (4.3) and (4.4) for the matrix A.

From Theorem 4.1 and equality (6.11) it follows that the matrices A and $\widetilde{A}$ are determined by the systems x_j, w_j, s_j, and t_j. This implies the uniqueness of the matrix A and the element a_{n+1}.

The theorem is proved. □

Theorem 6.2. *Let x_j and s_j $(j = 0, 1, \ldots, n)$ be given systems of elements in $\mathfrak{A}$ and the elements x_n and s_0 be invertible.*

For the existence of a Toeplitz matrix $A = \|a_{j-k}\|_{j,k=0}^n$ with elements in $\mathfrak{A}$ and an element $a_{n+1} \in \mathfrak{A}$ such that x_j and s_j are solutions of equations (4.1) *and* (4.3), *respectively, it is necessary and sufficient that the matrix*

$$K = \left\| \begin{array}{cc} S_{0,n-1} & X_{0,n-1} \\ S^{n1} & X^{n1} \end{array} \right\|$$

be invertible. If this condition is fulfilled, then the element a_{n+1} and the matrix A are uniquely determined, and solutions of equations (4.3) *and* (4.4) *are determined by equalities* (5.10).

Proof. The necessity of the hypotheses follows from Theorem 6.1. Let us prove the sufficiency. Assume that

$$\left\| \begin{array}{c} \chi_0 \\ \vdots \\ \chi_{n-1} \\ \chi_0' \\ \vdots \\ \chi_{n-1}' \end{array} \right\| = K^{-1} \left\| \begin{array}{c} 0 \\ \vdots \\ 0 \\ x_0 x_n^{-1} s_n - s_0 \\ \vdots \\ x_{n-1} x_n^{-1} s_n - s_{n-1} \end{array} \right\|. \tag{6.13}$$

Moreover, put

$$\chi_n = e, \quad \chi_n' = -x_n^{-1} s_n. \tag{6.14}$$

It is easy to see that from (6.13) and (6.14) it follows that

$$\widetilde{K} \chi = 0, \tag{6.15}$$

where $\chi = \{\chi_0, \ldots, \chi_n, \chi_0', \ldots, \chi_n'\}$. Put

$$w_k = x_n \chi_k, \quad t_k = -x_n \chi_k' \quad (k = 0, 1, \ldots, n).$$

In particular, $w_n = x_n$.

From (6.15) one can easily get the equalities

$$S_{0n} x_n^{-1} \left\| \begin{array}{c} w_0 \\ \vdots \\ w_n \end{array} \right\| = X_{0n} x_n^{-1} \left\| \begin{array}{c} t_0 \\ \vdots \\ t_n \end{array} \right\|$$

and

$$S_{n1}x_n^{-1}\left\|\begin{matrix} w_1 \\ \vdots \\ w_n \end{matrix}\right\| = X_{1n}x_n^{-1}\left\|\begin{matrix} t_1 \\ \vdots \\ t_n \end{matrix}\right\|.$$

This immediately implies condition 2) of Theorem 6.1. Thus all the conditions of Theorem 6.1 are fulfilled.

The theorem is proved. □

The following theorem is proved analogously.

Theorem 6.3. *Let w_j and t_j $(j = 0, 1, \ldots, n)$ be given systems of elements in $\mathfrak{A}$ and the elements w_n and t_0 be invertible.*

For the existence of a Toeplitz matrix $A = \|a_{j-k}\|_{j,k=0}^n$ with elements in $\mathfrak{A}$ and an element $a_{n+1} \in \mathfrak{A}$ such that w_j and t_j are solutions of equations (4.2) *and* (4.4), *respectively, it is necessary and sufficient that the matrix*

$$L = \left\|\begin{matrix} T^{0,n-1} & T_{n1} \\ W^{0,n-1} & W_{n1} \end{matrix}\right\|$$

be invertible. If this condition is fulfilled, then the element a_{n+1} and the matrix A are uniquely determined and the solutions of equations (4.1) *and* (4.3) *are given by equalities* (5.13).

References

[1] I.C. Gohberg and A.A. Semencul, *The inversion of finite Toeplitz matrices and their continual analogues.* Matem. Issled. **7** (1972), no. 2(24), 201–223 (in Russian). MR0353038 (50 #5524), Zbl 0288.15004.

[2] I.C. Gohberg and N.Ya. Krupnik, *A formula for the inversion of finite Toeplitz matrices.* Matem. Issled. **7** (1972), no. 2(24), 272–283 (in Russian). MR0353039 (50 #5525), Zbl 0288.15005.

[3] L.M. Kutikov, *The structure of matrices which are the inverse of the correlation matrices of random vector processes.* Zh. Vychisl. Matem. Matem. Fiz. **7** (1967), 764–773 (in Russian). English translation: U.S.S.R. Comput. Math. Math. Phys. **7** (1967), no. 4, 58–71. MR0217863 (36 #952), Zbl 0251.15023.

[4] I.I. Hirschman, Jr., *Matrix-valued Toeplitz operators.* Duke Math. J. **34** (1967), 403–415. MR0220002 (36 #3071), Zbl 0182.46203.

[5] M.A. Naimark, *Normed Rings,* (in Russian).

1st Russian edition: Gosudarstv. Izdat. Tehn.-Teor. Lit., Moscow, 1956. MR0090786 (19,870d), Zbl 0073.08902.

English translation of 1st Russian edition: P. Noordhoff N.V., Groningen, 1959. MR0110956 (22 #1824), Zbl 0089.10102.

German translation of 1st Russian edition: *Normierte Algebren.* VEB Deutscher Verlag der Wissenschaften, Berlin, 1959. Zbl 0089.10101.

English translation of 1st Russian edition, revised: P. Noordhoff N.V., Groningen, 1964. MR0205093 (34 #4928), Zbl 0137.31703. Reprinted by Wolters-Noordhoff Publishing, Groningen, 1970. MR0355601 (50 #8075), Zbl 0218.46042.

2nd Russian edition, revised: Nauka, Moscow, 1968. MR0355602 (50 #8076).

English translation of 2nd Russian edition: *Normed Algebras.* Wolters-Noordhoff Publishing, Groningen, 1972. MR0438123 (55 #11042).

German translation of 2nd Russian edition: *Normierte Algebren.* VEB Deutscher Verlag der Wissenschaften, Berlin, 1990. MR1034247 (90m:46004a), MR1038909 (90m:46004b), 0717.46039.

[6] L.M. Kutikov, *Inversion of correlation matrices.* Izv. Akad. Nauk SSSR Tehn. Kibernet. (1965), no. 5, 42–47 (in Russian). English translation: Engineering Cybernetics (1965), no. 5, 35–39. MR0203871 (34 #3718).

Operator Theory:
Advances and Applications, Vol. 206, 47–63

Matrix Integral Operators on a Finite Interval with Kernels Depending on the Difference of the Arguments

Israel Gohberg and Georg Heinig

By $L_p^n(0,\tau)$ $(1 \leq p \leq \infty,\ 0 < \tau < \infty)$ denote the Banach space of the vector functions $f = \{f_1, f_2, \dots, f_n\}$ with entries $f_j \in L_p(0,\tau)$ and the norm

$$\|f\|_{L_p^n} = \left(\sum_{j=1}^{n} \|f_j\|_{L_p}\right).$$

In this paper we consider integral operators of the form

$$((I-K)\varphi)(t) = \varphi(t) - \int_0^\tau k(t-s)\varphi(s)\,ds \quad (0 \leq t \leq \tau)$$

acting in the space $L_p^n(0,\tau)$, where $k(t) = \|k_{rj}(t)\|_{r,j=1}^n$ is a matrix function with entries in $L_1(-\tau,\tau)$.

The results of the paper [1] obtained there for $n = 1$ are generalized to the above operators. Theorems obtained below are continual analogues of the theorems appearing in the first three sections of the paper [2].

The paper consists of four sections. The first section has an auxiliary character. In the second section, a formula for $(I-K)^{-1}$ is constructed with the aid of solutions of the following four equations:

$$x(t) - \int_0^\tau k(t-s)x(s)\,ds = k(t), \tag{0.1}$$

$$w(t) - \int_0^\tau w(s)k(t-s)\,ds = k(t), \tag{0.2}$$

$$z(-t) - \int_0^\tau k(s-t)z(-s)\,ds = k(-t), \tag{0.3}$$

$$y(-t) - \int_0^\tau y(-s)k(s-t)\,ds = k(-t), \tag{0.4}$$

The paper was originally published as И.Ц. Гохберг, Г. Хайниг, О матричных интегральных операторах на конечном интервале с ядрами, зависящими от разности аргументов, Rev. Roumaine Math. Pures Appl. **20** (1975), 55–73.
MR0380495 (52 #1395), Zbl 0327.45009.

where $0 \le t \le \tau$. These equations are considered in the space $L_1^{n\times n}(0,\tau)$ of the matrix functions of order n with entries in $L_1(0,\tau)$.

In the third section main properties of solutions of equations (0.1)–(0.4) are studied. In the fourth section the problem of reconstructing a matrix function $k(t)$ from matrix functions $w(t)$, $x(t)$, $z(-t)$, and $y(-t)$ is considered.

1. Two lemmas

In this section two auxiliary statements are proved.

Lemma 1.1. *Let K be the operator defined in the space $L_p^n(0,\tau)$ $(1 \le p \le \infty)$ by the equality*

$$(K\varphi)(t) = \int_0^\tau k(t-s)\varphi(s)\,ds \quad (0 \le t \le \tau), \tag{1.1}$$

where $k(t) \in L_1^{n\times n}(-\tau,\tau)$. Then for every p the operator K is compact and the subspace $\mathrm{Ker}(I-K)$ *consists of absolutely continuous functions.*

Proof. It is easy to see that K is a bounded linear operator in the space $L_p^n(0,\tau)$ and

$$\|K\|_{L_p^n} \le \|k(t)\|_{L_1^{n\times n}(-\tau,\tau)}. \tag{1.2}$$

If the matrix function $k(t)$ has the form

$$k(t) = \sum_{j=-m}^{m} e^{i\pi jt/\tau} A_j, \tag{1.3}$$

where $A_j \in L(\mathbb{C}^n)$ [1], then, obviously, the corresponding operator K is of finite rank. It is known that the set of matrix functions of the form (1.3) is a dense set in the space $L_1^{n\times n}(-\tau,\tau)$. This fact and estimate (1.2) imply the compactness of the operator K. By $W^n(0,\tau)$ denote the space of all vector functions $f(t) = \{f_1(t),\dots,f_n(t)\}$ with absolutely continuous entries and the norm

$$\|f(t)\|_{W^n} = \|f(t)\|_{L_1^n} + \left\|\frac{d}{dt}f(t)\right\|_{L_1^n}.$$

Consider the restriction $\widehat{K}$ of the operator K to $W^n(0,\tau)$. The operator $\widehat{K}$ is a compact operator acting in the space $W^n(0,\tau)$. Indeed, for a vector $f \in W^n(0,\tau)$,

$$(\widehat{K}f)(t) = \int_{t-\tau}^{t} k(s)f(t-s)\,ds$$

and

$$\frac{d}{dt}(\widehat{K}f)(t) = \int_0^\tau k(t-s)\frac{d}{ds}f(s)\,ds + k(t)f(0) - k(t-\tau)f(\tau).$$

Therefore, the vector $\widehat{K}f$ belongs to $W^n(0,\tau)$ and

$$\|\widehat{K}f\|_{W^n} \le h\|k\|_{L_1^{n\times n}}\|f\|_{W^n},$$

[1] The algebra of all linear operators acting in the n-dimensional space $\mathbb{C}^n$ is denoted by $L(\mathbb{C}^n)$.

where h is a constant independent of the matrix $k(t)$ and the vector $f(t)$. The compactness of the operator $\widehat{K}$ is proved in the same way as the compactness of the operator K in the space $L_p^n(0,\tau)$.

Obviously, in the proof of the last statement of the lemma it is sufficient to confine oneself to the case $p=1$. Evidently, $\operatorname{Ker}(I-\widehat{K}) \subset \operatorname{Ker}(I-K)$, whence

$$\dim \operatorname{Ker}(I-\widehat{K}) \leq \dim \operatorname{Ker}(I-K).$$

Since the space $W^n(0,\tau)$ is dense in $L_1^n(0,\tau)$ and $\operatorname{Im}(I-\widehat{K}) \subset \operatorname{Im}(I-K)$, we have

$$\dim \operatorname{Coker}(I-\widehat{K}) \geq \dim \operatorname{Coker}(I-K).$$

Taking into account also that

$$\dim \operatorname{Ker}(I-\widehat{K}) = \dim \operatorname{Coker}(I-K)$$

and

$$\dim \operatorname{Ker}(I-\widehat{K}) = \dim \operatorname{Coker}(I-\widehat{K}),$$

we obtain

$$\dim \operatorname{Ker}(I-K) = \dim \operatorname{Ker}(I-\widehat{K}).$$

Hence $\operatorname{Ker}(I-K) = \operatorname{Ker}(I-\widehat{K}) \subset W^n(0,\tau)$. The lemma is proved. □

Lemma 1.2. *For a matrix function $k(t) \in L_1^{n\times n}(-\tau,\tau)$, the following statements are equivalent.*

1. *Equations* (0.1) *and* (0.3) *have solutions in the space $L_1^{n\times n}(0,\tau)$.*
2. *Equations* (0.2) *and* (0.4) *have solutions in the space $L_1^{n\times n}(0,\tau)$.*
3. *The operator $I-K$, where K is defined by equality* (1.1), *is invertible in every space $L_p^n(0,\tau)$ $(1 \leq p \leq \infty)$.*
4. *The operator $I-K$, where K is defined by equality* (1.1), *is invertible in some space $L_p^n(0,\tau)$ $(1 \leq p \leq \infty)$.*
5. *The operator $I-K$, where K is defined by equality* (1.1), *is invertible in every space $L_p^{n\times n}(0,\tau)$ $(1 \leq p \leq \infty)$ (or in some of them).*

Proof. According to Lemma 1.1, the subspace $\operatorname{Ker}(I-K)$ is the same in all spaces $L_p^n(0,\tau)$ $(1 \leq p \leq \infty)$. This immediately implies the equivalence of Statements 3 and 4.

The equivalence of Statements 3 and 5 is obvious.

Statement 2 implies Statement 3. Let $w(t)$ and $y(-t)$ be solutions of equations (0.2) and (0.4) in $L_1^{n\times n}(0,\tau)$. Assume that $f(t) \in \operatorname{Ker}(I-K)$. Then

$$\begin{aligned} 0 &= \int_0^\tau w(\tau-t)\left[\varphi(t) - \int_0^\tau k(t-s)\varphi(s)\,ds\right] dt \\ &= \int_0^\tau w(\tau-t)\varphi(t)\,dt - \int_0^\tau (w(\tau-t)-k(\tau-t))\varphi(t)\,dt = \varphi(\tau) \end{aligned} \tag{1.4}$$

and

$$\begin{aligned} 0 &= \int_0^\tau y(-t)\left[\varphi(t) - \int_0^\tau k(t-s)\varphi(s)\,ds\right]dt \\ &= \int_0^\tau y(-t)\varphi(t)\,dt - \int_0^\tau (y(-t)-k(t))\varphi(t)\,dt = \varphi(0). \end{aligned} \tag{1.5}$$

In view of Lemma 1.1, the function $\varphi(t)$ is absolutely continuous. Since

$$\begin{aligned} 0 &= \frac{d}{dt}\left(\varphi(t) - \int_0^\tau k(t-s)\varphi(s)\,ds\right) \\ &= \frac{d}{dt}\varphi(t) + k(t)\varphi(0) - k(t-\tau)\varphi(\tau) + \int_0^\tau k(t-s)\frac{d}{ds}\varphi(s)\,ds, \end{aligned}$$

we obtain according to (1.4) and (1.5) that

$$\frac{d}{dt}\varphi(t) - \int_0^\tau k(t-s)\frac{d}{ds}\varphi(s)\,ds = 0.$$

Therefore, $\frac{d}{dt}\varphi(t) \in \operatorname{Ker}(I-K)$.

From what has been proved above it also follows that

$$\frac{d^r}{dt^r}\varphi(t) \in \operatorname{Ker}(I-K), \quad \left(\frac{d^r}{dt^r}\varphi\right)(0) = \left(\frac{d^r}{dt^r}\varphi\right)(\tau) \quad \text{for} \quad r = 1, 2, \ldots.$$

Since the subspace $\operatorname{Ker}(I-K)$ is finite dimensional, we see that there exist numbers $m \in \mathbb{N}$ and $\alpha_j \in \mathbb{C}^1$ $(j = 0, 1, \ldots, m;\ \alpha_m \neq 0)$ such that

$$\sum_{j=0}^m \alpha_j \frac{d^j}{dt^j}\varphi(t) \equiv 0.$$

The last differential equation under the initial conditions

$$\left(\frac{d^r}{dt^r}\varphi\right)(0) = 0 \quad (r = 0, 1, \ldots, m-1)$$

has a unique solution, which is equal to zero. Thus $\operatorname{Ker}(I-K) = \{0\}$, whence the operator $I-K$ is invertible in all spaces $L_p^n(0,\tau)$ $(1 \le p \le \infty)$.

Let us show that Statement 3 implies Statement 2. Consider the operator K^* adjoint to the operator K. For $1 \le p < \infty$, the operator K^* has the form (1.1) and is determined by the matrix function $k^*(-t)$ adjoint to $k(-t)$. From Statement 3 it follows that the operator $I-K^*$ is invertible in the spaces $L_p^n(0,\tau)$ for $1 < p \le \infty$. In view of Lemma 1.1, from here it follows that this operator is invertible also in the space $L_1^n(0,\tau)$. The operator $I-K^*$ can be extended in a natural way to the space $L_1^{n\times n}(0,\tau)$ and it is invertible there. It is easily verified that the formulas

$$w(t) = [(I-K^*)^{-1}f]^*(\tau - t), \quad f(t) = k^*(\tau - t) \quad (0 \le t \le \tau) \tag{1.6}$$

and

$$y(-t) = [(I-K^*)^{-1}h]^*(t), \quad h(t) = k^*(-t) \quad (0 \le t \le \tau) \tag{1.7}$$

determine solutions of equations (0.2) and (0.4).

Analogously it is proved that Statement 3 implies Statement 1. It is easy to check that solutions $x(t)$ and $z(-t)$ of equations (0.1) and (0.3) are given by the formulas

$$x(t) = [(I-K)^{-1}k](t) \quad (0 \le t \le \tau) \tag{1.8}$$

and

$$z(-t) = [(I-K)^{-1}g](\tau - t), \quad g(t) = -k(t-\tau) \quad (0 \le t \le \tau). \tag{1.9}$$

It remains to show that Statement 1 implies Statement 3. Equations (0.1) and (0.3) can be rewritten as follows:

$$x^*(t) = -\int_0^\tau x^*(s)k^*(t-s)\,ds = k^*(t) \quad (0 \le t \le \tau) \tag{1.10}$$

and

$$z^*(-t) - \int_0^\tau z^*(-s)k^*(s-t)\,ds = k^*(-t) \quad (0 \le t \le \tau). \tag{1.11}$$

These equations can be considered as equations (0.4) and (0.2) for the function $k^*(-t)$. Thus, by what has been proved above, the solvability of the equations (1.10), (1.11) in $L_1^{n\times n}(0,\tau)$ implies the invertibility of the operator $I - K^*$, and whence it implies the invertibility of the operator $I - K$.

The lemma is proved. □

2. The inversion formula

2.1. The main result of this section is the following.

Theorem 2.1. *Let $k(t) \in L_1^{n\times n}(-\tau,\tau)$ and K be an operator*[2] *of the form* (1.1). *If the operator $I - K$ is invertible, then for its inverse the equality*

$$((I-K)^{-1}f)(t) = f(t) + \int_0^\tau \gamma(t,s)f(s)\,ds \quad (0 \le t \le \tau) \tag{2.1}$$

holds, where the kernel $\gamma(t,s)$ is determined from the solutions $x(t)$, $w(t)$, $z(-t)$, and $y(-t)$ of equations (0.1)–(0.4) *by the formula*

$$\begin{aligned}\gamma(t,s) &= x(t-s) + y(t-s)\\ &+ \int_0^{\min(t,s)} \big(x(t-r)y(r-s) - z(t-r-\tau)w(r-s+\tau)\big)\,dr.\end{aligned} \tag{2.2}$$

Here we follow the convention that $x(-t) = y(t) = 0$ for $t > 0$.

Proof. First, consider the case when the matrix function $k(t)$ has the form

$$k(t) = \sum_{j=-m}^{m} e^{i\pi jt/\tau} A_j, \tag{2.3}$$

[2] In what follows it is supposed that the operator K acts in one of the spaces $L_p^n(0,\tau)$ $(1 \le p \le \infty)$.

where $A_j \in L(\mathbb{C}^n)$. In this case the solution φ of the equation $(I-K)\varphi = f$ has the form

$$\varphi = f + \sum_{j=-m}^{m} e^{i\pi jt/\tau} A_j \xi_j, \quad \xi_j = \int_0^\tau e^{i\pi js/\tau} \varphi(s)\, ds.$$

The vectors ξ_j $(j = 0, \pm 1, \ldots, \pm m)$ are solutions of the system of equations

$$\xi_j = \sum_{k=-m}^{m} \alpha_{jk} A_k \xi_k = \int_0^\tau f(s) e^{-i\pi js/\tau}\, ds \quad (j = 0, \pm 1, \ldots, \pm m),$$

where $\alpha_{jk} = \tau((-1) - 1)/\pi i(j-k)$. Hence

$$\varphi(t) = f(t) + \int_0^\tau \left(\sum_{j,k=-m}^{m} e^{i\pi(jt-ks)/\tau} A_k \gamma_{jk} \right) f(s)\, ds,$$

where

$$\|\gamma_{jk}\|_{j,k=-m}^{m} = (\|\delta_{jk} - \alpha_{jk} A_k\|_{j,k=-m}^{m})^{-1}.$$

Thus, the operator $(I-K)^{-1}$ has the form

$$((I-K)^{-1} f)(t) = f(t) + \int_0^\tau \gamma(t,s) f(s)\, ds,$$

with kernel

$$\gamma(t,s) = \sum_{j,k=-m}^{m} e^{i\pi(jt-ks)/\tau} A_k \gamma_{jk}.$$

Obviously, the equalities

$$\gamma(t,s) - \int_0^\tau k(t-r)\gamma(r,s)\, dr = k(t-s), \tag{2.4}$$

$$\gamma(t,s) - \int_0^\tau \gamma(t,r) k(r-s)\, dr = k(t-s) \tag{2.5}$$

hold, where $0 \le s, t \le \tau$. In particular, these equalities imply the equalities

$$\gamma(t,0) = x(t), \quad \gamma(\tau,t) = w(\tau - t), \quad \gamma(t,\tau) = z(t-\tau), \quad \gamma(0,t) = y(-t). \tag{2.6}$$

For a sufficiently small $\varepsilon > 0$ we have

$$\begin{aligned}
\gamma(t,s) - \gamma(t-\varepsilon, s-\varepsilon) = & \int_0^\tau k(t-r)(\gamma(u,s) - \gamma(r, s-\varepsilon))\, dr \\
& - \int_0^\tau k(t-r)(\gamma(r,s) - \gamma(r-\varepsilon, s-\varepsilon))\, dr \\
& - \int_0^\varepsilon k(t-r)\gamma(r,-s)\, dr \\
& - \int_\tau^{\tau+\varepsilon} k(t-r)\gamma(r-\varepsilon, s-\varepsilon)\, dr.
\end{aligned}$$

From here it follows that

$$\begin{aligned}&\frac{1}{\varepsilon}(\gamma(t,s)-\gamma(t-\varepsilon,s-\varepsilon))-\int_0^\tau k(t-r)\frac{1}{\varepsilon}(\gamma(r,s)-\gamma(r-\varepsilon,s-\varepsilon))\,dr\\&=\frac{1}{\varepsilon}\int_0^\varepsilon k(t-r)\gamma(r,s)\,dr-\frac{1}{\varepsilon}\int_\tau^{\tau+\varepsilon} k(t-r)\gamma(r-\varepsilon,s-\varepsilon)\,dr.\end{aligned}\tag{2.7}$$

Put

$$\omega(t,s)=\lim_{\varepsilon\to 0}\frac{1}{\varepsilon}(\gamma(t,s)-\gamma(t-\varepsilon,s-\varepsilon)).$$

Then passing to the limit as $\varepsilon\to 0$ in equality (2.7) we get

$$\omega(t,s)-\int_0^\tau k(t-r)\omega(r,s)\,dr=k(t)\gamma(0,s)-k(t-\tau)\gamma(\tau,s)$$

or

$$\omega(t,s)-\int_0^\tau k(t-r)\omega(r,s)\,dr=k(t)y(-s)-k(t-\tau)w(t-s).\tag{2.8}$$

Let $g(t,s)=x(t)y(-s)-z(t-\tau)w(\tau-s)$. Then

$$g(t,s)-\int_0^\tau k(t-r)g(r,s)\,dr=k(t)y(-s)-z(t-\tau)w(\tau-s).$$

In view of the invertibility of the operator $I-K$, from here it follows that

$$\omega(t,s)=g(t,s)=x(t)y(-s)-z(t-\tau)w(\tau-s).\tag{2.9}$$

From the definition of the function $\omega(t,s)$ it follows that

$$\gamma(t,s)=\gamma(t-s,0)+\gamma(0,s-t)+\int_0^{\min(t,s)}\omega(t-r,s-r)\,dr.\tag{2.10}$$

This equality means that formulas (2.1) and (2.2) hold.

Now consider the case of an arbitrary matrix function $k(t)\in L_1^{n\times n}(-\tau,\tau)$. Suppose $k_m(t)$ is a sequence of matrix functions of the form (2.3) that converges to $k(t)$ in the norm of the space $L_1^{n\times n}(-\tau,\tau)$. By K_m denote the operators of the form (1.1) generated by the functions $k_m(t)$. According to (1.2) the operators K_m tend to the operator K in the norm as $m\to\infty$. Without loss of generality, one can assume that the operators $I-K_m$ are invertible. By $x_m(t)$, $y_m(-t)$, $z_m(-t)$, and $w_m(t)$ denote the solutions of equations (0.1)–(0.4) for the function $k_m(t)$, respectively. By what has been proved above, for the operators $(I-K_m)^{-1}$ formula (2.2) holds. Hence $(I-K_m)^{-1}=I+\Gamma_m$, where Γ_m is the integral operator determined by the corresponding kernel $\gamma_m(t,s)$ of the form (2.2):

$$\begin{aligned}\gamma_m(t,s)&=x_m(t-s)+y_m(t-s)\\&\quad+\int_0^{\min(t,s)}\big(x_m(t-r)y_m(r-s)-z_m(t-r-\tau)w_m(r-s+\tau)\big)\,dr.\end{aligned}$$

By Γ denote the integral operator with kernel $\gamma(t,s)$ defined by equality (2.2). We shall prove that

$$\lim_{m\to\infty}\|\Gamma_m-\Gamma\|=0.$$

Put

$$
\begin{aligned}
h_m(t) = {} & |x(t) - x_m(t)| + |y(t) - y_m(t)| \\
& + \int_0^\tau |x(t-r)y(-r) - x_m(t-r)y_m(-r)|\, dr \\
& + \int_0^\tau |z(t-r)w(r) - z_m(t-r)w_m(r)|\, dr,
\end{aligned} \tag{2.11}
$$

where $|\cdot|$ is a norm in the space $L(\mathbb{C}^n)$.

The matrix functions $x_m(t)$, $w_m(t)$, $z_m(-t)$, and $y_m(t)$ converge in the norm of the space $L_1^{n\times n}(0,\tau)$ as $m \to \infty$ to the matrix functions $x(t)$, $w(t)$, $z(-t)$, and $y(-t)$, respectively. Then from (2.11) one can easily derive that

$$\lim_{m\to\infty} \|h_m\|_{L_1} = 0.$$

Moreover, it is easy to see that the inequality

$$|\gamma(t,s) - \gamma_m(t,s)| \le h_m(t-s)$$

holds (cf. [3, Section 6]). Thus, for every function $\varphi(t) \in L_p^n(0,\tau)$, the inequality

$$
\begin{aligned}
& \left\| \int_0^\tau (\gamma(t,s) - \gamma_m(t,s))\varphi(s)\, ds \right\|_{L_p^n} \\
& \le \left(\int_0^\tau \left(\int_0^\tau |\gamma(t,s) - \gamma_m(t,s)|\, |\varphi(s)|\, ds \right)^p dt \right)^{1/p} \\
& \le \left\| \int_0^\tau h_m(t-s)|\varphi(s)|\, ds \right\|_{L_p} \le \|h_m(t)\|_{L_1} \|\varphi(t)\|_{L_p^n}
\end{aligned}
$$

holds. Therefore,

$$\|(\Gamma - \Gamma_m)\varphi\|_{L_p^n} \le \|h_m\|_{L_1} \|\varphi\|_{L_p^n} \quad \text{and} \quad \lim_{m\to\infty} \|\Gamma - \Gamma_m\| = 0.$$

Since

$$\lim_{m\to\infty} \|I + \Gamma_m - (I-K)^{-1}\|_{L_p^n} = 0,$$

from here it follows that

$$I + \Gamma = (I-K)^{-1}.$$

The theorem is proved. □

Corollary 2.1. *Let the hypotheses of Theorem* 2.1 *be fulfilled. Then for the resolvent kernel* $\gamma(t,s)$ *the formula*

$$
\begin{aligned}
\gamma(t,s) = {} & w(t-s) + z(t-s) \\
& + \int_{\max(t,s)}^\tau \big(z(t-r)w(r-s) - x(t-r+\tau)y(r-s-\tau)\big)\, dr
\end{aligned} \tag{2.12}
$$

also holds.

Proof. Indeed, from the definition of the function $\omega(t,s)$ it follows that

$$\gamma(t,s) = \gamma(t-s+\tau,\tau) + \gamma(\tau, s-t-\tau) - \int_0^{\min(\tau-s,\tau-t)} \omega(t+r, s+r)\, dr.$$

This implies that

$$\gamma(t,s) = w(t-s) + z(t-s) + \int_0^{\min(\tau-s,\tau-t)} \big(z(t+r-\tau)w(-s-r+\tau) - x(t+r)y(-r-s)\big)\, dr.$$

The last equality implies equality (2.12). □

2.2. Formula (2.2) can also be represented in a different form. By $\widetilde{\gamma}(\lambda,\mu)$ denote the Fourier transform of the function $\gamma(t,s)$:

$$\widetilde{\gamma}(\lambda,\mu) = \int_0^\tau \int_0^\tau \gamma(t,s) e^{i(\lambda t+\mu s)}\, dt\, ds.$$

The function $\widetilde{\gamma}(\lambda,\mu)$ is an entire function of μ and λ. We have the following.

Corollary 2.2. *For the matrix function $\widetilde{\gamma}(\lambda,\mu)$, the identity*

$$\widetilde{\gamma}(\lambda,\mu) = \frac{i}{\lambda+\mu}\Big(\big(1+\widetilde{x}(\lambda)\big)\big(1+\widetilde{y}(-\mu)\big) - e^{i\tau(\lambda+\mu)}\big(1+\widetilde{z}(\lambda)\big)\big(1+\widetilde{w}(-\mu)\big)\Big) \quad (2.13)$$

holds, where in each occurrence $\widetilde{u}(\lambda)$ denotes the Fourier transform defined for a function $u(t) \in L_1^{n\times n}(0,\tau)$ by

$$\widetilde{u}(\lambda) = \int_0^\tau u(t) e^{i\lambda t}\, dt.$$

Proof. Indeed, first assume that the functions $x(t)$, $z(-t)$, $w(t)$, and $y(-t)$ are continuously differentiable. Then the resolvent kernel $\gamma(t,s)$ is also continuously differentiable. For a sufficiently small $\varepsilon > 0$ we have

$$\begin{aligned}
\frac{e^{-i\varepsilon(\lambda+\mu)}-1}{\varepsilon}\widetilde{\gamma}(\lambda,\mu) &= \frac{1}{\varepsilon}\int_0^\tau\int_0^\tau \gamma(t,s)\left(e^{i(\lambda(t+\varepsilon)+\mu(s+\varepsilon))} - e^{i(\lambda t+\mu s)}\right) dt\, ds \\
&= \frac{1}{\varepsilon}\int_0^\tau\int_0^\tau \big(\gamma(t+\varepsilon,s+\varepsilon) - \gamma(t,s)\big) e^{i(\lambda t+\mu s)}\, dt\, ds \\
&\quad + \frac{1}{\varepsilon}\left(\int_0^\tau\int_{-\varepsilon}^0 + \int_{-\varepsilon}^0\int_{-\varepsilon}^\tau\right)\gamma(t+\varepsilon,s+\varepsilon)e^{i(\lambda t+\mu s)}\, dt\, ds \\
&\quad - \frac{1}{\varepsilon}\left(\int_{\tau-\varepsilon}^\tau\int_0^\tau + \int_0^{\tau-\varepsilon}\int_{\tau-\varepsilon}^\tau\right)\gamma(t,s)e^{i(\lambda t+\mu s)}\, dt\, ds.
\end{aligned}$$

Since the function

$$\frac{1}{\varepsilon}\big(\gamma(t+\varepsilon,s+\varepsilon) - \gamma(t,s)\big)$$

is uniformly bounded with respect to ε, passing to the limit as $\varepsilon \to \infty$, we obtain

$$\begin{aligned}-i(\lambda+\mu)\widetilde{\gamma}(\lambda,\mu) &= \int_0^\tau\int_0^\tau \omega(t,s)e^{i(\lambda t+\mu s)}\,dt\,ds \\ &+ \int_0^\tau \gamma(t,0)e^{i\lambda t}\,dt + \int_0^\tau \gamma(0,s)e^{i\mu s}\,ds \\ &+ \int_0^\tau \gamma(t,\tau)e^{i(\lambda t+\mu\tau)}\,dt + \int_0^\tau \gamma(\tau,s)e^{i(\lambda\tau+\mu s)}\,ds.\end{aligned}$$

It remains to take into account equalities (2.6) and (2.9).

In the general case formula (2.13) is proved by passing to the limit. □

3. Properties of the solutions of main equations

Solutions of all four equations (0.1)–(0.4) play a part in formulas (2.2) and (2.12). On the other hand, according to Lemma 1.2, the invertibility of the corresponding operator $I-K$ is equivalent to the existence of solutions for one pair of equations (0.1) and (0.3) or (0.2) and (0.4). It happens that solutions of equations (0.1) and (0.3) are related to solutions of equations (0.2) and (0.4). In the present section these relations are investigated and auxiliary results are obtained, which will be used for solving the inverse problem in the forthcoming section.

Everywhere in what follows we will assume that K is an operator of the form

$$(Kf)(t) = \int_0^\tau k(t-s)f(s)\,ds \quad (0 \le t \le \tau)$$

with $k(t) \in L_1^{n\times n}(0,\tau)$ and that the operator $I-K$ is invertible.

Proposition 3.1. *For the solutions of equations* (0.1)–(0.4) *the identities*

$$z(t-s-\tau)+\int_0^\tau z(t-r-\tau)w(r-s)\,dr = y(t-s-\tau)+\int_0^\tau x(t-r)y(r-s-\tau)\,dr \quad (3.1)$$

and[3]

$$\begin{aligned}w(\tau+t-s) &+ \int_0^\tau z(t-r)w(r-s+\tau)\,dr \\ &= x(\tau+t-s) + \int_0^\tau x(\tau+t-r)y(r-s)\,dr\end{aligned} \quad (3.2)$$

hold for $0 \le t,s \le \tau$.

Proof. Indeed, from formulas (2.2) and (2.6) it follows that

$$z(t-\tau) = \gamma(t,\tau) = y(t-\tau) + \int_0^t (x(t-r)y(r-\tau) - z(t-\tau-r)w(r))\,dr.$$

[3]Here and in what follows it is assumed that $x(t) = w(t) = y(-t) = z(-t) = 0$ for $t \notin [0,\tau]$.

This implies that

$$z(t-\tau)+\int_0^t z(t-\tau-r)w(r)\,dr = y(t-\tau)+\int_0^t x(t-r)y(r-\tau)\,dr. \tag{3.3}$$

By $g(t,s)$ and $h(t,s)$ denote the left-hand and right-hand sides of equality (3.1), respectively. It is easy to see that the functions $g(t,s)$ and $h(t,s)$ depend only on the difference of the arguments $t-s$. This implies that equality (3.1) holds if $g(t,0)=h(t,0)$ and $g(0,s)=h(0,s)$. The first of these equalities follows from (3.3) and the second equality is obvious because $g(0,s)=h(0,s)=0$.

Equality (3.2) is proved analogously.

The proposition is proved. □

Proposition 3.2. *For the solutions of equations* (0.1)–(0.4) *the identities*

$$z(t-s)+\int_0^\tau z(t-\tau-r)w(\tau-s-r)\,dr = y(t-s)+\int_0^\tau x(t-r)y(r-s)\,dr \tag{3.4}$$

and

$$x(t-s)+\int_0^\tau x(t-r)y(r-s)\,dr = w(t-s)+\int_0^\tau z(t-r-\tau)w(r-s-\tau)\,dr \tag{3.5}$$

hold for $0\le t,s\le\tau$.

Proof. Indeed, the left-hand side $g(t,s)$ and the right-hand side $h(t,s)$ of equality (3.4) depend only on the difference of the arguments $t-s$. Hence equality (3.4) holds if $g(t,\tau)=h(t,\tau)$ and $g(\tau,s)=h(\tau,s)$ $(0\le s,t\le\tau)$. The first of these equalities follows from (3.3) and the second is obvious.

Equality (3.5) is proved by analogy.

The proposition is proved. □

We define the following notation. Let $g(t)$ $(-\tau\le t\le\tau)$ be a matrix function in $L_1^{n\times n}(-\tau,\tau)$. Consider the following operators acting in the space $L_p^n(0,\tau)$:

$$(Gf)(t)=\int_0^\tau g(t-s)f(s)\,ds \quad (0\le t\le\tau),$$

$$(G_\pm f)(t)=\int_0^\tau g(t-s\pm\tau)f(s)\,ds \quad (0\le t\le\tau)$$

and

$$(G'f)(t)=\int_0^\tau f(s)g(t-s)\,ds \quad (0\le t\le\tau),$$

$$(G'_\pm f)(t)=\int_0^\tau f(s)g(t-s\pm\tau)\,ds \quad (0\le t\le\tau).$$

Note that if the function $g(t)$ $(\not\equiv 0)$ is equal to zero on one of the intervals $[-\tau,0]$ or $[0,\tau]$, then the operators G and G' are Volterra operators. Hence the operators $I+G$ and $I+G'$ are invertible. Besides that, if $g(t)=0$ for $\pm t\ge 0$, then $G_\pm G'_\pm=0$.

Equalities (3.1), (3.2), (3.4), and (3.5) can be rewritten in operator form. For instance, equalities (3.1) and (3.2) in operator notation have the form

$$Z_-(I+W) = (I+X)Y_-, \tag{3.1'}$$

$$(I+Z)W_+ = X_+(I+Y). \tag{3.2'}$$

Obviously, the equalities

$$Z'_-(I+W') = (I+X')Y'_-, \tag{3.1''}$$

$$(I+Z')W'_+ = X'_+(I+Y'). \tag{3.2''}$$

also hold.

Note also that formula (2.2) in operator notation has the form

$$(I-K)^{-1} = (I+X)(I+Y) - Z_-W_+. \tag{3.6}$$

In the sequel the following "paired" operators acting in the space $L_p^n(-\tau,\tau)$ or in the space $L_p^{n\times n}(-\tau,\tau)$ are considered:

$$(\widetilde{\Pi}\chi)(t) = \chi(t) + \int_{-\tau}^{0} x(t-s)\chi(s)\,ds + \int_0^{\tau} z(t-s-\tau)\chi(s)\,ds \quad (-\tau \le t \le \tau),$$

$$(\widetilde{\Lambda}\omega)(t) = \omega(t) + \int_{-\tau}^{0} \omega(s)w(t-s)\,ds + \int_0^{\tau} \omega(s)y(t-s-\tau)\,ds \quad (-\tau \le t \le \tau).$$

We also introduce the following operators:

$$(H_+f)(t) = f(t-\tau) \quad (0 \le t \le \tau),$$

$$(H_-f)(t) = f(t+\tau) \quad (-\tau \le t \le 0).$$

The operator H_+ maps $L_p^n(-\tau,0)$ onto $L_p^n(0,\tau)$ and the operator H_- is the inverse of the operator H_+. Considering the space $L_p^n(-\tau,\tau)$ as the direct sum of the spaces $L_p^n(0,\tau)$ and $L_p^n(-\tau,0)$, the operators $\widetilde{\Pi}$ and $\widetilde{\Lambda}$ can be represented in block form as

$$\widetilde{\Pi} = \begin{Vmatrix} H_-(I+X)H_+ & H_-Z_- \\ X_+H_+ & I+Z \end{Vmatrix}, \quad \widetilde{\Lambda} = \begin{Vmatrix} H_-(I+W')H_+ & H_-Y'_- \\ W'_+H_+ & I+Y' \end{Vmatrix}.$$

Obviously, these operators are similar to the operators

$$\Pi = \begin{Vmatrix} I+X & Z_- \\ X_+ & I+Z \end{Vmatrix}, \quad \Lambda = \begin{Vmatrix} I+W' & Y'_- \\ W'_+ & I+Y' \end{Vmatrix}$$

acting in the direct sum of two copies of the space $L_p^n(0,\tau)$.

Proposition 3.3. *The operators* Π *and* Λ *are invertible.*

Proof. Obviously, the operator $I+Z$ is invertible. It is straightforward to verify that

$$\Pi = \begin{Vmatrix} I & Z_-(I+Z)^{-1} \\ 0 & I \end{Vmatrix} \begin{Vmatrix} I+X-Z_-(I+Z)^{-1}X_+ & 0 \\ 0 & I+Z \end{Vmatrix} \times \begin{Vmatrix} I & 0 \\ (I+Z)^{-1}X_+ & I \end{Vmatrix}. \tag{3.7}$$

From this equality it follows that the operator Π is invertible if and only if the operator $C = I + X - Z_-(I+Z)^{-1}X_+$ is invertible. Since the operator $I+Y$ is invertible, from (3.2′) it follows that

$$C = ((I+X)(I+Y) - Z_-W_-)(I+Y)^{-1}.$$

In view of equality (3.6) this means that

$$C = (I-K)^{-1}(I+Y)^{-1}.$$

Hence the operator C, as well as the operator Π, is invertible.

The invertibility of the operator Λ is proved analogously.

The proposition is proved. □

Proposition 3.4. *The equalities*

$$\left\| \begin{matrix} y(t-\tau) \\ -w(t) \end{matrix} \right\| = \Pi^{-1} \left\| \begin{matrix} z(t-\tau) \\ -x(t) \end{matrix} \right\| \tag{3.8}$$

and

$$\left\| \begin{matrix} z(t-\tau) \\ -x(t) \end{matrix} \right\| = \Lambda^{-1} \left\| \begin{matrix} y(t-\tau) \\ -w(t) \end{matrix} \right\| \tag{3.9}$$

hold.

Proof. Indeed, from equality (3.1) for $s=0$ it follows that

$$y(t-\tau) = \int_0^\tau x(t-r)y(r-\tau)\,dr - \int_0^\tau z(t-r-\tau)w(r)\,dr = z(t-\tau),$$

and from equality (3.2) for $s=\tau$ it follows that

$$\int_0^\tau x(\tau+t-r)y(r-\tau)\,dr - w(t) - \int_0^\tau z(t-r)w(r)\,dr = -x(t).$$

The last equalities can be written in the form

$$\begin{aligned} (I+X)y(t-\tau) - Z_-w(t) &= z(t-\tau), \\ X_+y(t-\tau) - (I+Z)w(t) &= -x(t). \end{aligned}$$

This immediately implies equality (3.8).

From (3.1) for $t=\tau$ it follows that

$$z(s-t) + \int_0^\tau z(r-\tau)w(s-r)\,dr = y(s-\tau) + \int_0^\tau x(r)y(s-r-\tau)\,dr,$$

and from (3.2) for $t=0$ it follows that

$$w(s) + \int_0^\tau z(r-\tau)w(s-r+\tau)\,dr = x(s) + \int_0^\tau x(r)y(s-r)\,dr.$$

This immediately implies equality (3.9).

The proposition is proved. □

Proposition 3.5. *The operators defined by the equalities*

$$\Pi' = \begin{Vmatrix} I+X' & X'_+ \\ Z'_- & I+Z' \end{Vmatrix}, \quad \Lambda' = \begin{Vmatrix} I+W & W_+ \\ Y_- & I+Y \end{Vmatrix}$$

are invertible.

Proof. Indeed, since the operators Π and Λ are invertible, we see that their adjoint operators Π^* and Λ^* are also invertible. The operator Π^* is defined by the formula

$$\Pi^* = \begin{Vmatrix} \varphi_1(t) \\ \varphi_2(t) \end{Vmatrix} = \begin{Vmatrix} \psi_1(t) \\ \psi_2(t) \end{Vmatrix},$$

where

$$\psi_1(t) = \varphi_1(t) + \int_0^\tau x^*(s-t)\varphi_1(s)\,ds + \int_0^\tau x^*(s-t+\tau)\varphi_2(s)\,ds,$$

$$\psi_2(t) = \varphi_2(t) + \int_0^\tau z^*(s-t-\tau)\varphi_1(s)\,ds + \int_0^\tau z^*(s-t)\varphi_2(s)\,ds$$

for $0 \le t \le \tau$ and $\varphi_1(t), \varphi_2(t) \in L_p^{n\times n}(0,\tau)$.

Passing to the adjoint operators in the last two equalities, after a change of variables we obtain

$$\psi_1^*(t-\tau) = \varphi_1^*(t-\tau) + \int_0^\tau \varphi_1^*(s-\tau)x(t-s)\,ds + \int_0^\tau \varphi_2^*(s-\tau)x(t-s+\tau)\,ds,$$

$$\psi_2^*(t-\tau) = \varphi_2^*(t-\tau) + \int_0^\tau \varphi_1^*(s-\tau)z(t-s-\tau)\,ds + \int_0^\tau \varphi_2^*(s-\tau)z(t-s)\,ds.$$

These equalities can be written in the form

$$\begin{Vmatrix} \psi_1^*(t-\tau) \\ \psi_2^*(t-\tau) \end{Vmatrix} = \Pi' \begin{Vmatrix} \varphi_1^*(t-\tau) \\ \varphi_2^*(t-\tau) \end{Vmatrix}.$$

Thus the operator Π' is invertible. The invertibility of the operator Λ' is proved analogously. □

4. Inverse problem

Theorem 4.1. *Let four matrix functions be given: $x(t), w(t) \in L_1^{n\times n}(0,\tau)$ and $y(t), z(t) \in L_1^{n\times n}(-\tau,0)$. For the existence of a matrix function $k(t) \in L_1^{n\times n}(-\tau,\tau)$ such that $x(t)$, $w(t)$, $z(-t)$, and $y(-t)$ are solutions of equations* (0.1)–(0.4), *respectively, it is necessary and sufficient that the following conditions be fulfilled:*

1) *$Z_-(I+W) = (I+X)Y_-$ and $(I+Z)W_+ = X_+(I+Y)$;*
2) *at least one of the operators*

$$\Pi = \begin{Vmatrix} I+X & Z_- \\ X_+ & I+Z \end{Vmatrix}, \quad \Pi' = \begin{Vmatrix} I+X' & X'_+ \\ Z'_- & I+Z' \end{Vmatrix},$$

$$\Lambda' = \begin{Vmatrix} I+W & W_+ \\ Y_- & I+Y \end{Vmatrix}, \quad \Lambda = \begin{Vmatrix} I+W' & Y'_- \\ W'_+ & I+Y' \end{Vmatrix}$$

is invertible.

If conditions 1) *and* 2) *are fulfilled, then all the operators* Π, Π', Λ, *and* Λ' *are invertible and the matrix function* $k(t)$ *is uniquely determined by the relations*

$$\left\| \begin{array}{c} k_+(t) \\ k_-(t) \end{array} \right\| = (\Pi')^{-1} \left\| \begin{array}{c} x(t) \\ z(t-\tau) \end{array} \right\| = (\Lambda')^{-1} \left\| \begin{array}{c} w(t) \\ y(t-\tau) \end{array} \right\|, \tag{4.1}$$

where $k_+(t) = k(t)$ *for* $0 \le t \le \tau$ *and* $k_-(t) = k(t-\tau)$ *for* $-\tau \le t \le 0$.

Proof. The necessity of conditions 1) and 2) is established in Propositions 3.1, 3.3, and 3.5. Let us prove formula (4.1). From equalities (0.1) and (0.3), we obtain after simple transformations that

$$k(t) + \int_{t-\tau}^{t} k(s)x(t-s)\,ds = x(t) \quad (0 \le t \le \tau)$$

and

$$k(t-\tau) + \int_{t}^{t-\tau} k(s-t)z(t-s)\,ds = z(t-\tau) \quad (0 \le t \le \tau).$$

From here it follows that

$$k(t) + \int_0^\tau k(s)x(t-s)\,ds + \int_0^\tau k(s-t)x(t-s+\tau)\,ds = x(t)$$

and

$$k(t-\tau) + \int_0^\tau k(s-t)z(t-s)\,ds + \int_0^\tau k(s)z(t-s-\tau)\,ds = z(t-\tau).$$

The last equalities are equivalent to the first equality in (4.1). The second equality in (4.1) is proved analogously.

Let us show the sufficiency of conditions 1) and 2). For instance, let the operator Π be invertible. From the proof of Proposition 3.5 it follows that the operator Π' is also invertible. Put

$$\left\| \begin{array}{c} k_+(t) \\ k_-(t) \end{array} \right\| = (\Pi')^{-1} \left\| \begin{array}{c} x(t) \\ z(t-\tau) \end{array} \right\| \tag{4.2}$$

and $k(t) = k_+(t)$ for $0 \le t \le \tau$ and $k(t) = k_-(t)$ for $-\tau \le t \le 0$. From (4.2) we obtain after simple transformations that

$$x(t) - \int_0^\tau k(t-s)x(s)\,ds = k(t) \quad (0 \le t \le \tau)$$

and

$$z(t-\tau) - \int_0^\tau k(t-s)z(s-t)\,ds = k(t-\tau) \quad (0 \le t \le \tau).$$

This means that for the function $k(t)$, the equations (0.1) and (0.3) have solutions. In view of Theorem 2.1, from here it follows that the operator $I - K$ is invertible.

It is obvious that condition 1) for the matrix functions $y(-t)$ and $w(t)$ can be written in the form

$$y(t-\tau) + \int_0^\tau x(t-s)y(s-\tau)\,ds = z(t-\tau) + \int_0^\tau z(t-s-\tau)w(s)\,ds$$

and

$$x(t) + \int_0^\tau x(\tau + t - s)y(s - \tau)\,ds = w(t) + \int_0^\tau z(t - s)w(s)\,ds$$

or, equivalently, in the form

$$(I + X)y(t - \tau) - Z_- w(t) = z(t - \tau),$$
$$X_+ y(t - \tau) - (I + Z)w(t) = -x(t).$$

Thus, for the functions $y(-t)$ and $w(t)$ the equality

$$\Pi \left\| \begin{matrix} y(t-\tau) \\ -w(t) \end{matrix} \right\| = \left\| \begin{matrix} z(t-\tau) \\ -x(t) \end{matrix} \right\|$$

holds.

On the other hand, in view of Proposition 3.4 (see (3.8)), the solutions of equations (0.2) and (0.4) are also determined by the last equality. Taking into account the invertibility of the operator Π, we conclude that the functions $y(-t)$ and $w(t)$ are solutions of equations (0.2) and (0.4), respectively.

The theorem is proved analogously in the case when all other operators Π', Λ', and Λ are invertible.

The theorem is proved. □

Theorem 4.2. *Let matrix functions* $x(t) \in L_1^{n\times n}(0, \tau)$ *and* $z(t) \in L_1^{n\times n}(-\tau, 0)$ *be given. For the existence of a matrix function* $k(t) \in L_1^{n\times n}(-\tau, \tau)$ *such that* $x(t)$ *and* $z(-t)$ *are solutions of equations* (0.1) *and* (0.3), *respectively, it is necessary and sufficient that the operator* Π *be invertible.*

Proof. The necessity of the hypotheses follows from Proposition 3.3. For the proof of the sufficiency we introduce the matrix functions $y(-t)$ and $w(t)$ by the equality

$$\left\| \begin{matrix} y(t-\tau) \\ -w(t) \end{matrix} \right\| = (\Pi^*)^{-1} \left\| \begin{matrix} z(t-\tau) \\ -x(t) \end{matrix} \right\|.$$

From here it follows that

$$(I + X)y(t - \tau) = Z_- w(t) + z(t - \tau), \quad x(t) + X_+ y(t - \tau) = (I + Z)w(t).$$

With the aid of the arguments from the proof of Proposition 3.1 from here we get the equalities

$$(I + X)Y_- = Z_-(W + I), \quad X_+(I + Y) = (I + Z) = W_+.$$

Thus the conditions of Theorem 4.1 are fulfilled. The theorem is proved. □

The following theorem is proved analogously.

Theorem 4.3. *Let matrix functions* $w(t) \in L_1^{n\times n}(0, \tau)$ *and* $y(t) \in L_1^{n\times n}(-\tau, 0)$ *be given. For the existence of a matrix function* $k(t) \in L_1^{n\times n}(-\tau, \tau)$ *such that* $w(t)$ *and* $y(-t)$ *are solutions of equations* (0.2) *and* (0.4), *respectively, it is necessary and sufficient that the operator* Λ' *be invertible.*

References

[1] I.C. Gohberg and A.A. Semencul, *The inversion of finite Toeplitz matrices and their continual analogues.* Matem. Issled. **7** (1972), no. 2(24), 201–223 (in Russian). MR0353038 (50 #5524), Zbl 0288.15004.

[2] I.C. Gohberg and G. Heinig, *Inversion of finite Toeplitz matrices consisting of elements of a noncommutative algebra.* Rev. Roumaine Math. Pures Appl. **19** (1974), 623–663 (in Russian). English translation: **this volume.** MR0353040 (50 #5526), Zbl 0337.15005.

[3] M.G. Krein, *Integral equations on a half-line with kernel depending upon the difference of the arguments.* Uspehi Mat. Nauk **13** (1958), no. 5 (83), 3–120 (in Russian). English translation: Amer. Math. Soc. Transl. (2) **22** (1962), 163–288. MR0102721 (21 #1507), Zbl 0088.30903.

Operator Theory:
Advances and Applications, Vol. 206, 65–88

The Resultant Matrix and its Generalizations. I. The Resultant Operator for Matrix Polynomials

Israel Gohberg and Georg Heinig

Let $a(\lambda) = a_0 + a_1\lambda + \cdots + a_n\lambda^n$ and $b(\lambda) = b_0 + b_1\lambda + \cdots + b_m\lambda^m$ be two polynomials with coefficients in $\mathbb{C}^1$. The determinant of the following matrix

$$R(a,b) = \left[\begin{array}{ccccccc} a_0 & a_1 & \dots & a_n & & & \\ & a_0 & a_1 & \dots & a_n & & \\ & & \ddots & \ddots & \ddots & \ddots & \\ & & & a_0 & a_1 & \dots & a_n \\ b_0 & b_1 & \dots & b_m & & & \\ & b_0 & b_1 & \dots & b_m & & \\ & & \ddots & \ddots & \ddots & \ddots & \\ & & & b_0 & b_1 & \dots & b_m \end{array}\right] \begin{array}{l} \left.\vphantom{\begin{array}{c}a\\a\\a\\a\end{array}}\right\} m \\ \left.\vphantom{\begin{array}{c}a\\a\\a\\a\end{array}}\right\} n \end{array}$$

is said to be the resultant of these polynomials.

It is known that

$$\det R(a,b) = a_n^n b_m^m \prod_{j=1}^{n} \prod_{k=1}^{m} (\lambda_j(a) - \lambda_k(b)),$$

where $\lambda_j(a)$ $(j = 1, 2, \dots, n)$ is the complete collection of the roots of the polynomial $a(\lambda)$. In particular, the polynomials $a(\lambda)$ and $b(\lambda)$ have at least one common root if and only if the resultant $\det R(a,b)$ is equal to zero (see, e.g., [1]). A more complete result is known (see, e.g., [2, 3]). It says that the number of common roots of the polynomials $a(\lambda)$ and $b(\lambda)$ (taking into account their multiplicities) is equal to the number

$$m + n - \operatorname{rank} R(a,b).$$

This relation immediately follows from the next statement.

The paper was originally published as И.Ц. Гохберг, Г. Хайниг, Результантная матрица и её обобщения. I. Результантный оператор матричных полиномов, Acta Sci. Math. (Szeged) **37** (1975), 41–61. MR0380471 (52 #1371), Zbl 0298.15013.

Theorem 0.1. *Let λ_l $(l = 1, 2, \dots, l_0)$ be all distinct common roots of the polynomials $a(\lambda)$ and $b(\lambda)$ and k_l be the multiplicity of the common root λ_l.*

Then the system of vectors

$$H_k(\lambda_l) = \left(\binom{p}{k} \lambda_l^{p-k} \right)_{p=0}^{m+n-1} \in \mathbb{C}^{m+n} \quad (l = 1, 2, \dots, l_0, \ k = 0, 1, \dots, k_l - 1)$$

forms a basis of the subspace $\operatorname{Ker} R(a, b)$. *In particular,*

$$\sum_{l=0}^{l_0} k_l = \dim \operatorname{Ker} R(a, b). \tag{0.1}$$

In the present paper various generalizations of this theorem are established.

In the first part of the paper the resultant operator for matrix polynomials is studied. In contrast with the scalar case, this operator is defined, as a rule, by a rectangular matrix.

This part consists of five sections. The first two sections have an auxiliary character. In the third section the main theorem is proved, from which, in particular, Theorem 0.1 follows. In fourth and fifth sections examples of applications of the main theorem are presented.

In the second part of the paper continual analogues of the results of this paper will be exposed.

The authors have started this investigation under the influence of conversations with M.G. Krein. He kindly drew the authors' attention to relations of this circle of questions with results on inversion of finite Toeplitz matrices and their continual analogues [4, 5].

The authors wish to express their sincere gratitude to M.G. Krein.

1. Lemma on multiple extensions of systems of vectors

1.1. Let L denote some linear space and let L^m be the linear space of all vectors of the form $f = (f_j)_{j=0}^{m-1}$ with components $f_j \in L$. Let

$$\mathfrak{F} = \{\varphi_{jk} : \ k = 0, 1, \dots, k_j - 1; \ j = 1, 2, \dots, j_0\}$$

be a system of vectors in L and λ_0 be a complex number. We say that the system of vectors

$$\mathfrak{F}^m(\lambda_0) = \{\Phi_{jk}(\lambda_0) : \ k = 0, 1, \dots, k_j - 1; \ j = 1, 2, \dots, j_0\}$$

in L^m, where

$$\Phi_{jk}(\lambda_0) = \left(\varphi_{jk}^p(\lambda_0)\right)_{p=0}^{m-1}, \quad \varphi_{jk}^p(\lambda_0) = \sum_{s=0}^{p} \binom{p}{s} \lambda_0^{p-s} \varphi_{j,k-s} \tag{1.1}$$

is the extension of multiplicity m of the system $\mathfrak{F}$ with respect to λ_0. It is easy to see that for the vectors $\Phi_{jk}(\lambda_0)$ the equalities

$$\varphi^p_{jk}(\lambda_0) = \frac{d^p}{dt^p} e^{\lambda_0 t}\left(\varphi_{jk} + \frac{t}{1!}\varphi_{j,k-1} + \cdots + \frac{t^k}{k!}\varphi_{j0}\right)\Big|_{t=0} \tag{1.2}$$

hold.

The following recurrent formula

$$\varphi^{p+1}_{jk}(\lambda_0) = \lambda_0 \varphi^p_{jk}(\lambda_0) + \varphi^p_{jk-1}(\lambda_0) \quad (k = 0, 1, \ldots, k_j - 1;\ j = 1, 2, \ldots, j_0) \tag{1.3}$$

also holds, where it is assumed that $\varphi^p_{j,-1}(\lambda_0) = 0$. Indeed, from equality (1.1) it follows that

$$\begin{aligned}
&\varphi^{p+1}_{jk}(\lambda_0) - \lambda_0 \varphi^p_{jk}(\lambda_0) \\
&= \sum_{s=0}^{p+1}\left(\binom{p+1}{s}\lambda_0^{p-s+1}\varphi_{j,k-s}(\lambda_0) - \binom{p}{s}\lambda_0^{p-s+1}\varphi_{j,k-s}(\lambda_0)\right) \\
&= \sum_{s=0}^{p+1}\binom{p}{s-1}\lambda_0^{p-s+1}\varphi_{j,k-s}(\lambda_0).
\end{aligned}$$

Therefore,

$$\varphi^{p+1}_{jk}(\lambda_0) - \lambda_0 \varphi^p_{jk}(\lambda_0) = \sum_{s=0}^{p}\binom{p}{s}\lambda_0^{p-s}\varphi_{j,k-1-s}(\lambda_0) = \varphi^p_{j,k-1}(\lambda_0).$$

From formula (1.3) one can get without difficulty a more general formula

$$\varphi^{p+r}_{jk}(\lambda_0) = \sum_{s=0}^{p}\binom{r}{s}\lambda_0^{r-s}\varphi^p_{j,k-s}(\lambda_0) \quad (r = 1, 2, \ldots). \tag{1.4}$$

Indeed, for $r = 1$ formula (1.4) coincides with (1.3). Assume that equality (1.4) holds. Then

$$\begin{aligned}
\varphi^{p+r+1}_{jk}(\lambda_0) &= \lambda_0 \varphi^{p+r}_{jk}(\lambda_0) + \varphi^{p+r}_{j,k-1}(\lambda_0) \\
&= \sum_{s=0}^{r}\left(\binom{r}{s}\lambda_0^{r+1-s}\varphi^p_{j,k-s}(\lambda_0) - \binom{r}{s-1}\lambda_0^{r+1-s}\right)\varphi_{j,k-s}(\lambda_0),
\end{aligned}$$

whence

$$\varphi^{p+r+1}_{jk}(\lambda_0) = \sum_{s=0}^{r+1}\binom{r+1}{s}\lambda_0^{r+1-s}\varphi^p_{j,k-s}(\lambda_0).$$

1.2. The next statement plays an important role in what follows.

Lemma 1.1. *Let $\Lambda = \{\lambda_1, \lambda_2, \ldots, \lambda_l\}$ be a system of distinct complex numbers and $\mathfrak{F}_l$ $(l = 1, 2, \ldots, l_0)$ be systems of vectors in L:*

$$\mathfrak{F}_l = \{\varphi_{jk,l} :\ k = 0, 1, \ldots, k_{jl} - 1;\ j = 1, 2, \ldots, j_l\}.$$

If for every $l = 1, 2, \ldots, l_0$ *the system of vectors* $\varphi_{j0,l}$ $(j = 1, 2, \ldots, j_l)$ *is linearly independent and a number* m *satisfies the condition*

$$m \geq \sum_{l=1}^{l_0} \max_{j=1,2,\ldots,j_l} k_{jl}, \tag{1.5}$$

then the system of vectors[1]

$$\mathfrak{F}^m(\Lambda) = \bigcup_{l=1}^{l_0} \mathfrak{F}_l^m(\lambda_l)$$

is also linearly independent.

Proof. Obviously, it is sufficient to confine oneself to the case

$$m = \sum_{l=1}^{l_0} \max_{j=1,2,\ldots,j_l} k_{jl}.$$

Consider the operator matrices

$$\Pi_r(\lambda) = \|\pi_{st}^r I\|_{s,t=1}^{m-1} \quad (r = 0, 1, \ldots, m-1),$$

where

$$\pi_{st}^r = \begin{cases} \delta_{st} & (s \leq r), \\ (-\lambda)^{s-t}\binom{s-r}{t-r} & (s > r) \end{cases}$$

and I is the identity operator in the space L. Now we elucidate how the operator $\Pi_r(\lambda)$ acts on the vectors $\Phi_{jk}(\lambda_l) = (\varphi_{jk,l}^p(\lambda_l))_{p=0}^{m-1}$ of the system $\mathfrak{F}^m(\Lambda)$. First of all, note that the operator $\Pi_r(\lambda)$ does not change the first $r+1$ components of the vector $\Phi_{jk,l}(\lambda_l)$.

Consider the systems of vectors

$$\mathfrak{G}_r^m(\lambda, \lambda_l) = \{P_r\Pi_r(\lambda)\Phi_{jk}(\lambda_l) : \ k = 0, 1, \ldots, k_{jl}-1; \ j = 1, 2, \ldots, j_l\}, \tag{1.6}$$

where

$$P_r = \underbrace{\begin{bmatrix} 0 & \ldots & 0 & I & & 0 \\ \vdots & & \vdots & & \ddots & \\ 0 & \ldots & 0 & 0 & & I \end{bmatrix}}_{m} \Bigg\} m - r,$$

$l = 1, 2, \ldots, l_0$ and $r = 0, 1, \ldots, m-1$.

---

[1]Here and in what follows by $\bigcup_{l=1}^{l_0} \mathfrak{F}_l^m(\lambda_l)$ we denote the system

$$\{\Phi_{jk,l}(\lambda_l) : \ k = 0, 1, \ldots, k_{jl}-1; \ j = 1, 2, \ldots, j_l; \ l = 1, 2, \ldots, l_0\}.$$

We shall prove that each of these systems of vectors is the extension of multiplicity $m-r$ with respect to $\lambda_l - \lambda$ of the corresponding system of vectors

$$\mathfrak{S}_r(\lambda_l) = \left\{\varphi^r_{jk,l}(\lambda_l): \ k=0,1,\dots,k_{jl}-1; \ j=1,2,\dots,j_l\right\}.$$

Put

$$(\psi^p_{jk,l})^{m-1-r}_{p=0} = P_r\Pi_r(\lambda)\Phi_{jk,l}(\lambda_l).$$

Then

$$\psi^p_{jk,l} = \sum_{s=r}^{p+r}(-\lambda)^{p+r-s}\binom{p}{s-r}\varphi^s_{jk,l}(\lambda_l) = \sum_{u=0}^{p}(-\lambda)^u\binom{p}{u}\varphi^{p+k-u}_{jk,l}(\lambda_l).$$

In view of (1.4),

$$\varphi^{p-u+r}_{jk,l}(\lambda_l) = \sum_{s=0}^{p-u}\lambda_l^{p-u-s}\binom{p-u}{s}\varphi^r_{j,k-s,l}(\lambda_l),$$

hence

$$\psi^p_{jk,l} = \sum_{u=0}^{p}\sum_{s=0}^{p-u}\lambda_l^{p-u-s}(-\lambda)^u\binom{p}{u}\binom{p-u}{s}\varphi^r_{j,k-s,l}.$$

Taking into account that

$$\binom{p-u}{s}\binom{p}{u} = \binom{p-s}{u}\binom{p}{s},$$

we get

$$\psi^p_{jk,l} = \sum_{s=0}^{p}\sum_{u=0}^{p-s}\lambda_l^{p-u-s}(-\lambda)^u\binom{p-s}{u}\binom{p}{s}\varphi^r_{j,k-s,l}(\lambda_l).$$

Thus,

$$\psi^p_{jk,l} = \sum_{s=0}^{p}(\lambda_l-\lambda)^{p-s}\binom{p}{s}\varphi^r_{j,k-s,l}(\lambda_l).$$

The last equality means that the system of vectors $\mathfrak{S}^m_r(\lambda,\lambda_l)$ is the extension of multiplicity $m-r$ of the system $\mathfrak{S}^m_r(\lambda_l)$ with respect to $\lambda_l-\lambda$.

Consider the system $\mathfrak{F}^{(1)}_m(\lambda_l)$ of vectors

$$\Phi^{(1)}_{jk}(\lambda_l) = \Pi_0(\lambda)\Phi_{jk,l}(\lambda_l) \quad (k=0,1,\dots,k_{jl}-1; \ j=1,2,\dots,j).$$

From the above arguments it follows that the system $\mathfrak{F}^{(1)}_m(\lambda_l)$ is the extension of multiplicity m of the system $\mathfrak{F}_l$ $(l=1,2,\dots,l_0)$ with respect to $\lambda_l-\lambda_1$. In particular, the vectors $\Phi_{jk}^{(1)}(\lambda_1)$ have the form

$$\Phi_{jk}^{(1)}(\lambda_1) = \begin{bmatrix} \varphi_{jk,1} \\ \vdots \\ \varphi_{j0,1} \\ 0 \\ \vdots \\ 0 \end{bmatrix} \quad (k = 0, 1, \ldots, k_{j1} - 1;\ j = 1, 2, \ldots, j_1).$$

Now let us form the system $\mathfrak{F}_m^{(2)}(\lambda_l)$ of vectors

$$\Phi_{jk}^{(2)}(\lambda_l) = \Pi_{\kappa_1}(\lambda_2 - \lambda_1)\Phi_{jk}^{(1)}(\lambda_1) \quad (k = 0, 1, \ldots, k_{jl} - 1;\ j = 1, 2, \ldots, l_0),$$

where

$$\kappa_1 = \max_{j=1,2,\ldots,j_1} k_{j1}.$$

Then the system of vectors $P_{\kappa_1}\Phi_{jk}^{(2)}(\lambda_l)$ is the extension of multiplicity $m - \kappa_1$ of the system of vectors $\varphi_{jk,l}^{\kappa_1}(\lambda_l - \lambda_1)$ with respect to

$$\lambda_l - \lambda_1 - (\lambda_2 - \lambda_1) = \lambda_l - \lambda_2.$$

In particular, $P_{\kappa_1}\Phi_{jk}^{(2)}(\lambda_1) = 0$ and the vectors $P_{\kappa_1}\Phi_{jk}^{(2)}(\lambda_2)$ have the form

$$P_{\kappa_1}\Phi_{jk}^{(2)}(\lambda_2) = \begin{bmatrix} * \\ \vdots \\ * \\ (\lambda_2 - \lambda_1)^{\kappa_1}\varphi_{jk,2} \\ 0 \\ \vdots \\ 0 \end{bmatrix} \begin{matrix} \\ \\ \\ \leftarrow k \\ \\ \\ \\ \end{matrix}$$

where the vectors not playing any role in what follows are replaced by $*$.

We shall repeat this process: from the system $\mathfrak{F}_m^{(s)}(\lambda_l)$ of vectors

$$\Phi_{jk}^{(s)}(\lambda_l) \quad (k = 0, 1, \ldots, k_{j,l} - 1;\ j = 1, \ldots, j_l)$$

we construct the system $\mathfrak{F}_m^{(s+1)}(\lambda_l)$ of vectors

$$\Phi_{jk}^{(s+1)}(\lambda_l) = \Pi_{\kappa_2}(\lambda_{s+1} - \lambda_s)\Phi_{jk}^{(s)},$$

where

$$\kappa_s = \sum_{l=1}^{s} \max_{j=1,2,\ldots,j_s} k_{jl}.$$

For the vectors $P_{\kappa_s}\Phi_{jk}^{(s+1)}(\lambda_l)$, the equalities

$$P_{\kappa_s}\Phi_{jk}^{(s+1)}(\lambda_l) = 0 \quad (l = 1, 2, \ldots, s)$$

hold. and the vectors $P_{\kappa_s}\Phi_{jk}^{(s+1)}(\lambda_{s+1})$ have the form

$$\begin{bmatrix} * \\ \vdots \\ * \\ (\lambda_{s+1}-\lambda_s)^{\kappa_s}\varphi_{jk,s+1} \\ 0 \\ \vdots \\ 0 \end{bmatrix} \leftarrow k \quad (k=0,1,\dots,k_{j,s+1}-1;\ j=1,2,\dots,j_{s+1}).$$

Finally, we construct the systems $\mathfrak{F}_m^{l_0}(\lambda_l)$ $(l=1,2,\dots,l_0)$. The vectors of this system $P_{\kappa_{l_0}-1}\Phi_{jk}^{(l_0)}(\lambda_l)=0$ for $l\neq l_0$, and the vectors $P_{\kappa_{l_0}-1}\Phi_{jk}^{(l_0)}(\lambda_{l_0})$ have the form

$$\begin{bmatrix} * \\ \vdots \\ * \\ (\lambda_{l_0}-\lambda_{l_0-1})^{\kappa_{l_0}-1}\varphi_{jk,l_0} \\ 0 \\ \vdots \\ 0 \end{bmatrix} \leftarrow k. \tag{1.7}$$

Now let us prove that the system of vectors

$$\mathfrak{F}_m^{(l_0)}(\Lambda)=\bigcup_{l=1}^{l_0}\mathfrak{F}_m^{(l_0)}(\lambda_l)$$

is linearly independent. Let

$$\sum_{l=1}^{l_0}\sum_{j=1}^{j_l}\sum_{k=0}^{k_{jl}-1}\alpha_{jkl}\Phi_{jk}^{(l_0)}(\lambda_l)=0.$$

Then

$$\sum_{j,k,l}\alpha_{jkl}P_{\kappa_{l_0}-1}\Phi_{jk}^{(l_0)}(\lambda_l)=\sum_{j=1}^{j_{l_0}}\sum_{k=0}^{k_{jl}-1}\alpha_{jkl_0}P_{\kappa_{l_0}-1}\Phi_{jk}^{(l_0)}(\lambda_{l_0})=0.$$

Since the vectors $P_{\kappa_{l_0}-1}\Phi_{jk}^{(l_0)}(\lambda_{l_0})$ have the form (1.7) and the vectors φ_{j0,l_0} $(j=1,2,\dots,j_l)$ are linearly independent, we have

$$\alpha_{jkl_0}=0 \quad (k=0,1,\dots,k_{jl_0}-1;\ j=1,2,\dots,j_{l_0}).$$

Thus,

$$\sum_{l=1}^{l_0-1}\sum_{j=1}^{j_l}\sum_{k=0}^{k_{jl}-1}\alpha_{jkl}\Phi_{jk}^{(l_0)}(\lambda_l)=0.$$

Applying the operator $P_{\kappa_{l_0}-2}$ to this equality, we obtain that

$$\alpha_{jk,l_0-1} = 0 \quad (k = 0, 1, \ldots, k_{jl_0-1} - 1;\ j = 1, \ldots, j_{l_0} - 1).$$

Continuing this process analogously, we obtain that

$$\alpha_{jk_s} = 0 \quad (k = 0, 1, \ldots, k_{j_s} - 1;\ j = 1, 2, \ldots, j_s)$$

for $s = l_0-2, l_0-3, \ldots, 1$. The obtained linear independency of the system $\mathfrak{F}_m^{(l_0)}(\Lambda)$ implies the linear independency of the system $\mathfrak{F}^m(\Lambda)$.

The lemma is proved. □

It is easily seen that condition (1.5) is essential in the formulation of Lemma 1.1 even in the case $L = \mathbb{C}^1$.

1.3. The next lemma is derived from Lemma 1.1.

Lemma 1.2. *Let* $\Lambda = \{\lambda_1, \lambda_2, \ldots, \lambda_{l_0}\}$ *be a set of complex numbers and let* $\mathfrak{F}_l$ $(l = 1, 2, \ldots, l_0)$ *be a system of vectors in* L*:*

$$\mathfrak{F}_l = \{\varphi_{jk,l} :\ k = 0, 1, \ldots, k_{jl} - 1;\ j = 1, 2, \ldots, j_l\}.$$

If a number m *satisfies condition* (1.5) *and*

$$\sum_{l=1}^{l_0} \sum_{j=1}^{j_l} \sum_{k=0}^{k_{jl}-1} \alpha_{jkl} \Phi_{jk,l}(\lambda_l) = 0 \quad (\alpha_{jkl} \in \mathbb{C}^1), \tag{1.8}$$

then

$$\Phi_{k,l}(\lambda_l) = \sum_j \alpha_{jkl} \Phi_{jk,l}(\lambda_l) = 0 \quad (k = 0, 1, \ldots, \max_j k_{jl} - 1;\ l = 1, 2, \ldots, l_0),$$

where the summation spreads over all indices j *such that* $k_{jl} \geq k$.

Proof. Let us form the system

$$\mathfrak{G}_l = \left\{ \varphi_{k,l} = \sum_j \alpha_{jkl} \varphi_{jk,l} :\ \varphi_{0,l} \neq 0;\ k = 0, 1, \ldots, \max_j k_{jl} - 1 \right\},$$

where $l = 1, 2, \ldots, l_0$, and assume that it is not empty. Obviously, then the system

$$\mathfrak{G}_l^m = \left\{ \Phi_{k,l}(\lambda_l) : k = 0, 1, \ldots, \max_j k_{jl} - 1 \right\}$$

is the extension of multiplicity m of the system $\mathfrak{G}_l$. Applying Lemma 1.1 to the system $\mathfrak{G} = \bigcup_{l=1}^{l_0} \mathfrak{G}_l$, we obtain that the system of vectors $\Phi_{k,l}(\lambda_l)$ is linearly independent. The last fact contradicts equality (1.7). Thus $\mathfrak{G}_l = \emptyset$, whence

$$\varphi_{0,l} = 0 \quad (l = 1, 2, \ldots, l_0).$$

Now we form the systems

$$\mathfrak{G}_{l,1} = \left\{ \varphi_{k,l}^{(1)} = \varphi_{k+1,l} :\ \varphi_{1,l} \neq 0,\ k = 0, 1, \ldots, \max_j k_{jl} - 2 \right\} \quad (l = 1, 2, \ldots, l_0).$$

It is easy to see that the extension of multiplicity m of the system $\mathfrak{S}_{l,1}$ consists of the vectors

$$\Phi_{k+1,l}(\lambda_l) = \Phi_{k,l,1}(\lambda_l) \quad (\varphi_{1,l} \neq 0;\ k = 0, \dots, \max_j k_{jl} - 2).$$

Hence, applying Lemma 1.1 again, we obtain that $\mathfrak{S}_{1,l} = \emptyset$, whence

$$\varphi_{1,l} = 0 \quad (l = 1, 2, \dots, l_0).$$

Continuing this process, we get

$$\varphi_{r,l} = 0 \quad (l = 1, 2, \dots, l_0) \quad \text{for} \quad r = 2, 3, \dots, \max_{j,l} k_{jl} - 1.$$

The lemma is proved. □

2. Auxiliary propositions

2.1. Let d be a natural number and $A(\lambda) = A_0 + \lambda A_1 + \dots + \lambda^n A_n$ be a matrix pencil with coefficients $A_k \in L(\mathbb{C}^d)$ [2]. Everywhere in what follows we will suppose that the matrix A_n is invertible.

A number $\lambda_0 \in \mathbb{C}^1$ is said to be an eigenvalue of the pencil $A(\lambda)$ if

$$\det A(\lambda_0) = 0.$$

If for some vector $\varphi_0 \in \mathbb{C}^d$ ($\varphi_0 \neq 0$) the equality $A(\lambda_0)\varphi_0 = 0$ holds, then the vector φ_0 is said to be an eigenvector of the pencil $A(\lambda)$ at λ_0. A chain of vectors $\varphi_0, \varphi_1, \dots, \varphi_r$ is called a Jordan chain (a chain of an eigenvector and associated vectors) of length $r + 1$ if the equalities

$$A(\lambda_0)\varphi_k + \frac{1}{1!}\left(\frac{d}{d\lambda}A\right)(\lambda_0)\varphi_{k-1} + \dots + \frac{1}{k!}\left(\frac{d^k}{d\lambda^k}A\right)(\lambda_0)\varphi_0 = 0 \qquad (2.1)$$

hold for $k = 0, 1, \dots, r$.

Let λ_0 be an eigenvalue of the pencil $A(\lambda)$. It is easy to prove that one can construct a basis $\varphi_{10}, \varphi_{20}, \dots, \varphi_{r0}$ in the kernel of the matrix $A(\lambda_0)$ with the following property. For every vector φ_{j0} there exists a chain of associated vectors $\varphi_{j1}, \varphi_{j2}, \dots, \varphi_{j,k_j-1}$, where $k_1 \geq k_2 \geq \dots \geq k_r$ and the number $k_1 + k_2 + \dots + k_r$ is equal to the multiplicity of the zero of the function $\det A(\lambda)$ at the point λ_0.

The numbers k_j ($j = 1, 2, \dots, r$) are called partial multiplicities of the eigenvalue λ_0, and the system

$$\varphi_{j0}, \varphi_{j1}, \dots, \varphi_{j,k_j-1} \quad (j = 1, \dots, r)$$

is said to be a canonical system of Jordan chains for the pencil $A(\lambda)$ at the eigenvalue λ_0.

---

[2]The space of quadratic matrices of order d is denoted by $L(\mathbb{C}^d)$.

Lemma 2.1. *Let* $\Lambda = \{\lambda_1, \lambda_2, \dots, \lambda_l\}$ *be the complete collection of all distinct eigenvalues of a pencil* $A(\lambda) = A_0 + \lambda A_1 + \cdots + \lambda^n A_n$ *with coefficients* $A_k \in L(\mathbb{C}^d)$ *and let*

$$\mathfrak{F} = \{\varphi_{jk,l} : \ k = 0, 1, \dots, k_{jl} - 1; \ j = 1, 2, \dots, j_l\} \quad (l = 1, 2, \dots, q)$$

be a canonical system of Jordan chains of the pencil $A(\lambda)$ *at the eigenvalue* λ_l.

Then for every natural number m *the system*[3]

$$\begin{aligned}\mathfrak{F}^{m+n}(\Lambda) &= \bigcup_{l=1}^{l_0} \mathfrak{F}_l^{m+n}(\lambda_l) \\ &= \{\Phi_{jk,l}(\lambda_l) = (\varphi_{jk,l}^p(\lambda_l))_{p=0}^{m-1} : \ j = 1, \dots, j_l; \ k = 0, \dots, k_{jl} - 1\}\end{aligned}$$

is a basis of the kernel of the operator

$$\mathcal{A}_m = \left[\begin{array}{cccccccc} A_0 & A_1 & \dots & A_n & \dots & \dots & 0 \\ \vdots & A_0 & \dots & A_{n-1} & A_n & & \vdots \\ \vdots & & \ddots & \ddots & \ddots & \ddots & \vdots \\ 0 & \dots & \dots & A_0 & \dots & A_{n-1} & A_n \end{array}\right] \Bigg\} m$$

acting from $\mathbb{C}^{(m+n)d}$ *to* $\mathbb{C}^{md}$.

Proof. First, we shall show that the equality

$$\sum_{p=r}^{n+r} A_{p-r} \varphi_{jk,l}^p(\lambda_l) = 0 \quad (j = 1, 2, \dots, j_l) \tag{2.2}$$

holds for $k = 0$ and every r. Indeed, since $\varphi_{j0,l}^p(\lambda_l) = \lambda_l^p \varphi_{j0,l}$, we have

$$\sum_{p=r}^{n+r} A_{p-r} \varphi_{j0,l}^p(\lambda_l) = \sum_{p=r}^{n+r} \lambda_l^p A_{p-r} \varphi_{j0,l} = \lambda_l^r A(\lambda_l) \varphi_{j0,l} = 0.$$

Assume that equality (2.2) is true for every r and $k = k_0$. Then it is also true for $k = k_0 + 1$ and every r. Indeed, by the assumption, we have

$$\begin{aligned} 0 = \sum_{p+r}^{n+r} A_{p-r} \varphi_{jk_0,l}^p(\lambda_l) &= \lambda_l \sum_{p=r}^{n+r} \sum_{s=0}^{p} \binom{p}{s} \lambda_l^{p-s} A_{p-r} \varphi_{j,k_0-s,l} \\ &= \sum_{p=r}^{n+r} \sum_{s=0}^{p} \binom{p}{s-1} \lambda_l^{p-s} A_{p-r} \varphi_{j,k_0+1-s,l}. \end{aligned}$$

[3]Recall that the system $\mathfrak{F}^{m+n}(\lambda_l)$ is the extension of multiplicity $m + n$ of the system $\mathfrak{F}$ with respect to λ_l (for the definition, see Section 1).

This implies that

$$\sum_{p=r}^{n+r} A_{p-r}\varphi^{p}_{j,k_0+1}(\lambda_0) = \sum_{p=r}^{n+r}\sum_{s=0}^{p}\left(\binom{p}{s}\lambda_l^{p-s}A_{p-r}\varphi_{j,k_0+1-s,l} - \binom{p}{s-1}\lambda_l^{p-s}A_{p-r}\varphi_{j,k_0+1-s,l}\right)$$

$$= \sum_{p=r}^{n+r}\sum_{s=1}^{p-1}\binom{p-1}{s-1}\lambda_l^{p-s}A_{p-r}\varphi_{j,k_0+1-s,l}$$

$$= \sum_{p=r-1}^{n+r-1} A_{p-(r-1)}\varphi^{p}_{jk_0,l}(\lambda_l) = 0.$$

Thus it has been proved that equality (2.2) holds for every $k = 0, 1, \dots, k_{jl} - 1$ and every $r = 0, 1, \dots, m-1$. This implies that $\Phi_{jk,l}(\lambda_l) \in \operatorname{Ker}\mathcal{A}_m$.

The system $\mathfrak{F}^{m+n}(\Lambda)$ consists of nd vectors. According to Lemma 1.1, this system is linearly independent. On the other hand, obviously,

$$\dim\operatorname{Ker}\mathcal{A}_m = (m+n)d - md = nd.$$

Thus $\operatorname{Ker}\mathcal{A}_m = \operatorname{lin}\mathfrak{F}^{m+n}(\Lambda)$. [4]

The lemma is proved. □

2.2.

Lemma 2.2. *Let*

$$\mathfrak{F}_l = \{\varphi_{jk,l} : \ k = 0, 1, \dots, k_{jl} - 1; \ j = 1, 2, \dots, j_l\} \quad (l = 1, 2, \dots, q)$$

be a system of vectors in the space $\mathbb{C}^d$ and $\Lambda = \{\lambda_1, \lambda_2, \dots, \lambda_q\}$ be a finite set of complex numbers. Suppose $\mathfrak{F}_l^{m+n}(\lambda_l)$ is the extension of multiplicity $m+n$ of the system $\mathfrak{F}_l$ with respect to λ_l and the vector

$$\Omega = \sum_{l=1}^{q}\sum_{j=1}^{j_l}\sum_{k=0}^{k_{jl}-1} \alpha_{jkl}\Phi_{jk,l}(\lambda_l) \in \mathbb{C}^{(m+n)d},$$

where $\alpha_{jkl} \in \mathbb{C}^1$, belongs to the kernel of the operator $\mathcal{A}_m$. If a number m satisfies the condition[5]

$$m \geq \sum_{l=1}^{q}\max_j k_{jl}$$

then all vectors

$$\Omega_{k,l} = \sum_j \alpha_{jkl}\Phi_{jk,l}(\lambda_l) \quad (k = 0, 1, \dots, \max k_{jl} - 1; \ l = 1, \dots, q) \tag{2.3}$$

where the summation runs over all indices j such that $k_{jl} \geq k$, belong to the kernel of the operator $\mathcal{A}_m$.

---

[4] The linear hull of a system of vectors $\mathfrak{F}$ is denoted by $\operatorname{lin}\mathfrak{F}$.

[5] Notice that this condition is automatically fulfilled if $m \geq nd$.

Proof. Put

$$\psi_{jk,l} = \sum_{p=0}^{n} A_p \varphi_{jk,l}^{p}(\lambda_l),$$

$$\mathfrak{S}_l = \{\psi_{jk,l} : j = 1, 2, \ldots, j_l;\ k = 0, 1, \ldots, k_{jl-1}\} \quad (l = 1, 2, \ldots, q),$$

and

$$\widetilde{\Psi}_{jk,l} = (\psi_{jk,l}^{p})_{p=0}^{m-1} \stackrel{\text{def}}{=} \mathcal{A}_m \Phi_{jk,l}(\lambda_l).$$

Let us prove that the system

$$\widetilde{\mathfrak{S}}_l = \{\widetilde{\Psi}_{jk,l} : k = 0, 1, \ldots, k_{jl} - 1;\ l = 1, 2 \ldots, q\}$$

is the extension of multiplicity m of the system $\mathfrak{F}_l$ with respect to λ_l, that is, $\mathfrak{F}_l^m(\lambda_l) = \widetilde{\mathfrak{S}}_l$.

The equality

$$\psi_{jk,l}^{s} = \sum_{p=s}^{s+n} A_{p-s} \varphi_{jk,l}^{p}(\lambda_l) = \sum_{p=0}^{n} A_p \varphi_{jk,l}^{p+s}(\lambda_l)$$

holds. In view of (1.4),

$$\varphi_{jk,l}^{s} = \sum_{p=0}^{n} \sum_{u=0}^{s} \binom{s}{u} \lambda_l^{s-u} \varphi_{j,k-u,l}^{p}(\lambda_l).$$

Therefore,

$$\psi_{jk,l}^{s} = \sum_{p=0}^{n} \sum_{u=0}^{s} \binom{s}{u} \lambda_l^{s-u} A_p \varphi_{j,k-u,l}^{p}(\lambda_l) = \sum_{u=0}^{s} \binom{s}{u} \lambda_l^{s-u} \psi_{j,k-u,l}.$$

From here it follows that $\widetilde{\psi}_{jk,l}^{s} = \psi_{jk,l}^{s}(\lambda_l)$ and $\widetilde{\Psi}_{jk,l} = \Psi_{jk,l}(\lambda_l)$.

Let

$$\Omega = \sum_{l=1}^{q} \sum_{j=1}^{j_l} \sum_{k=0}^{k_{jl}-1} \Phi_{jk,l}(\lambda_l) \in \operatorname{Ker} \mathcal{A}_m.$$

Then

$$\mathcal{A}_m \Omega = \sum_{l=1}^{q} \sum_{j=1}^{j_l} \sum_{k=0}^{k_{jl}-1} \alpha_{jkl} \widetilde{\Psi}_{jk,l}.$$

In view of Lemma 1.2, from here it follows that

$$\sum_{j} \alpha_{jkl} \widetilde{\Psi}_{jk,l} = 0 \quad (k = 0, 1, \ldots, k_{jl} - 1;\ l = 1, 2, \ldots, q),$$

where the summation runs over all j such that $k_{jl} \geq k$.

The lemma is proved. □

Note that it can happen that, under the hypotheses of Lemma 2.2, all vectors $\Phi_{jk,l}(\lambda_l)$ do not belong to the kernel of the operator $\mathcal{A}_m$ for every m. One can verify this by the following example.

For $d = 2$ consider the pencil $A(\lambda) = A_0 + \lambda I$, where

$$A_0 = \left\| \begin{matrix} -1 & 0 \\ 0 & 1 \end{matrix} \right\|.$$

Put $\Lambda = \{1, -1\}$, $\mathfrak{F}_1\{\varphi_{10}\}$, $\mathfrak{F}_2 = \{\varphi_{20}\}$, where $\varphi_{10} = (0; 1)$ and $\varphi_{20} = (1; -1)$. Let m be an arbitrary natural number. Obviously, then the system $\mathfrak{F}_1^m(1)$ consists of a unique vector

$$\Phi_{10,1}(1) = (\underbrace{0, 1, 0, 1, \ldots, 0, 1}_{2m}),$$

and the system $\mathfrak{F}_2^m(-1)$ consists of the vector

$$\Phi_{20,2}(-1) = (\underbrace{-1, 1, -1, 1, \ldots, -1, 1}_{2m}).$$

The operator $\mathcal{A}_m$ is defined in the case under consideration by the equality

$$\mathcal{A}_m = \left.\begin{bmatrix} A_0 & I & \ldots & \ldots & 0 \\ \vdots & A_0 & I & & \vdots \\ \vdots & & \ddots & \ddots & \vdots \\ 0 & \ldots & \ldots & A_0 & I \end{bmatrix}\right\} m$$

It is easy to see that $\mathcal{A}_m(\Phi_{10,1}(1) + \Phi_{20,2}(-1)) = 0$, while $\mathcal{A}_m\Phi_{10,1}(1) \neq 0$ and $\mathcal{A}_m\Phi_{20,2}(-1) \neq 0$.

2.3.

Lemma 2.3. *Let $\mathfrak{F} = \{\varphi_k : k = 0, 1, \ldots, k_0\}$ [6] be a system of vectors in $\mathbb{C}^d$ and let $\mathfrak{F}^m = \{\Phi_k(\lambda_0) = (\varphi_k^p(\lambda_0))_{p=0}^{m-1}\}$ be its extension of multiplicity m with respect to $\lambda_0 \in \mathbb{C}^1$. If $m > k_0$ and the vector $\Phi_{k_0}(\lambda_0)$ belongs to the kernel of the operator $\mathcal{A}_m$, then λ_0 is an eigenvalue of the pencil $A(\lambda)$ and $\varphi_0, \varphi_1, \ldots, \varphi_{k_0}$ is a Jordan chain.*

Proof. Let

$$\psi_k = \sum_{p=0}^{n} A_p \varphi_k^p.$$

Repeating the corresponding argument from the proof of Lemma 2.2, we show that the system

$$\widetilde{\mathfrak{S}} = \{\mathcal{A}_m\Phi_k(\lambda_0) : \ k = 0, 1, \ldots, k_0\}$$

is the extension of multiplicity m of the system

$$\mathfrak{S} = \{\psi_k : k = 0, 1, \ldots, k_0\}.$$

[6]Here the index k corresponds to the second index of vectors from the definition of an extension of multiplicity m.

Therefore, according to (1.3), the equality

$$\psi_{p+1}^{k} = \lambda_0 \psi_k^p + \psi_{k-1}^p$$

holds, where $\mathcal{A}_m \Phi_k(\lambda_0) = (\psi_k^p)_{p=0}^{m-1}$.

Since $\psi_{k_0}^p = 0$ $(p = 0, 1, \dots, m-1)$ and $m > k_0$, from here it follows that $\psi_k = 0$ for $k = 0, 1, \dots, k_0$. The last fact means that

$$0 = \sum_{p=0}^{n} A_p \varphi_k^p = \sum_{p=0}^{n} \sum_{r=0}^{p} \binom{p}{r} \lambda_0^{p-r} A_p \varphi_{k-r}.$$

Since

$$\left(\frac{d^r}{d\lambda^r} A\right)(\lambda) = r! \sum_{p=0}^{n} \binom{p}{r} \lambda^{p-r} A_p,$$

we have

$$0 = \sum_{r=0}^{k} \frac{1}{r!} \left(\frac{d^r}{d\lambda^r} A\right)(\lambda_0) \varphi_{k-r} \quad (k = 0, 1, \dots, k_0).$$

This means that $\varphi_0, \varphi_1, \dots, \varphi_{k_0}$ is a Jordan chain of the pencil $A(\lambda)$ at the eigenvalue λ_0.

The lemma is proved. □

3. Main theorem

3.1. Let

$$A(\lambda) = A_0 + \lambda A_1 + \cdots + \lambda^n A_n, \quad B(\lambda) = B_0 + \lambda B_1 + \cdots + \lambda^m B_m$$

be two matrix pencils with coefficients $A_j, B_k \in \mathbb{C}^d$, where $j = 0, 1, \dots, n$ and $k = 0, 1, \dots, m$. Let λ_0 be a common eigenvalue of the pencils $A(\lambda)$ and $B(\lambda)$, and let

$$\mathfrak{R} = \operatorname{Ker} A(\lambda_0) \cap \operatorname{Ker} B(\lambda_0).$$

Let $\varphi_0, \varphi_1, \dots, \varphi_r$ be a Jordan chain simultaneously for the pencils $A(\lambda)$ and $B(\lambda)$ corresponding to the eigenvalue λ_0. The number $r+1$ is called the length of this chain. The largest length of such a chain starting with the vector φ_0 is called the rank of the common eigenvector φ_0 and is denoted by $\operatorname{rank}(\lambda_0, \varphi_0)$.

We choose a basis $\varphi_{10}, \varphi_{20}, \dots, \varphi_{j_0 0}$ in the subspace $\mathfrak{R}$ such that the ranks of its vectors have the following properties: k_1 is the maximum of the numbers $\operatorname{rank}(\lambda_0, \varphi)$ $(\varphi \in \mathfrak{R})$ and k_j $(j = 1, 2, \dots, j_0)$ is the maximum of the numbers $\operatorname{rank}(\lambda_0, \varphi)$ for all vectors of the direct complement to $\operatorname{lin}\{\varphi_{10}, \varphi_{20}, \dots, \varphi_{j-1,0}\}$ in $\mathfrak{R}$ that contains φ_{j0}.

It is easy to see that the number $\operatorname{rank}(\lambda_0, \varphi_0)$ for every vector $\varphi_0 \in \mathfrak{R}$ is equal to one of the numbers k_j $(j = 1, 2, \dots, j_0)$. Therefore the numbers k_j $(j = 1, 2, \dots, j_0)$ are uniquely determined by the pencils $A(\lambda)$ and $B(\lambda)$.

By $\varphi_{j1}, \varphi_{j2}, \dots, \varphi_{j,k_j-1}$ denote the chain of associated vectors to the eigenvector φ_{j0} $(j = 1, 2, \dots, j_0)$ common for $A(\lambda)$ and $B(\lambda)$.

The system

$$\varphi_{j0}, \varphi_{j1}, \dots, \varphi_{j,k_j-1} \quad (j = 1, 2, \dots, j_0)$$

is said to be a canonical system of common Jordan chains for the pencils $A(\lambda)$ and $B(\lambda)$ at the eigenvalue λ_0, and the number

$$\nu(A, B, \lambda_0) \overset{\text{def}}{=} \sum_{j=1}^{j_0} k_j$$

is called the common multiplicity for the eigenvalue λ_0 for the pencils $A(\lambda)$ and $B(\lambda)$.

3.2. To the pencils $A(\lambda)$ and $B(\lambda)$ and an integer number $w > \max\{n, m\}$ we assign the operator

$$R_w(A,B) = \begin{bmatrix} A_0 & A_1 & \dots & A_n & & & \\ & A_0 & \dots & A_{n-1} & A_n & & \\ & & \ddots & \vdots & \ddots & \ddots & \\ & & & A_0 & \dots & A_{n-1} & A_n \\ B_0 & B_1 & \dots & B_m & \dots & \dots & \\ & B_0 & \dots & B_{m-1} & B_m & & \\ & & \ddots & \vdots & \ddots & \ddots & \\ & & & B_0 & \dots & B_{m-1} & B_m \end{bmatrix} \begin{matrix} \left.\vphantom{\begin{matrix}a\\a\\a\\a\end{matrix}}\right\} w-n \\ \left.\vphantom{\begin{matrix}a\\a\\a\\a\end{matrix}}\right\} w-m \end{matrix}$$

acting from the space $\mathbb{C}^{wd}$ to $\mathbb{C}^{(2w-m-n)d}$.

We call $R_w(A, B)$ the resultant operator or the resultant matrix of the pencils $A(\lambda)$ and $B(\lambda)$.

Theorem 3.1. *Let*

$$A(\lambda) = A_0 + \lambda A_1 + \dots + \lambda^n A_n, \quad B(\lambda) = B_0 + \lambda B_1 + \dots + \lambda^m B_m \tag{3.1}$$

be two matrix pencils ($A_j, B_k \in \mathbb{C}^d$) *with the invertible leading coefficients* A_n *and* B_m, *let* $\Lambda = \{\lambda_1, \lambda_2, \dots, \lambda_l\}$ *be the set of all (distinct) common eigenvalues of the pencils* $A(\lambda)$ *and* $B(\lambda)$, *and let*

$$\mathfrak{F}_l = \{\varphi_{jk,l} : \ k = 0, 1, \dots, k_{jl} - 1; \ j = 1, 2, \dots, j_l\} \quad (l = 1, 2, \dots, q)$$

be a canonical system of common Jordan chains for the pencils $A(\lambda)$ *and* $B(\lambda)$ *at the eigenvalue* λ_l.

If the condition[7]

$$w \geq \min\{n + md, m + nd\} \tag{3.2}$$

[7]It is easy to see that the presented proof remains valid if one replaces condition (3.2) by a weaker condition

$$w \geq \min\left\{n + \sum_{l=1}^{q} \max_j k_{jl}(B); m + \sum_{l=1}^{q} \max_j k_{jl}(A)\right\} \tag{3.2'}$$

where $\{k_{jl}(C) : j = 1, \dots, j_l(C)\}$ is the collection of the partial multiplicities of a pencil $C(\lambda)$.

holds, then the system

$$\mathfrak{F}^w(\Lambda) = \bigcup_{l=1}^{q} \mathfrak{F}_l^w(\lambda_l)$$

is a basis of the subspace $\operatorname{Ker} R_w(A,B)$.

In particular, under condition (3.1) *the equality*

$$\nu(A,B) = \dim \operatorname{Ker} R_w(A,B),$$

holds, where

$$\nu(A,B) = \sum_{l=1}^{q} \nu(A,B,\lambda_l).$$

Proof. Without loss of generality one can assume that $m \geq n$.

From Lemma 2.1 it follows that $\mathfrak{F}^w(\Lambda) \subset \operatorname{Ker} \mathcal{A}_{w-n}$ and $\mathfrak{F}^w(\Lambda) \subset \operatorname{Ker} \mathcal{B}_{w-m}$. Hence $\mathfrak{F}^w(\Lambda) \subset \operatorname{Ker} R_w(A,B) = \operatorname{Ker} \mathcal{A}_{w-n} \cap \operatorname{Ker} \mathcal{B}_{w-m}$.

Now let $\Omega \in \operatorname{Ker} R_w(A,B)$. Then $\Omega \in \operatorname{Ker} \mathcal{A}_{w-n}$ and, in view of Lemma 2.1, Ω can be represented in the form

$$\Omega = \sum_{l=1}^{q} \sum_{j=1}^{j_l} \sum_{k=0}^{k_{jl}-1} \alpha_{jkl} \Phi_{jk,l}(\lambda_l),$$

where

$$\mathfrak{F}_l^w(\Lambda) = \{\Phi_{jk,l}(\lambda_l) : j = 1,2,\dots,j_l;\ k = 0,1,\dots,k_{jl}-1\}.$$

By Lemma 2.2, this implies that

$$\Omega_k(\lambda_l) = \sum_j \alpha_{jkl} \Phi_{jk,l}(\lambda_l) \in \operatorname{Ker} \mathcal{B}_{w-m}$$

for $k = 0,1,\dots,k_{jl}-1$ and $l = 1,2,\dots,q$, where the summation runs over all j such that $k_{jl} \geq k$. According to Lemma 2.3, the vectors

$$\omega_{kr}(\lambda_l) = \sum_j \alpha_{jkl} \varphi_{jr}(\lambda_l) \quad (r = 0,1,\dots,k;\ k = 0,1,\dots,k_l - 1;\ k_l = \max_j k_{jl})$$

form a Jordan chain for the pencil $B(\lambda)$. Since, besides that, the vectors $\omega_{kr}(\lambda_l)$ $(r = 0,1,\dots,k)$ also represent a Jordan chain for the pencil $A(\lambda)$, we arrive at the conclusion that the vectors $\Omega_k(\lambda_l)$ can be represented as a linear combination of vectors from $\mathfrak{F}^w(\Lambda)$. Therefore the vector Ω is a linear combination of vectors from $\mathfrak{F}^w(\Lambda)$.

The theorem is proved. □

Corollary 3.1. *The pencils* $A(\lambda)$ *and* $B(\lambda)$ *from* (3.1) *have a common eigenvalue* λ_0 *and a common eigenvector at* λ_0 *if and only if the rank of the resultant matrix* $R_w(A,B)$ *for* $w \geq \min\{md+n; nd+m\}$ *is less than maximal.*

3.3. Theorem 0.1 formulated in the introduction is easily derived from Theorem 3.1. Obviously, the classical resultant matrix $R(a, b)$ presented in the introduction coincides with $R_{m+n}(a, b)$. Clearly, in this case (for $d = 1$), condition (3.2) is satisfied. Therefore $\operatorname{Ker} R(a, b)$ consists of the linear span of the vectors

$$\Phi_k(\lambda_l) = (\varphi_k^p(\lambda_l))_{p=0}^{m+n-1}, \quad \text{where} \quad \varphi_k^p(\lambda_l) = \sum_{s=0}^{k} \binom{p}{s} \lambda_l^{p-s}.$$

Since

$$\Phi_k(\lambda_l) = \sum_{s=0}^{k} H_s(\lambda_l), \quad H_k(\lambda_l) = \Phi_k(\lambda_l) - \Phi_{k-1}(\lambda_l),$$

this immediately yields Theorem 0.1.

In the case $d > 1$ for the classical resultant matrix, that is, for $R_{m+n}(A, B)$ one can claim only that

$$\operatorname{Ker} R_{m+n}(A, B) = \operatorname{Ker} \mathcal{A}_m \cap \operatorname{Ker} \mathcal{B}_n$$

and

$$\nu(A, B) \leq \dim \operatorname{Ker} R_{m+n}(A, B).$$

These relations follow straightforwardly from Lemma 2.1.

We present one more statement for the classical resultant matrix for $d \neq 1$.

Let $\mathfrak{F}(\lambda_l, A)$ (resp. $\mathfrak{F}(\mu_l, B)$) be a canonical system of Jordan chains for the pencil $A(\lambda)$ (resp. $B(\lambda)$) at the eigenvalue λ_l (resp. μ_l). Then the number $\dim \operatorname{Ker} R_{m+n}(A, B)$ is equal to the codimension of the subspace

$$\operatorname{lin} \mathfrak{F}^{m+n}(A) \cup \mathfrak{F}^{m+n}(B),$$

where

$$\mathfrak{F}^{m+n}(A) = \bigcup_l \mathfrak{F}^{m+n}(\lambda_l, A), \quad \mathfrak{F}^{m+n}(B) = \bigcup_l \mathfrak{F}^{m+n}(\mu_l, B).$$

In particular, the operator $R_{m+n}(A, B)$ is invertible if and only if the system $\mathfrak{F}^{m+n}(A) \cup \mathfrak{F}^{m+n}(B)$ is complete in $\mathbb{C}^{(m+n)d}$.

It is easy to see that for $w = m+n$ and $d > 1$ condition (3.2′) is fulfilled only in some particular cases. Condition (3.2) (and corresponding condition (3.2′)) of Theorem 3.1 is essential. For the classical resultant matrix, Theorem 3.1 is not true, in general. One can demonstrate this by the following example.

Let

$$A(\lambda) = \begin{bmatrix} 1+\lambda & 0 \\ -1 & -1+\lambda \end{bmatrix}, \quad B(\lambda) = \begin{bmatrix} 1+\lambda & 1 \\ 1 & 1+\lambda \end{bmatrix}.$$

Then the kernel of the resultant matrix $R_2(A, B)$ consists of the set of vectors $(-1, 1, 1, 0)t$ ($t \in \mathbb{C}^1$). On the other hand, ± 1 are the eigenvalues of the pencil $A(\lambda)$ and $0, -2$ are the eigenvalues of the pencil $B(\lambda)$. Thus, $\nu(A, B) = 0$, while $\dim \operatorname{Ker} R_2(A, B) = 1$.

4. Applications

We shall present two applications of the results of Section 3.

4.1. We start with a generalization of the method of elimination of an unknown from a system of equations with two unknowns (see, e.g., [1, Chap. 11, Section 54]).

Let $A(\lambda,\mu)$ and $B(\lambda,\mu)$ be matrix pencils of two variables

$$A(\lambda,\mu)=\sum_{j=0}^{n}\sum_{k=0}^{m}\lambda^j\mu^k A_{jk},\quad B(\lambda,\mu)=\sum_{j=0}^{q}\sum_{k=0}^{p}\lambda^j\mu^k B_{jk},$$

where $\lambda,\mu\in\mathbb{C}^1$ and $A_{jk},B_{jk}\in L(\mathbb{C}^d)$.

Consider the following system of equations

$$A(\lambda,\mu)\varphi=0,\quad B(\lambda,\mu)\varphi=0 \tag{4.1}$$

with unknown numbers λ and μ and an unknown vector $\varphi\in\mathbb{C}^d$ ($\varphi\neq 0$). We make the following assumptions:

a) for some $\mu\in\mu_0\in\mathbb{C}^1$ the system (4.1) does not have a solution;

b) the determinants

$$\det\sum_{k=0}^{m}\mu^k A_{nk},\quad \det\sum_{k=0}^{p}\mu^k B_{qk}$$

are not equal identically to zero;

c) the determinants

$$\det\sum_{j=0}^{n}\lambda^j A_{jm},\quad \det\sum_{j=0}^{q}\lambda^j B_{jp}$$

are not equal identically to zero;

Let conditions a) and b) be fulfilled. We write the pencils $A(\lambda,\mu)$ and $B(\lambda,\mu)$ in powers of the variable λ:

$$\begin{aligned}A(\lambda,\mu)&=A_0(\mu)+\lambda A_1(\mu)+\cdots+\lambda^n A_n(\mu),\\ B(\lambda,\mu)&=B_0(\mu)+\lambda B_1(\mu)+\cdots+\lambda^q B_q(\mu),\end{aligned}$$

where

$$A_j(\mu)=\sum_{k=0}^{m}\mu^k A_{jk},\quad B_j(\mu)=\sum_{k=0}^{p}\mu^k B_{jk}.$$

Put

$$R_w(\mu) = \begin{bmatrix} A_0(\mu) & A_1(\mu) & \dots & A_n(\mu) & & & \\ & A_0(\mu) & \dots & A_{n-1}(\mu) & A_n(\mu) & & \\ & & \ddots & \vdots & \ddots & \ddots & \\ & & & A_0(\mu) & \dots & A_{n-1}(\mu) & A_n(\mu) \\ B_0(\mu) & B_1(\mu) & \dots & B_m(\mu) & & & \\ & B_0(\mu) & \dots & B_{m-1}(\mu) & B_m(\mu) & & \\ & & \ddots & \vdots & \ddots & \ddots & \\ & & & B_0(\mu) & \dots & B_{m-1}(\mu) & B_m(\mu) \end{bmatrix}.$$

Let M_0 be a (finite) set of zeros of the function $\det A_n(\mu)B_q(\mu)$. If w has the property

$$w \geq \min\{nd+q; qd+n\},$$

then for all $\mu \notin M_0$ one can apply Corollary 3.1. Therefore the set of points μ, for which system (4.1) has a solution, consists of the points of the set M_1 of the numbers $\mu_0 \notin M_0$ such that

$$\operatorname{rank} R_w(\mu_0) < w, \tag{4.2}$$

and, maybe, some points of the set M_0.

Substituting in (4.1) the points μ_0 of the set $M_0 \cup M_1$ instead of μ, we get

$$\left.\begin{aligned} A(\lambda,\mu_0)\varphi = 0 \\ B(\lambda,\mu_0)\varphi = 0 \end{aligned}\right\} \quad (\mu_0 \in M_0 \cup M_1). \tag{4.3}$$

In this way solving system (4.1) is reduced to solving the system with one unknown number and one unknown vector.

Now assume that, in addition to conditions a) and b), condition c) is also fulfilled. Then the described process can be repeated interchanging the variables μ and λ. Let Λ_0 be the set of all zeros of the function

$$\det\left(\sum_{j=0}^{n}\lambda^j A_{jm}\right)\left(\sum_{j=0}^{p}\lambda^j B_{jp}\right).$$

The set of the numbers λ, for which system (4.1) has a solution, consists of the set Λ_1 of all points λ_0 such that[8]

$$\operatorname{rank} \widetilde{R}_w(\lambda_0) < w \tag{4.4}$$

and, maybe, some points of the set Λ_0.

[8] $\widetilde{R}_w(\lambda)$ is defined similarly to $R_w(\lambda)$.

Thus, it remains to solve the systems of equations

$$\left.\begin{array}{c} A(\lambda_0,\mu_0)\varphi = 0 \\ B(\lambda_0,\mu_0)\varphi = 0 \end{array}\right\} \tag{4.5}$$

where λ_0 runs through the set $A_0 \cup A_1$ and μ_0 runs through the set $M_0 \cup M_1$.

4.2. Consider the homogeneous differential equations

$$A_n\left(\frac{d^n}{dt^n}\varphi\right)(t) + \cdots + A_1\left(\frac{d}{dt}\varphi\right)(t) + (A_0\varphi)(t) = 0, \tag{4.6}$$

$$B_m\left(\frac{d^m}{dt^m}\psi\right)(t) + \cdots + B_1\left(\frac{d}{dt}\psi\right)(t) + (B_0\psi)(t) = 0, \tag{4.7}$$

where $A_j, B_k \in L(\mathbb{C}^d)$ and the matrices A_n and B_m are invertible. The matrix pencils

$$A(\lambda) = \sum_{k=0}^{n} \lambda^k A_k, \quad B(\lambda) = \sum_{k=0}^{m} \lambda^k B_k$$

correspond to these equations, respectively. There is a close relation between the solutions of equation (4.6) and Jordan chains for the pencil $A(\lambda)$. The general solution of the equation is a linear combination of vector functions of the form

$$\varphi(t) = e^{\lambda_0 t}\left(\frac{t^k}{k!}\varphi_0 + \cdots + \frac{t}{1!}\varphi_{k-1} + \varphi_k\right), \tag{4.8}$$

where $\varphi_0, \varphi_1, \ldots, \varphi_k$ run over all Jordan chains for the pencil $A(\lambda)$. Thus Theorem 3.1 immediately yields the following.

Theorem 4.1. *Let $\mathfrak{N}$ be the subspace of the common solutions of equations* (4.6) *and* (4.7). *If a number w satisfies the condition*

$$w \geq \min\{nd + m, md + n\},$$

then the equality

$$\dim \mathfrak{N} = \dim \operatorname{Ker} R_w(A, B)$$

holds.

With the aid of this theorem and the method of Section 4.1 one can indicate a method for solving the system of differential equations depending on a parameter μ of the form

$$\left.\begin{array}{c} A_n(\mu)\left(\frac{d^n}{dt^n}\varphi\right)(t) + \cdots + A_1(\mu)\left(\frac{d}{dt}\varphi\right)(t) + A_0(\mu)\varphi(t) = 0 \\ B_m(\mu)\left(\frac{d^m}{dt^m}\varphi\right)(t) + \cdots + B_1(\mu)\left(\frac{d}{dt}\varphi\right)(t) + B_0(\mu)\varphi(t) = 0 \end{array}\right\},$$

where $A_k(\mu)$ and $B_k(\mu)$ are matrix pencils.

Consider also one more problem. For given vectors χ_k $(k = 0, 1, \ldots, m+n-1)$ in the space $\mathbb{C}^d$, we will look for all pairs of functions $(\varphi(t), \psi(t))$, where $\varphi(t)$ is a solution of equation (4.6) and $\psi(t)$ is a solution of equation (4.7) that satisfy the initial conditions

$$\frac{d^k}{dt^k}(\varphi(t)+\psi(t))|_{t=0} = \chi_k \quad (k=0,1,\dots,m+n-1). \tag{4.9}$$

This problem has a unique solution for every collection of vectors χ_k ($k = 0,1,\dots,m+n-1$) if and only if the classical resultant matrix $R_{m+n}(A,B)$ is invertible.

Indeed, for every solution $\varphi(t)$ of equation (4.6) (resp. every solution $\psi(t)$ of equation (4.7)), we construct the vector function

$$\Phi(t) = \left(\left(\frac{d^k}{dt^k}\varphi\right)(t)\right)_{k=0}^{m+n-1} \quad \left(\text{resp. } \Psi(t) = \left(\left(\frac{d^k}{dt^k}\psi\right)(t)\right)_{k=0}^{m+n-1}\right).$$

Setting $X = (\chi_k)_{k=0}^{m+n-1}$, we see that initial conditions (4.9) take the form

$$\Phi(0) + \Psi(0) = X.$$

If the vector function $\varphi(t)$ (resp. $\psi(t)$) runs over all solutions of equation (4.6) (resp. (4.7)), then from equality (1.2) it follows immediately that the vectors $\Phi(0)$ (resp. $\Psi(0)$) run over the system of vectors of the extension $\mathfrak{F}^{m+l}$ (resp. $\mathfrak{G}^{m+l}$) of multiplicity $m+n$ of Jordan chains for the pencil $A(\lambda)$ (resp. $B(\lambda)$). Therefore, for every $X \in \mathbb{C}^{(m+n)d}$ the problem has a solution if and only if the union of the systems $\mathfrak{F}^{m+l}$ and $\mathfrak{G}^{m+l}$ is complete in $\mathbb{C}^{(m+n)d}$. As it is noticed in Section 3, the last fact holds if and only if the matrix $R_{m+n}(A,B)$ is invertible. It is easy to see that in this case the solution is unique.

5. Kernel of Bezoutian

5.1. As yet another application of Theorem 0.1, we present a description of the kernel of the Bezoutian of two polynomials in the case $d=1$.

Let $a(\lambda) = a_0 + a_1\lambda + \cdots + a_n\lambda^n$ and $b(\lambda) = b_0 + b_1\lambda + \cdots + b_m\lambda^m$ ($m \le n$) be two polynomials ($a_k, b_k \in \mathbb{C}^1$; $a_n \neq 0$). Consider the polynomial in two variables

$$B(\lambda,\mu) = \frac{a(\lambda)b(\mu) - a(\mu)b(\lambda)}{\lambda - \mu} = \sum_{p,q=0}^{n-1} b_{pq}\lambda^p\mu^q.$$

The quadratic matrix $\mathcal{B}(a,b) = \|b_{pq}\|_{p,q=0}^{n-1}$ is said to be the Bezoutian of the polynomials $a(\lambda)$ and $b(\lambda)$. It is known (see, e.g., [2, 3]) that the defect of the Bezoutian is equal to the degree of the greatest common divisor of the polynomials $a(\lambda)$ and $b(\lambda)$. This statement admits the following refinement.

Theorem 5.1. *The kernel of the Bezoutian $\mathcal{B}(a,b)$ of the polynomials $a(\lambda)$ and $b(\lambda)$ consists of the linear span of the vectors*

$$\varphi_{jk} = \left(\binom{p}{k}\lambda_j^p\right)_{p=0}^{n-1} \quad (k = 0,1,\dots,\nu_j - 1;\ j = 1,2,\dots,l),$$

where λ_j $(j = 1,2,\dots,l)$ are all common zeros of the polynomials $a(\lambda)$ and $b(\lambda)$, and ν_j is the common multiplicity of the zero λ_j.

Proof. For the Bezoutian $\mathcal{B}(a,b)$ the equality

$$\begin{aligned}\mathcal{B}(a,b) = &\begin{bmatrix} a_1 & \dots & a_{n-1} & a_n \\ a_2 & \dots & a_n & 0 \\ \vdots & \iddots & \vdots & \vdots \\ a_n & \dots & 0 & 0 \end{bmatrix}\begin{bmatrix} b_0 & b_1 & \dots & b_{n-1} \\ 0 & b_0 & \dots & b_{n-2} \\ \vdots & \vdots & \ddots & \vdots \\ 0 & 0 & \dots & b_0 \end{bmatrix} \\ &- \begin{bmatrix} b_1 & \dots & b_{n-1} & b_n \\ b_2 & \dots & b_n & 0 \\ \vdots & \iddots & \vdots & \vdots \\ b_n & \dots & 0 & 0 \end{bmatrix}\begin{bmatrix} a_0 & a_1 & \dots & a_{n-1} \\ 0 & a_0 & \dots & a_{n-2} \\ \vdots & \vdots & \ddots & \vdots \\ 0 & 0 & \dots & a_0 \end{bmatrix}\end{aligned} \tag{5.1}$$

holds (see [3]). We form the matrices

$$\widetilde{A}_n = \begin{bmatrix} a_n & 0 & \dots & 0 \\ a_{n-1} & a_n & \dots & 0 \\ \vdots & \vdots & \ddots & \vdots \\ a_1 & a_2 & \dots & a_n \end{bmatrix}, \quad A_n = \begin{bmatrix} a_0 & a_1 & \dots & a_{n-1} \\ 0 & a_0 & \dots & a_{n-2} \\ \vdots & \vdots & \ddots & \vdots \\ 0 & 0 & \dots & a_0 \end{bmatrix},$$

$\Delta = [\delta_{j,n-k-1}]_{j,k=0}^{n-1}$, and the matrices $\widetilde{B_n}$ and B_n in the same fashion. Then equality (5.1) takes the form

$$\mathcal{B}(a,b) = \Delta(\widetilde{A}_n B_n - \widetilde{B}_n A_n). \tag{5.2}$$

Therefore the equation $\mathcal{B}(a,b)\varphi = 0$ is equivalent to the equation

$$(\widetilde{A_n} B_n - \widetilde{B}_n A_n)\varphi = 0. \tag{5.3}$$

Obviously, in the presented notation,

$$R(b,a) = \begin{bmatrix} B_n & \widetilde{B}_n \\ A_n & \widetilde{A}_n \end{bmatrix}.$$

It is easy to see that

$$\begin{bmatrix} B_n & \widetilde{B}_n \\ A_n & \widetilde{A}_n \end{bmatrix} = \begin{bmatrix} I & \widetilde{B}_n\widetilde{A}_n^{-1} \\ 0 & I \end{bmatrix}\begin{bmatrix} B_n - \widetilde{B}_n\widetilde{A}_n^{-1}A_n & 0 \\ 0 & \widetilde{A}_n \end{bmatrix}\begin{bmatrix} I & 0 \\ \widetilde{A}_n^{-1}A_n & I \end{bmatrix}.$$

Obviously, the matrices $\widetilde{B}_n$ and $\widetilde{A}_n$ commute. Therefore,

$$B_n - \widetilde{B}_n\widetilde{A}_n^{-1}A_n = \widetilde{A}_n^{-1}(\widetilde{A}_n B_n - \widetilde{B}_n A_n). \tag{5.4}$$

Let

$$\begin{bmatrix} B_n & \widetilde{B}_n \\ A_n & \widetilde{A}_n \end{bmatrix} \begin{bmatrix} f \\ g \end{bmatrix} = 0. \tag{5.5}$$

Then from equalities (5.2)–(5.4) it follows that $\mathcal{B}(a,b)f = 0$.

Conversely, if $\mathcal{B}(a,b)f = 0$, then equality (5.5) holds for $g = \widetilde{A}_n^{-1} A_n f$. It remains to apply Theorem 0.1.

The theorem is proved. □

5.2. Given two polynomials of the form

$$x(\lambda) = x_0 + x_1\lambda + \cdots + x_n\lambda^n, \quad y(\lambda) = y_0 + y_{-1}\lambda^{-1} + \ldots y_{-n}\lambda^{-n}, \tag{5.6}$$

where $x_n \neq 0$ or $y_{-n} \neq 0$. The quadratic matrix $\widetilde{\mathcal{B}}(x,y) = \|b_{pq}\|_{p,q=0}^{n-1}$, where

$$\sum_{p,q=0}^{n-1} b_{pq}\lambda^p\mu^q = \frac{x(\lambda)y(\mu^{-1}) - (\lambda\mu)^n x(\mu^{-1})y(\lambda)}{1-\lambda\mu}$$

is said to be the Bezoutian of the polynomials $x(\lambda)$ and $y(\lambda)$. By a straightforward verification we obtain

$$\widetilde{\mathcal{B}}(x,y) = \begin{bmatrix} x_0 & 0 & \ldots & 0 \\ x_1 & x_0 & \ldots & 0 \\ \vdots & \vdots & \ddots & \vdots \\ x_{n-1} & x_{n-2} & \ldots & x_0 \end{bmatrix} \begin{bmatrix} y_0 & y_{-1} & \ldots & y_{1-n} \\ 0 & y_0 & \ldots & y_{2-n} \\ \vdots & \vdots & \ddots & \vdots \\ 0 & 0 & \ldots & y_0 \end{bmatrix}$$
$$- \begin{bmatrix} y_{-n} & 0 & \ldots & 0 \\ y_{1-n} & y_{-n} & \ldots & 0 \\ \vdots & \vdots & \ddots & \vdots \\ y_{-1} & y_{-2} & \ldots & y_{-n} \end{bmatrix} \begin{bmatrix} x_n & x_{n-1} & \ldots & x_1 \\ 0 & x_n & \ldots & x_2 \\ \vdots & \vdots & \ddots & \vdots \\ 0 & 0 & \ldots & x_n \end{bmatrix}.$$

For the Bezoutian $\widetilde{\mathcal{B}}(x,y)$ an analogue of Theorem 5.1 holds. For completeness we present its formulation.

Theorem 5.2. *The kernel of the Bezoutian of the polynomials $x(\lambda)$ and $y(\lambda)$ of the form* (5.6) *consists of the linear span of the vectors*

$$\varphi_{jk} = \left(\lambda_j^p \binom{p}{k}\right)_{p=0}^{n-1} \quad (k = 0, 1, \ldots, k_j - 1),$$

where λ_j $(j = 1, 2, \ldots, l)$ are all common zeros of the functions $x(\lambda)$ and $y(\lambda)$, and k_j is the common multiplicity of the zero λ_j.

References

[1] A.G. Kurosh, *A Course in Higher Algebra*, 9th edition, Nauka, Moscow, 1968 (in Russian).

English translation from the 10th Russian edition: *Higher Algebra.* Mir Publishers, Moscow, 1975. MR0384363 (52 #5240).

[2] M.G. Krein and M.A. Naimark, *The method of symmetric and Hermitian forms in the theory of the separation of the roots of algebraic equations.* Khar'kov, 1936 (in Russian). English translation: Linear and Multilinear Algebra **10** (1981), no. 4, 265–308. MR0638124 (84i:12016), Zbl 0584.12018.

[3] F.I. Lander, *The Bezoutian and the inversion of Hankel and Toeplitz matrices.* Matem. Issled. **9** (1974), no. 2(32), 69–87 (in Russian). MR0437559 (55 #10483), Zbl 0331.15017.

[4] I.C. Gohberg and A.A. Semencul, *The inversion of finite Toeplitz matrices and their continual analogues.* Matem. Issled. **7** (1972), no. 2(24), 201–223 (in Russian). MR0353038 (50 #5524), Zbl 0288.15004.

[5] I.C. Gohberg and G. Heinig, *Inversion of finite Toeplitz matrices consisting of elements of a noncommutative algebra.* Rev. Roumaine Math. Pures Appl. **19** (1974), 623–663 (in Russian). English translation: **this volume.** MR0353040 (50 #5526), Zbl 0337.15005.

[6] I.C. Gohberg and G. Heinig, *Matrix integral operators on a finite interval with kernels depending on the difference of the arguments.* Rev. Roumaine Math. Pures Appl. **20** (1975), 55–73 (in Russian). English translation: **this volume.** MR0380495 (52 #1395), Zbl 0327.45009.

[7] M.G. Krein, *Distribution of roots of polynomials orthogonal on the unit circle with respect to a sign-alternating weight.* Teor. Funkts., Funkts. Anal. Prilozh. (Khar'kov) **2** (1966), 131–137 (in Russian). MR0201702 (34 #1584), Zbl 0257.30002.

[8] I.C. Gohberg and M.G. Krein, *Introduction to the Theory of Linear Nonselfadjoint Operators in Hilbert space.* Nauka, Moscow, 1965 (in Russian). MR0220070 (36 #3137), Zbl 0138.07803.

English translation: *Introduction to the Theory of Linear Nonselfadjoint Operators.* Amer. Math. Soc., Providence, R.I. 1969. MR0246142 (39 #7447), Zbl 0181.13504.

French translation: *Introduction à la Théorie des Opérateurs Linéaires non Auto-Adjoints Dans un Espace Hilbertien.* Dunod, Paris, 1971. MR0350445 (50 #2937).

[9] I.C. Gohberg and E.I. Sigal, *An operator generalization of the logarithmic residue theorem and the theorem of Rouché.* Matem. Sbornik, New Ser. **84(126)** (1971), 607–629 (in Russian). English translation: Math. USSR Sbornik **13** (1971), 603–625. MR0313856 (47 #2409), Zbl 0254.47046.

Operator Theory:
Advances and Applications, Vol. 206, 89–109

The Resultant Matrix and its Generalizations. II. The Continual Analogue of the Resultant Operator

Israel Gohberg and Georg Heinig

Let $\mathcal{A}(\lambda)$ and $\mathcal{B}(\lambda)$ ($\lambda \in \mathbb{C}^1$) be entire functions of the form

$$\mathcal{A}(\lambda) = a_0 + \int_0^\tau a(t)e^{i\lambda t}\,dt, \quad \mathcal{B}(\lambda) = b_0 + \int_{-\tau}^0 b(t)e^{i\lambda t}\,dt, \tag{0.1}$$

where $a_0, b_0 \in \mathbb{C}^1$, $a(t) \in L_1(0,\tau)$, $b(t) \in L_1(-\tau,0)$, and τ is some positive number.

To this pair of functions assign a bounded linear operator $R_0(\mathcal{A},\mathcal{B})$ acting in the space $L_1(-\tau,\tau)$ by the formula

$$(R_0(\mathcal{A},\mathcal{B})f)(t) = \begin{cases} f(t) + \displaystyle\int_{-\tau}^{\tau} a(t-s)f(s)\,ds & (0 \le t \le \tau), \\ f(t) + \displaystyle\int_{-\tau}^{\tau} b(t-s)f(s)\,ds & (-\tau \le t < 0), \end{cases}$$

where we assume that $a(t) = 0$ for $t \neq [0,\tau]$ and $b(t) = 0$ for $t \notin [-\tau, 0]$. It is natural to consider this operator as a continual analogue of the resultant operator for two polynomials (see [1]). The operator $R_0(\mathcal{A},\mathcal{B})$ is said to be the resultant operator of the functions $\mathcal{A}(\lambda)$ and $\mathcal{B}(\lambda)$.

The following theorem holds.

Theorem 0.1. *Let $\mathcal{A}(\lambda)$ and $\mathcal{B}(\lambda)$ be entire functions of the form* (0.1). *Suppose $\lambda_1, \lambda_2, \dots, \lambda_l$ is the complete collection of distinct common zeros of the functions $\mathcal{A}(\lambda)$ and $\mathcal{B}(\lambda)$, and k_j is the multiplicity of a common zero λ_j. Then the system of functions*

$$\psi_{jk}(t) = t^k e^{-i\lambda_j t} \quad (k = 0, 1, \dots, k_j - 1;\ j = 1, 2, \dots, l)$$

The paper is a translation of the paper И.Ц. Гохберг, Г. Хайниг, Результантная матрица и её обобщения. II. Континуальный аналог результантного оператора, Acta Math. Acad. Sci. Hungar. **28** (1976), no. 3-4, 189–209. MR0425652 (54 #13606), Zbl 0341.15011.

forms a basis of the subspace $\operatorname{Ker} R_0(\mathcal{A},\mathcal{B})$. *In particular,*

$$\sum_{j=1}^{l} k_j = \dim \operatorname{Ker} R_0(\mathcal{A},\mathcal{B}).$$

This theorem is a continual analogue of Theorem 0.1 in [1].

The present paper is a continuation of the authors' paper [1], here continual analogues of other theorems from [1] are obtained as well.

Note that in the continual matrix case the definition of the resultant operator becomes more involved. By analogy with the discrete case, the continual resultant operator for a matrix function acts from one space of vector functions to another space of vector functions with wider support.

There is an essential difference between the discrete and continual resultant operators for matrix functions. It consists in the fact that the choice of spaces, in which the resultant operator acts in the continual case, does not depend on the matrices itself and their orders.

The paper consists of six sections. In Section 1 the main theorem is formulated. It is proved in Section 3. In Section 4 Theorem 0.1 is proved. Section 2 has an auxiliary character. In the last two sections some applications of the main theorem are presented.

The authors express their sincere gratitude to M.G. Krein for extremely useful discussions.

1. Formulation of the main theorem

1.1. Let α and β ($-\infty < \alpha < \beta < \infty$) be a pair of real numbers and d be a natural number. By $L_1^d(\alpha,\beta)$ denote the Banach space of all vector functions $f(t) = (f_j(t))_{j=1}^d$ with entries in $L_1(\alpha,\beta)$. Analogously, by $L_1^{d\times d}(\alpha,\beta)$ denote the space of matrix functions $a(t) = \|a_{jk}(t)\|_{j,k=1}^d$ of order d with entries in $L_1(\alpha,\beta)$.

By $F^{d\times d}(\alpha,\beta)$ denote the space of all matrix functions of the form

$$\mathcal{A}(\lambda) = a_0 + \int_\alpha^\beta a(t)e^{i\lambda t}\,dt, \tag{1.1}$$

where $a_0 \in L(\mathbb{C}^d)$ [1] and $a(t) \in L_1^{d\times d}(\alpha,\beta)$, and by $F_0^{d\times d}(\alpha,\beta)$ denote the subspace of $F^{d\times d}(\alpha,\beta)$ with invertible first summands.

The space $F^{d\times d}(\alpha,\beta)$ consists of entire matrix functions.

Let τ be a positive number. To each pair of matrix functions of the form

$$\mathcal{A}(\lambda) = a_0 + \int_0^\tau a(t)e^{i\lambda t}\,dt \in F^{d\times d}(0,\tau)$$

and

$$\mathcal{B}(\lambda) = a_0 + \int_{-\tau}^0 b(t)e^{i\lambda t}\,dt \in F^{d\times d}(-\tau,0)$$

[1] The space of the quadratic matrices of order d is denoted by $L(\mathbb{C}^d)$.

and to each number $\varepsilon > 0$ we assign the operator $R_\varepsilon(\mathcal{A},\mathcal{B})$ acting from the space $L_1^d(-\tau,\tau+\varepsilon)$ to the space $L_1^d(-\tau-\varepsilon,\tau+\varepsilon)$ by the rule[2]

$$(R_\varepsilon(\mathcal{A},\mathcal{B})\phi)(t) = \begin{cases} \phi(t) + \displaystyle\int_{-\tau}^{\tau+\varepsilon} a(t-s)\phi(s)\,ds & (0 \le t \le \tau+\varepsilon), \\ \phi(t+\varepsilon) + \displaystyle\int_{-\tau}^{\tau+\varepsilon} b(t+\varepsilon-s)\phi(s)\,ds & (-\tau-\varepsilon \le t < 0). \end{cases} \tag{1.2}$$

By analogy with the discrete case, we refer to the operator $R_\varepsilon(\mathcal{A},\mathcal{B})$ as the resultant operator of the matrix functions $\mathcal{A}(\lambda)$ and $\mathcal{B}(\lambda)$. The operator $R_0(\mathcal{A},\mathcal{B})$ acting in the space $L_1^d(-\tau,\tau)$ is called the classical resultant operator.

It is easy to see that the resultant operator $R_\varepsilon(\mathcal{A},\mathcal{B})$ is closely connected to the operator $\widetilde{R}_\varepsilon(\mathcal{A},\mathcal{B})$ acting from the space $L_1^d(-\tau,\tau+\varepsilon)$ to the space $L_1^d(-\tau,\varepsilon)\dot{+}L_1^d(0,\tau+\varepsilon)$ by the formula

$$(\widetilde{R}_\varepsilon(\mathcal{A},\mathcal{B})\phi)(t) = \begin{cases} \phi(t) + \displaystyle\int_{-\tau}^{\tau+\varepsilon} a(t-s)\phi(s)\,ds & (0 \le t \le \tau+\varepsilon), \\ \phi(t) + \displaystyle\int_{-\tau}^{\tau+\varepsilon} b(t-s)\phi(s)\,ds & (-\tau \le t \le \varepsilon). \end{cases}$$

Indeed, if $\widetilde{R}_\varepsilon(\mathcal{A},\mathcal{B})\phi = (f_1,f_2)$ where $f_1 \in L_1^d(-\tau,\varepsilon)$ and $f_2 \in L_1^d(0,\tau+\varepsilon)$, then

$$(R_\varepsilon(\mathcal{A},\mathcal{B})\phi)(t) = \begin{cases} f_1(t+\varepsilon) & (-\tau-\varepsilon \le t < 0), \\ f_2(t) & (0 \le t \le \tau+\varepsilon). \end{cases}$$

Obviously, the equality

$$\operatorname{Ker} R_\varepsilon(\mathcal{A},\mathcal{B}) = \operatorname{Ker} \widetilde{R}_\varepsilon(\mathcal{A},\mathcal{B})$$

holds.

1.2. Recall some definitions and notation from [1][3]. Let $\mathcal{A}(\lambda)$ be an entire matrix function. A number $\lambda_0 \in \mathbb{C}^1$ is said to be an eigenvalue of the matrix function $\mathcal{A}(\lambda)$ if $\det \mathcal{A}(\lambda_0) = 0$. A vector $\phi_0 \in \mathbb{C}^d$ is said to be an eigenvector at the eigenvalue λ_0 if $\mathcal{A}(\lambda_0)\phi_0 = 0$. A collection of vectors $\phi_0, \phi_1, \dots, \phi_r$ is said to be a Jordan chain a chain of an eigenvector and associated vectors (or a Jordan chain) of length $r+1$ if the equalities

$$\mathcal{A}(\lambda_0)\phi_k + \frac{1}{1!}\left(\frac{d}{d\lambda}\mathcal{A}\right)(\lambda_0)\phi_{k-1} + \cdots + \frac{1}{k!}\left(\frac{d^k}{d\lambda^k}\mathcal{A}\right)(\lambda_0)\phi_0 = 0 \tag{1.3}$$

hold for $k = 0, 1, \dots, r$.

Let λ_0 be an eigenvalue of the matrix function $\mathcal{A}(\lambda)$. It is easy to prove that in the kernel of the operator $\mathcal{A}(\lambda_0)$ one can construct a basis $\phi_{10}, \phi_{20}, \dots, \phi_{r0}$ with the following property: for each vector there exists a Jordan chain $\phi_{j0}, \phi_{j1}, \dots, \phi_{j,k_j-1}$,

[2]Here and in what follows it is set $a(t) = 0$ for $t \notin [0,\tau]$ and $b(t) = 0$ for $t \notin [-\tau, 0]$.
[3]See also [2] and [3].

where $k_1 \geq k_2 \geq \cdots \geq k_r$ and $\sum_j k_j$ is equal to the multiplicity of zero of the function $\det \mathcal{A}(\lambda)$ at the point λ_0. The numbers k_j $(j = 1, 2, \ldots, r)$ are said to be partial multiplicities of the eigenvalue λ_0 and the system $\phi_{j0}, \phi_{j1}, \ldots, \phi_{j,k_j-1}$ $(j = 1, 2, \ldots, r)$ is said to be a canonical system of Jordan chains for the matrix function $\mathcal{A}(\lambda)$ at the eigenvalue λ_0.

Consider two entire matrix functions $\mathcal{A}(\lambda)$ and $\mathcal{B}(\lambda)$. Let λ_0 be a common eigenvalue of the matrix functions $\mathcal{A}(\lambda)$ and $\mathcal{B}(\lambda)$ and let

$$\mathfrak{N} = \operatorname{Ker} \mathcal{A}(\lambda_0) \cap \operatorname{Ker} \mathcal{B}(\lambda_0).$$

Let $\phi_0, \phi_1, \ldots, \phi_r$ be a Jordan chain simultaneously for the pencils $\mathcal{A}(\lambda)$ and $\mathcal{B}(\lambda)$ at the eigenvalue λ_0. The number $r+1$ is called the length of this common chain. The greatest length of such a common chain starting with the vector ϕ_0 is said to be the rank of the common eigenvector ϕ_0 and is denoted by $\operatorname{rank}_{\lambda_0} \phi_0$.

In the subspace $\mathfrak{N}$ we choose a basis $\phi_{10}, \phi_{20}, \ldots, \phi_{j0}$ such that the ranks k_j of its vectors have the following properties: k_1 is the greatest of the numbers $\operatorname{rank}_{\lambda_0} \phi$ $(\phi \in \mathfrak{N})$, and k_j $(j = 2, 3, \ldots, l)$ is the greatest of the numbers $\operatorname{rank}_{\lambda_0} \phi$ for all vectors ϕ from the direct complement to $\operatorname{lin}\{\phi_{10}, \phi_{20}, \ldots, \phi_{j-1,0}\}$ in $\mathfrak{N}$ that contains ϕ_{j0}.

It is easy to see that the number $\operatorname{rank}_{\lambda_0} \phi_0$ for every vector $\phi_0 \in \mathfrak{N}$ is equal to one of the numbers k_j $(j = 1, 2, \ldots, l)$. Hence the numbers k_j $(j = 1, 2, \ldots, l)$ are determined uniquely by the pencils $\mathcal{A}(\lambda)$ and $\mathcal{B}(\lambda)$. By $\phi_{j1}, \phi_{j2}, \ldots, \phi_{j,k_j-1}$ denote the corresponding common for $\mathcal{A}(\lambda)$ and $\mathcal{B}(\lambda)$ chain of vectors associated with the common eigenvector ϕ_{j0} $(j = 1, 2, \ldots, l)$.

The system

$$\phi_{j0}, \phi_{j1}, \ldots, \phi_{j,k_j-1} \quad (j = 1, 2, \ldots, l)$$

is called a canonical system of common Jordan chains for the matrix functions $\mathcal{A}(\lambda)$ and $\mathcal{B}(\lambda)$ at the common eigenvalue λ_0, and the number

$$\nu(\mathcal{A}, \mathcal{B}, \lambda_0) \stackrel{\text{def}}{=} \sum_{j=1}^{l} k_j$$

is called the common multiplicity of the eigenvalue λ_0 of the matrix functions $\mathcal{A}(\lambda)$ and $\mathcal{B}(\lambda)$.

Let us also agree on the following notation:

$$\nu(\mathcal{A}, \mathcal{B}) \stackrel{\text{def}}{=} \sum_{l} \nu(\mathcal{A}, \mathcal{B}, \lambda_l)$$

where λ_l runs over all common eigenvalues of the matrix functions $\mathcal{A}(\lambda)$ and $\mathcal{B}(\lambda)$.

Note that for every pair of matrix functions $\mathcal{A}(\lambda) \in F_0^{d\times d}(0, \tau)$ and $\mathcal{B}(\lambda) \in F_0^{d\times d}(-\tau, 0)$ the number $\nu(\mathcal{A}, \mathcal{B})$ is finite. Indeed, it is easy to see that the matrix function $\mathcal{A}(\lambda)$ is bounded in the upper half-plane and

$$\lim_{\operatorname{Im}\lambda \geq 0, \lambda \to \infty} \mathcal{A}(\lambda) = a_0,$$

and the matrix function $\mathcal{B}(\lambda)$ is bounded in the lower half-plane and

$$\lim_{\operatorname{Im}\lambda\le 0,\lambda\to\infty} \mathcal{B}(\lambda) = b_0.$$

This implies that the function $\det \mathcal{A}(\lambda)$ has at most a finite number of zeros in the upper half-plane and the function $\det \mathcal{B}(\lambda)$ has at most a finite number of zeros in the lower half-plane.

1.3. The main result of this paper is the following.

Theorem 1.1. *Let $\mathcal{A}(\lambda)$ and $\mathcal{B}(\lambda)$ be two matrix functions in $F_0^{d\times d}(0,\tau)$ and $F_0^{d\times d}(-\tau,0)$, respectively, let $\Lambda = \{\lambda_1, \lambda_2, \dots, \lambda_l\}$ be the complete collection of distinct common eigenvalues of the matrix functions $\mathcal{A}(\lambda)$ and $\mathcal{B}(\lambda)$, and let*

$$\mathfrak{F}_l = \left\{\phi_{jk,l} :\ k = 0, 1, \dots, k_{jl} - 1;\ j = 1, 2, \dots, j_l\right\} \quad (l = 1, 2, \dots, l_0)$$

be a canonical system of common Jordan chains at the eigenvalue λ_l.

Then for every $\varepsilon > 0$ the system of functions

$$\phi_{jk,l}(t) = e^{-i\lambda_l t}\left(\frac{(-it)^k}{k!}\phi_{j0,l} + \frac{(-it)^{k-1}}{(k-1)!}\phi_{j1,l} + \cdots + \phi_{jk,l}\right) \tag{1.4}$$

where $-\tau \le t \le \tau + \varepsilon$ and $k = 0, 1, \dots, k_{jl} - 1$; $j = 1, 2, \dots, j_l$; $l = 1, 2, \dots, l_0$, forms a basis of the kernel of the resultant operator $R_\varepsilon(\mathcal{A}, \mathcal{B})$. In particular, the equality

$$\nu(\mathcal{A}, \mathcal{B}) = \dim \operatorname{Ker} R_\varepsilon(\mathcal{A}, \mathcal{B}) \tag{1.5}$$

holds.

1.4. It turns out that finding the kernel of the resultant operator can be reduced to finding the kernels of two operators acting in the same space (in contrast to the operator $R_\varepsilon(\mathcal{A}, \mathcal{B})$). We introduce the operators $R'_\varepsilon(\mathcal{A}, \mathcal{B})$ and $R''_\varepsilon(\mathcal{A}, \mathcal{B})$ acting in the space $L_1^d(-\tau, \tau + \varepsilon)$ by the formulas

$$(R'_\varepsilon(\mathcal{A}, \mathcal{B})\phi)(t) = \begin{cases} a_0\phi(t) + \displaystyle\int_{-\tau}^{\tau+\varepsilon} a(t-s)\phi(s)\,ds & (0 \le t \le \tau + \varepsilon), \\ b_0\phi(t) + \displaystyle\int_{-\tau}^{\tau+\varepsilon} b(t-s)\phi(s)\,ds & (-\tau \le t < 0), \end{cases} \tag{1.6}$$

and

$$(R''_\varepsilon(\mathcal{A}, \mathcal{B})\phi)(t) = \begin{cases} a_0\phi(t) + \displaystyle\int_{-\tau}^{\tau+\varepsilon} a(t-s)\phi(s)\,ds & (\varepsilon \le t \le \tau + \varepsilon), \\ b_0\phi(t) + \displaystyle\int_{-\tau}^{\tau+\varepsilon} b(t-s)\phi(s)\,ds & (-\tau \le t < \varepsilon). \end{cases}$$

It is easy to see that the equality

$$\operatorname{Ker} R_\varepsilon(\mathcal{A}, \mathcal{B}) = \operatorname{Ker} R'_\varepsilon(\mathcal{A}, \mathcal{B}) \cap \operatorname{Ker} R''_\varepsilon(\mathcal{A}, \mathcal{B}) \tag{1.7}$$

holds.

Note also that the role of the resultant operator can be played also by the operator $\widehat{R}_\varepsilon(\mathcal{A},\mathcal{B})$ acting from the space $L_1^d(-\tau-\varepsilon,\tau)$ to the space $L_1^d(-\tau-\varepsilon,\tau+\varepsilon)$ by the rule

$$(\widehat{R}_\varepsilon(\mathcal{A},\mathcal{B})\phi)(t)=\begin{cases} a_0\phi(t-\varepsilon)+\displaystyle\int_{-\tau-\varepsilon}^{\tau} a(t-\varepsilon-s)\phi(s)\,ds & (0\le t\le \tau+\varepsilon),\\ b_0\phi(t)+\displaystyle\int_{-\tau-\varepsilon}^{\tau} b(t-s)\phi(s)\,ds & (-\tau-\varepsilon\le t<0).\end{cases}$$

In fact, the operator $\widehat{R}_\varepsilon(\mathcal{A},\mathcal{B})$ coincides with the operator $R_\varepsilon(\mathcal{A},\mathcal{B})$.

1.5. We give a simple example showing that Theorem 1.1 does not remain true in the matrix case for $\varepsilon=0$, that is, Theorem 0.1 does not admit a direct generalization to the matrix case.

Let

$$a(t)=\left\|\begin{matrix}1 & 1\\ 1 & -1\end{matrix}\right\| \quad (0\le t\le 1);\qquad a(t)=0\quad (t\notin[0,1]),$$

$$b(t)=\left\|\begin{matrix}1 & 1\\ -1 & -1\end{matrix}\right\| \quad (-1\le t\le 0);\quad b(t)=0\quad (t\notin[-1,0]),$$

and $a_0=b_0=I$, where I is the identity matrix.

Then

$$\mathcal{A}(\lambda)=a_0+\int_0^1 a(t)e^{i\lambda t}\,dt=\frac{1}{i\lambda}\left\|\begin{matrix} i\lambda+e^{i\lambda}-1 & e^{i\lambda}-1\\ e^{i\lambda}-1 & i\lambda-e^{i\lambda}+1\end{matrix}\right\|$$

and

$$\mathcal{B}(\lambda)=b_0+\int_{-1}^0 a(t)e^{i\lambda t}\,dt=\frac{1}{i\lambda}\left\|\begin{matrix} i\lambda+1-e^{-i\lambda} & 1-e^{-i\lambda}\\ e^{i\lambda}-1 & i\lambda-1+e^{-i\lambda}\end{matrix}\right\|$$

for $\lambda\neq 0$ and

$$\mathcal{A}(0)=\left\|\begin{matrix}2 & 1\\ 1 & 0\end{matrix}\right\|,\quad \mathcal{B}(0)=\left\|\begin{matrix}2 & 1\\ -1 & 0\end{matrix}\right\|.$$

We have

$$\det\mathcal{A}(\lambda)=1-2\left(\frac{e^{i\lambda}-1}{i\lambda}\right)^2\quad(\lambda\neq 0);\quad \det\mathcal{A}(0)=-1$$

and

$$\det\mathcal{B}(\lambda)\equiv 1.$$

Thus $\nu(\mathcal{A},\mathcal{B})=0$.

The operator $R_0(\mathcal{A},\mathcal{B})$ is determined by the equality

$$(R_0(\mathcal{A},\mathcal{B})\phi)(t)=\phi(t)+\begin{cases}\left\|\begin{matrix}1 & 1\\ 1 & -1\end{matrix}\right\|\displaystyle\int_0^1\phi(t-s)\,ds & (0\le t\le 1),\\ \left\|\begin{matrix}1 & 1\\ -1 & -1\end{matrix}\right\|\displaystyle\int_{-1}^0\phi(t-s)\,ds & (-1\le t<0).\end{cases}$$

It is easy to see that for the vector function

$$\phi(t) = \left\| \begin{array}{c} 1+t-\theta(t) \\ -1-t \end{array} \right\|,$$

where

$$\theta(t) = \begin{cases} 0 & (t<0), \\ 1 & (t\geq 0), \end{cases}$$

the equality $R_0(\mathcal{A},\mathcal{B})\phi = 0$ holds. Therefore $1 \leq \dim \operatorname{Ker} R_0(\mathcal{A},\mathcal{B})$ and

$$\nu(\mathcal{A},\mathcal{B}) \neq \dim \operatorname{Ker} R_0(\mathcal{A},\mathcal{B}).$$

2. A lemma

The statement obtained in this section will be used in Section 3 in the proof of the main theorem.

Lemma 2.1. *Let $\mathcal{A}(\lambda)$ and $\mathcal{B}(\lambda)$ be entire functions in $F_0^{d\times d}(0,\tau)$ and $F_0^{d\times d}(-\tau,0)$, respectively. Then for every $\varepsilon > 0$ the kernel of the resultant operator $R_\varepsilon(\mathcal{A},\mathcal{B})$ consists of absolutely continuous functions only.*

Proof. First, we show that the kernel of the operator $R'_\varepsilon(\mathcal{A},\mathcal{B})$ consists of vector functions that are absolutely continuous on the intervals $[-\tau,0)$ and $[0,\tau+\varepsilon]$ only. Consider the operator K'_ε acting in the space $L_1^d(-\tau,\tau+\varepsilon)$ by the rule

$$(K'_\varepsilon f)(t) = \begin{cases} \displaystyle\int_{-\tau}^{\tau+\varepsilon} a(t-s)\phi(s)\,ds & (0\leq t\leq \tau+\varepsilon), \\ \displaystyle\int_{-\tau}^{\tau+\varepsilon} b(t-s)\phi(s)\,ds & (-\tau\leq t<0). \end{cases}$$

Obviously, the equality

$$(R'_\varepsilon(\mathcal{A},\mathcal{B})f)(t) = (K'_\varepsilon f)(t) + \begin{cases} a_0 f(t) & (0\leq t\leq \tau+\varepsilon), \\ b_0 f(t) & (-\tau\leq t<0) \end{cases} \tag{2.1}$$

holds. It is easy to see that K'_ε is a bounded linear operator in $L_1^d(-\tau,\tau+\varepsilon)$ and the estimate

$$\|K'_\varepsilon\|_{L_1^d} \leq \|a(t)\|_{L_1^{d\times d}} + \|b(t)\|_{L_1^{d\times d}} \quad [4] \tag{2.2}$$

holds.

[4]Editor's remark. Here and elsewhere the following notations are used: If A is a bounded linear operator acting on a space $L_1^d(\alpha,\beta)$ whose support (α,β) is evident from the context we just write $\| A \|_d$ for the norm of A; similarly, if a is a $d\times d$ matrix function whose support is evident from the context we write $\| a \|_{L_1^{d\times d}}$.

We will prove that the operator K'_ε is compact in the space $L_1^d(-\tau, \tau+\varepsilon)$. Indeed, the functions $a(t)$ and $b(t)$ can be approximated in the norm of the space $L_1^d(-\tau, \tau+\varepsilon)$ to any required degree of accuracy by matrix functions of the form

$$\sum_{j=-m}^{m} e^{2\pi ijt/(2\tau+\varepsilon)} a_j. \tag{2.3}$$

Let $\widetilde{a}(t)$ and $\widetilde{b}(t)$ be the functions of the form (2.3) such that $\|\widetilde{a}(t) - a(t)\|_{L_1^{d\times d}} < \delta$ and $\|\widetilde{b}(t) - b(t)\|_{L_1^{d\times d}} < \delta$, where $\delta > 0$ is a given number.

Obviously, the operator $\widetilde{K}$ defined by the equality

$$(\widetilde{K}f)(t) = \begin{cases} \displaystyle\int_{-\tau}^{\tau+\varepsilon} \widetilde{a}(t-s)\phi(s)\,ds & (0 \le t \le \tau+\varepsilon), \\ \displaystyle\int_{-\tau}^{\tau+\varepsilon} \widetilde{b}(t-s)\phi(s)\,ds & (-\tau \le t < 0) \end{cases}$$

is of finite rank. From estimate (4) it follows that

$$\|K'_\varepsilon - \widetilde{K}\|_{L_1^d} \le \|\widetilde{a}(t) - a(t)\|_{L_1^{d\times d}} + \|\widetilde{b}(t) - b(t)\|_{L_1^{d\times d}} < 2\delta.$$

Thus the operator K can be approximated (in the norm) to any required degree of accuracy by finite rank operators. Hence the operator K'_ε is compact. In view of (2.1), from the compactness of the operator K'_ε it follows that

$$\dim \operatorname{Ker} R'_\varepsilon(\mathcal{A}, \mathcal{B}) = \dim \operatorname{Coker} R'_\varepsilon(\mathcal{A}, \mathcal{B}) < \infty. \tag{2.4}$$

By $\widehat{W}_0^d$ denote the Banach space of the vector functions $\phi(t) \in L_1^d(-\tau, \tau+\varepsilon)$ that are absolutely continuous on the intervals $[-\tau, 0)$ and $[0, \tau+\varepsilon]$ and have the limits

$$\phi(-0) \left(= \lim_{t\to -0} \phi(t) \right) \in \mathbb{C}^1.$$

Obviously, the space $\widehat{W}_0^d$ is a direct sum of the space of all absolutely continuous vector functions on $[-\tau, \tau]$ and the space of the vector functions of the form $c\theta(t)$, where $c \in \mathbb{C}^d$ and

$$\theta(t) = \begin{cases} 0 & (t < 0), \\ 1 & (t \ge 0). \end{cases}$$

A norm in the space $\widehat{W}_0^d$ is defined by the equality

$$\|\phi(t)\|_{\widehat{W}_0^d} = \|\phi(t)\|_{L_1^d} + \|(D\phi)(t)\|_{L_1^d},$$

here and in the sequel we set $(D\phi)(t) = \left(\frac{d}{dt}\phi\right)(t)$ for $t \neq 0$ (almost everywhere) and

$$(D\phi)(0) = \lim_{t\to +0} (D\phi)(t)$$

(if this limit exists).

Let $\widehat{K}'_\varepsilon$ be the restriction of the operator K'_ε to the space $\widehat{W}_0^d$. The operator $\widehat{K}'_\varepsilon$ is a compact operator in the space $\widehat{W}_0^d$. Indeed, for a vector function $f(t) \in \widehat{W}_0^d$ one has

$$(K'_\varepsilon f)(t) = \begin{cases} f(t) + \int_0^t a(r)f(t-r)\,dr + \int_t^\tau a(r)f(t-r)\,dr & (0 \le t \le \tau+\varepsilon), \\ f(t) + \int_{-\tau}^t b(r)f(t-r)\,dr + \int_t^0 b(r)f(t-r)\,dr & (-\tau \le t < 0). \end{cases}$$

Obviously, the right-hand side of the last equality is differentiable almost everywhere and

$$\left(\frac{d}{dt}\widehat{K}'_\varepsilon f\right)(t) = \begin{cases} (Df)(t) + \int_0^t a(r)Df(t-r)\,dr \\ \qquad + \int_t^\tau a(r)Df(t-r)\,dr & (0 \le t \le \tau), \\ \qquad + a(t)(f(0) - f(-0)) \\ (Df)(t) + \int_{-\tau}^t b(r)Df(t-r)\,dr \\ \qquad + \int_t^0 b(r)Df(t-r)\,dr & (-\tau \le t < 0). \\ \qquad + b(t)(f(0) - f(-0)) \end{cases}$$

Since $Df(t) \in L_1^d(-\tau, \tau+\varepsilon)$ and the operator K'_ε maps the space $L_1^d(-\tau, \tau+\varepsilon)$ into itself, from the last equality it follows that $K'_\varepsilon f \in \widehat{W}_0^d$ and the estimate

$$\|\widehat{K}'_\varepsilon f\|_{\widehat{W}_0^d} \le \varrho\big(\|a(t)\|_{L_1^{d\times d}} + \|b(t)\|_{L_1^{d\times d}}\big)\|f\|_{\widehat{W}_0^d}$$

holds, where ϱ is some constant that does not depend on $f(t)$, $a(t)$, and $b(t)$. With the aid of this estimate, the compactness of the operator $\widehat{K}'_\varepsilon$ in the space $\widehat{W}_0^d$ is proved in the same way as the compactness of the operator K'_ε in the space $L_1^d(-\tau, \tau+\varepsilon)$.

In view of the compactness of the operator $\widehat{K}_\varepsilon$ and equality (2.1), we get

$$\dim \operatorname{Ker} \widehat{R}'_\varepsilon = \dim \operatorname{Coker} \widehat{R}'_\varepsilon \tag{2.5}$$

where $\widehat{R}'_\varepsilon$ is the restriction of the operator $R'_\varepsilon(\mathcal{A}, \mathcal{B})$ to the space $\widehat{W}_0^d$. Since

$$\dim \operatorname{Ker} R'_\varepsilon \le \dim \operatorname{Ker} R'_\varepsilon(\mathcal{A}, \mathcal{B}), \quad \dim \operatorname{Coker} \widehat{R}'_\varepsilon \ge \dim \operatorname{Coker} R'_\varepsilon(\mathcal{A}, \mathcal{B}),$$

from (2.4) and (2.5) it follows that

$$\dim \operatorname{Ker} \widehat{R}'_\varepsilon = \dim \operatorname{Ker} R'_\varepsilon(\mathcal{A}, \mathcal{B}) < \infty.$$

Therefore, $\operatorname{Ker} \widehat{R}'_\varepsilon = \operatorname{Ker} R'_\varepsilon(\mathcal{A}, \mathcal{B})$ and

$$\operatorname{Ker} R'_\varepsilon(\mathcal{A}, \mathcal{B}) \subset \widehat{W}_0^d. \tag{2.6}$$

Now consider the operator K''_ε acting in the space $L_1^d(-\tau,\tau+\varepsilon)$ by the rule

$$(K''_\varepsilon\phi)(t) = \begin{cases} a_0\phi(t) + \int_0^\tau a(s)\phi(t-s)\,ds & (\varepsilon \le t \le \tau+\varepsilon), \\ b_0\phi(t) + \int_{-\tau}^0 b(s)\phi(t-s)\,ds & (-\tau \le t < \varepsilon). \end{cases}$$

Then the equality

$$(R''_\varepsilon\phi)(t) = (K''_\varepsilon\phi)(t) + \begin{cases} a_0\phi(t) & (\varepsilon \le t \le \tau+\varepsilon), \\ b_0\phi(t) & (-\tau \le t < \varepsilon) \end{cases}$$

holds.

By $\widehat{W}^d_\varepsilon$ denote the space of vector functions $\phi(t) \in L_1^d(-\tau,\tau+\varepsilon)$ that are absolutely continuous on the intervals $[-\tau,\varepsilon)$ and $[\varepsilon,\tau+\varepsilon]$ and have the limit

$$\phi(\varepsilon-0)\ \left(= \lim_{h\to\varepsilon-0}\phi(h)\right) \in \mathbb{C}^1.$$

With the aid of the previous arguments one can prove that the operator K''_ε maps the space $\widehat{W}^d_\varepsilon$ into itself and the restriction of the operator K''_ε to the space $\widehat{W}^d_\varepsilon$ is a compact operator in $\widehat{W}^d_\varepsilon$. Hence, the embedding

$$\operatorname{Ker} R''_\varepsilon(\mathcal{A},\mathcal{B}) \subset \widehat{W}^d_\varepsilon$$

holds. According to equality (1.7), from here and relation (2.6) it follows that

$$\operatorname{Ker} R''_\varepsilon(\mathcal{A},\mathcal{B}) \subset \widehat{W}^d_\varepsilon \cap \widehat{W}^d_0.$$

Since $\widehat{W}^d_\varepsilon \cap \widehat{W}^d_0$ coincides with the set of absolutely continuous vector functions on the interval $[-\tau,\tau+\varepsilon]$, this immediately implies the statement.

The lemma is proved. □

3. Proof of the main theorem

Let us introduce the operators $R_\varepsilon(\mathcal{A})$ and $R_\varepsilon(\mathcal{B})$ setting

$$(R_\varepsilon(\mathcal{A})\phi)(t) = \phi(t) + \int_{-\tau}^{\tau+\varepsilon} a(t-s)\phi(s)\,ds \quad (0 \le t \le \tau+\varepsilon)$$

and

$$(R_\varepsilon(\mathcal{B})\phi)(t) = \phi(t) + \int_{-\tau}^{\tau+\varepsilon} b(t-s)\phi(s)\,ds \quad (-\tau \le t < \varepsilon).$$

The operator $R_\varepsilon(\mathcal{A})$ acts from the space $L_1^d(-\tau,\tau+\varepsilon)$ to the space $L_1^d(0,\tau+\varepsilon)$ and the operator $R_\varepsilon(\mathcal{B})$ acts from $L_1^d(-\tau,\tau+\varepsilon)$ to $L_1^d(-\tau,\varepsilon)$. Obviously, the equality

$$\operatorname{Ker} R_\varepsilon(\mathcal{A},\mathcal{B}) = \operatorname{Ker} R_\varepsilon(\mathcal{A}) \cap \operatorname{Ker} R_\varepsilon(\mathcal{B}) \tag{3.1}$$

holds.

Let ϕ_{0k} $(k = 0, 1, \dots, k_0 - 1)$ be a common Jordan chain for the matrix functions $\mathcal{A}(\lambda)$ and $\mathcal{B}(\lambda)$ at the eigenvalue λ_0. Let us prove that then all functions

$$\phi_{0k}(t) = e^{-i\lambda_0 t}\left(\frac{(-it)^k}{k!}\phi_{00} + \dots + \frac{-it}{1!}\phi_{0,k-1} + \phi_{0k}\right) \tag{3.2}$$
$$(-\tau \le t \le \tau + \varepsilon;\ k = 0, 1, \dots, k_0 - 1)$$

belong to $\operatorname{Ker} R_\varepsilon(\mathcal{A}, \mathcal{B})$. Indeed, for $k = 0, 1, \dots, k_0 - 1$ and $0 \le t \le \tau + \varepsilon$,

$$\begin{aligned}(R_\varepsilon(\mathcal{A})\phi_{0k})(t) &= e^{-i\lambda_0 t}\sum_{r=0}^{k}\left(\frac{(-it)^r}{r!} + \int_0^\tau a(s)e^{i\lambda_0 s}\frac{[i(t-s)]^r}{r!}\,ds\right)\phi_{0,k-r}\\ &= e^{-i\lambda_0 t}\sum_{r=0}^{k}\sum_{p=0}^{r}\frac{-i^r t^p}{r!}\left(\delta_{rp} + \int_0^\tau \binom{r}{p} a(s)e^{i\lambda_0 s}(is)^{r-p}\,ds\right)\phi_{0,k-r}.\end{aligned}$$

Since

$$\left(\frac{d^p}{d\lambda^p}\mathcal{A}\right)(\lambda) = a_0\delta_{0p} + (-i)^p\int_0^\tau (-s)^p a(s)e^{i\lambda s}ds,$$

we have

$$\begin{aligned}R_\varepsilon(\mathcal{A})\phi_{0k}(t) &= e^{-i\lambda_0 t}\sum_{r=0}^{k}\sum_{p=0}^{r}\frac{(-i)^p}{p!(r-p)!}t^p\left(\frac{d^{r-p}}{d\lambda^{r-p}}\mathcal{A}\right)(\lambda_0)\phi_{0,k-r}\\ &= e^{-\lambda_0 t}\sum_{p=0}^{k}\frac{(-it)^p}{p!}\sum_{r=p}^{k}\frac{1}{(r-p)!}\left(\frac{d^{r-p}}{d\lambda^{r-p}}\mathcal{A}\right)(\lambda_0)\phi_{0,k-r}.\end{aligned} \tag{3.3}$$

Due to the definition of a Jordan chain, this implies that $R_\varepsilon(\mathcal{A})\phi_{0k}(t) = 0$. Analogously it is proved that $R_\varepsilon(\mathcal{B})\phi_{0k}(t) = 0$. Therefore $\phi_{0k}(t) \in \operatorname{Ker} R_\varepsilon(\mathcal{A}, \mathcal{B})$. From what has been proved above, in particular, it follows that

$$\dim \operatorname{Ker} R_\varepsilon(\mathcal{A}, \mathcal{B}) \ge \nu(\mathcal{A}, \mathcal{B}).$$

Now assume that the vector function $\phi(t)$ belongs to the kernel of the operator $R_\varepsilon(\mathcal{A}, \mathcal{B})$. Then

$$\phi(t) + \int_0^\tau a(s)\phi(t-s)\,ds = 0 \quad (0 \le t \le \tau + \varepsilon) \tag{3.4}$$

and

$$\phi(t) + \int_{-\tau}^0 b(s)\phi(t-s)\,ds = 0 \quad (-\tau \le t < \varepsilon). \tag{3.5}$$

In view of Lemma 2.1, the vector function $\phi(t)$ is absolutely continuous. From here and equalities (3.4)–(3.5) it follows that for every $r = 0, 1, \dots, k$ the function $\frac{d^r}{dt^r}\phi(t)$ belongs to $\operatorname{Ker} R_\varepsilon(\mathcal{A}, \mathcal{B})$. Since $\dim \operatorname{Ker} R_\varepsilon(\mathcal{A}, \mathcal{B}) < \infty$, this implies that there exist numbers α_j $(j = 1, 2, \dots, m_0)$ such that

$$\sum_{j=0}^{m_0}\alpha_j\frac{d^j}{dt^j}\phi(t) = 0 \quad (-\tau \le t \le \tau + \varepsilon).$$

Hence, the vector function $\phi(t)$ has the form

$$\phi(t) = \sum_{j=1}^{l} p_j(t)e^{-i\lambda_j t},$$

where $p_j(t)$ are polynomials with vector coefficients.

Let us show that all summands $p_j(t)e^{-i\lambda_j t}$ $(j = 1, 2, \dots, l)$ also belong to the kernel of the operator $R_\varepsilon(\mathcal{A}, \mathcal{B})$. If $\phi(t) \in \operatorname{Ker} R_\varepsilon(\mathcal{A}, \mathcal{B})$, then, in particular,

$$0 = \phi(t) + \int_0^\tau a(s)\phi(t-s)\,ds$$

for $0 \le t \le \tau + \varepsilon$. According to (3.5) we obtain

$$\begin{aligned} 0 &= \sum_{j=1}^{l} \left(p_j(t)e^{-i\lambda_j t} + \int_0^\tau a(s)p_j(t-s)e^{i\lambda_j(t-s)}\,ds \right) \\ &= \sum_{j=1}^{l} e^{-i\lambda_j t} \left(p_j(t) + \int_0^\tau a(s)p_j(t-s)e^{-i\lambda_j s}\,ds \right). \end{aligned} \tag{3.6}$$

It is easy to see that the vector function

$$q_j(t) = p_j(t) + \int_0^\tau a(s)p_j(t-s)e^{i\lambda_j s}\,ds$$

is a polynomial with vector coefficients.

It is known that a system of scalar functions of the form $e^{\mu_j t}r_j(t)$ $(j = 1, 2, \dots, l)$, where $r_j(t)$ are polynomials and μ_j are pairwise distinct complex numbers, is linearly independent. Hence the system of vector functions $q_j(t)e^{-i\lambda_j t}$ is linearly independent. This fact and (3.5) imply that

$$0 = e^{-i\lambda_j t}q_j(t) = e^{-i\lambda_j t}p_j(t) + \int_0^\tau a(s)p_j(t-s)e^{-i\lambda_j(t-s)}\,ds.$$

The last equality means that $e^{-i\lambda_j t}p_j(t) \in \operatorname{Ker} R_\varepsilon(\mathcal{A})$. Analogously it is proved that $e^{-i\lambda_j t}p_j(t) \in \operatorname{Ker} R_\varepsilon(\mathcal{B})$. According to equality (3.1), this implies that

$$e^{-i\lambda_j t}p_j(t) \in \operatorname{Ker} R_\varepsilon(\mathcal{A}, \mathcal{B}) \quad (j = 1, 2, \dots, l).$$

Let

$$p_j(t) = \frac{(-it)^{k_j}}{k_j!}\phi_{j0} + \frac{(-it)^{k_j-1}}{(k_j-1)!}\phi_{j1} + \dots + \phi_{jk_j} \quad (j = 1, 2, \dots, l).$$

Then

$$0 = R_\varepsilon(\mathcal{A})q_j(t) = e^{-i\lambda_j t}\sum_{k=0}^{k_j} \left(\frac{(-it)^k}{k!} + \int_0^\tau a(s)e^{i\lambda_j s}\frac{[-i(t-s)]^k}{k!}i^k\,ds \right) \phi_{j,k_j-k}.$$

Taking into account that

$$\left(\frac{d^k}{d\lambda^k}\mathcal{A}\right)(\lambda) = a_0\delta_{k0} + \int_0^\tau (is)^k a(s)e^{i\lambda s}\,ds,$$

we obtain

$$0 = e^{-i\lambda_j t} \sum_{p=0}^{k_j} \frac{(-it)^p}{p!} \sum_{k=p}^{k_j} \frac{1}{(k-p)!} \left(\frac{d^{k-p}}{d\lambda^{k-p}} \mathcal{A} \right) (\lambda_j) \phi_{j,k_j-k}.$$

Thus,

$$\sum_{k=p}^{k_j} \frac{1}{(k-p)!} \left(\frac{d^{k-p}}{d\lambda^{k-p}} \mathcal{A} \right) (\lambda_j) \phi_{j,k_j-k} = 0 \quad (0 \le t \le \tau + \varepsilon)$$

for $p = 0, 1, \ldots, k_j$ and $j = 1, 2, \ldots, l$. This means that for every $j = 1, 2, \ldots, l$ the vectors $\phi_{j0}, \phi_{j1}, \ldots, \phi_{jk}$ $(k = 0, 1, \ldots, k_j)$ form a Jordan chain for the matrix function $\mathcal{A}(\lambda)$ at the eigenvalue λ_j. Analogously it is proved that $\phi_{j0}, \phi_{j1}, \ldots, \phi_{jk}$ is also a Jordan chain for the matrix function $\mathcal{B}(\lambda)$.

Hence it is proved that every vector function in $\operatorname{Ker} R_\varepsilon(\mathcal{A}, \mathcal{B})$ is a linear combination of vector functions of the form (1.4). In particular, this implies equality (1.5). The theorem is proved. □

Theorem 1.1 and equality (1.7) imply the following.

Corollary 3.1. *Entire matrix functions $\mathcal{A}(\lambda) \in F_0^{d\times d}(0, \tau)$ and $\mathcal{B}(\lambda) \in F_0^{d\times d}(-\tau, 0)$ do not have any common eigenvector corresponding to the same eigenvalue if and only if for some $\varepsilon > 0$ the operator $R'_\varepsilon(\mathcal{A}, \mathcal{B})$ or the operator $R''_\varepsilon(\mathcal{A}, \mathcal{B})$ is invertible in the space $L_1^d(-\tau, \tau + \varepsilon)$.*

Corollary 3.2. *Under the hypotheses of Theorem 1.1, the inequality*

$$\nu(\mathcal{A}, \mathcal{B}) \le \dim \operatorname{Ker} R_0(\mathcal{A}, \mathcal{B})$$

holds.

Indeed, this follows from the obvious inclusion

$$\operatorname{Ker} R_\varepsilon(\mathcal{A}, \mathcal{B}) \subset \operatorname{Ker} R_0(\mathcal{A}, \mathcal{B}).$$

4. Scalar case

In this section Theorem 0.1 is proved.

4.1. We will need the following lemma.

Lemma 4.1. *Let $a(t) \in L_1(0, \tau)$, $b(t) \in L_1(-\tau, 0)$, and*

$$\mathcal{A}(\lambda) = 1 + \int_0^\tau a(s) e^{i\lambda s}\, ds, \quad \mathcal{B}(\lambda) = 1 + \int_{-\tau}^0 b(s) e^{i\lambda s}\, ds.$$

If the system of equations

$$\begin{aligned} a(t) - \int_0^\tau a(s)\omega(t-s)\, ds &= \omega(t) \quad (0 \le t \le \tau), \\ b(t) - \int_{-\tau}^0 b(s)\omega(t-s)\, ds &= \omega(t) \quad (-\tau \le t < 0) \end{aligned} \tag{4.1}$$

has a solution $\omega(t) \in L_1(-\tau,\tau)$*, then the classical resultant operator* $R_0(\mathcal{A},\mathcal{B})$ *is invertible.*

This lemma is obtained in [4] (see also [5, Proposition 3.5]).

With the aid of Lemma 4.1 we prove the following.

Lemma 4.2. *Let* $\mathcal{A}(\lambda) \in F_0^{1\times 1}(0,\tau)$ *and* $\mathcal{B}(\lambda) \in F_0^{1\times 1}(-\tau,0)$*. Then the kernel of the classical resultant operator* $R_0(\mathcal{A},\mathcal{B})$ *consists of absolutely continuous functions only.*

Proof. With the aid of the arguments from the proof of Lemma 2.1 one can show that every function $\phi(t) \in \operatorname{Ker} R_0(\mathcal{A},\mathcal{B})$ can be represented in the form $\phi(t) = \phi_0(t) - c\theta(t)$, where $\phi_0(t)$ is an absolutely continuous function and $c \in \mathbb{C}^1$.

Let

$$\mathcal{A}(\lambda) = a_0 + \int_0^\tau a(t)e^{i\lambda t}\,dt, \quad \mathcal{B}(\lambda) = b_0 + \int_{-\tau}^0 b(t)e^{i\lambda t}\,dt.$$

Assume that

$$\phi(t) = \phi_0(t) - c\theta(t) \in \operatorname{Ker} R_0(\mathcal{A},\mathcal{B}) \tag{4.2}$$

and $c \neq 0$. Since

$$R_0(\mathcal{A},\mathcal{B})\theta(t) = \begin{cases} a_0 + \int_0^t a(s)\,ds & (0 \le t \le \tau), \\ \int_{-\tau}^t b(s)\,ds & (-\tau \le t < 0), \end{cases}$$

we have

$$\phi_0(t) + \int_0^\tau a(s)\phi(t-s)\,ds = \left(a_0 + \int_0^t a(s)\,ds\right)c \quad (0 \le t \le \tau),$$

$$\phi_0(t) + \int_{-\tau}^0 b(s)\phi(t-s)\,ds = \left(\int_{-\tau}^t b(s)\,ds\right)c \quad (-\tau \le t < 0).$$

Differentiating the last equalities and setting

$$\omega(t) = \frac{1}{c}\frac{d}{dt}\phi(t) \quad (t \neq 0),$$

we obtain

$$a(t) - \int_0^\tau a(s)\omega(t-s)\,ds = \omega(t) \quad (0 \le t \le \tau),$$

$$b(t) - \int_{-\tau}^0 b(s)\omega(t-s)\,ds = \omega(t) \quad (-\tau \le t < 0).$$

In view of Lemma 4.1 this implies that the operator $R_0(\mathcal{A},\mathcal{B})$ is invertible. This contradicts the assumption $\phi(t) \in \operatorname{Ker} R_0(\mathcal{A},\mathcal{B})$ $(\phi \neq 0)$. Hence we have $c = 0$ in representation (4.2). This means that the function $\phi(t)$ is absolutely continuous.

The lemma is proved. □

Note that Lemma 4.1 does not remain valid for matrix functions. One can demonstrate this with the example presented in Section 1.

4.2. *Proof of Theorem* 0.1. For the proof of Theorem 0.1 we repeat the proof of Theorem 1.1, where we apply Lemma 4.2 instead of Lemma 2.1. We obtain that the kernel of the operator $R_0(\mathcal{A},\mathcal{B})$ consists of the linear hull of the functions of the form

$$\phi_{jk}(t) = e^{-i\lambda_j t}\left(\frac{(-it)^k}{k!}\phi_{j0} + \cdots + \frac{-it}{1!}\phi_{j,k-1} + \phi_{jk}\right)$$
$$(k = 0,1,\ldots,k_j;\ j = 1,2,\ldots,l)$$

where ϕ_{jk} are complex numbers and $\phi_{j0} \neq 0$.

The equality

$$\phi_{jk}(t) = \sum_{r=0}^{k} \frac{i^r}{r!}\phi_{jr}\psi_{jr}(t)$$

holds. On the other hand,

$$\phi_{jk}(t) - \phi_{j,k-1}(t) = \frac{-i^k}{k!}\phi_{j0}t^k.$$

From the last two equalities it follows that the linear hull of the functions $\phi_{jk}(t)$ coincides with the linear hull of the functions $\psi_{jk}(t)$.

The theorem is proved. □

Corollary 4.1. *Entire functions $\mathcal{A}(\lambda) \in F_0^{1\times 1}(0,\tau)$ and $\mathcal{B}(\lambda) \in F_0^{1\times 1}(-\tau,0)$ do not have any common zero if and only if the operator $R_0(\mathcal{A},\mathcal{B})$ is invertible.*

5. Applications

We present an example of application of theorems on a continual analogue of the resultant operator to the problem of elimination of an unknown variable from a system of two (in general, transcendental) equations with two unknowns.

5.1. First, consider the scalar case. Let $\mathcal{A}(\lambda,\mu)$ and $\mathcal{B}(\lambda,\mu)$ be entire functions in λ and μ of the form

$$\begin{aligned}\mathcal{A}(\lambda,\mu) &= a_0 + \int_0^\tau\int_0^\tau a(t,s)e^{i(\lambda t+\mu s)}\,ds\,dt,\\ \mathcal{B}(\lambda,\mu) &= b_0 + \int_{-\tau}^0\int_{-\tau}^0 b(t,s)e^{i(\lambda t+\mu s)}\,ds\,dt,\end{aligned} \tag{5.1}$$

where $a_0, b_0 \in \mathbb{C}^1$; $a_0, b_0 \neq 0$; $0<\tau<\infty$; and

$$a(t,s) \in L_1([0,\tau]\times[0,\tau]), \quad b(t,s) \in L_1([-\tau,0]\times[-\tau,0]).$$

Consider the system of equations

$$\begin{cases} \mathcal{A}(\lambda,\mu) = 0,\\ \mathcal{B}(\lambda,\mu) = 0 \end{cases} \tag{5.2}$$

with unknowns $\lambda \in \mathbb{C}^1$ and $\mu \in \mathbb{C}^1$. Consider the functions

$$a_\mu(t) = \int_0^\tau a(t,s)e^{i\mu s}\,ds \quad (\mu \in \mathbb{C}^1,\ t \in [0,\tau])$$

and

$$b_\mu(t) = \int_{-\tau}^0 b(t,s)e^{i\mu s}\,ds \quad (\mu \in \mathbb{C}^1,\ t \in [-\tau,0]).$$

Obviously, for every fixed μ the function $a_\mu(t)$ (resp. $b_\mu(t)$) belongs to the space $L_1(0,\tau)$ (resp. $L_1(-\tau,0)$). Hence the functions $a_\mu(t)$ and $b_\mu(t)$ can be considered as vector functions in μ with values in the spaces $L_1(0,\tau)$ and $L_1(-\tau,0)$, respectively. These functions are entire. Indeed, let

$$a'_\mu(t) = \int_0^\tau a(t,s)(is)e^{i\mu s}\,ds.$$

Then for $h \in \mathbb{C}^1$ we have

$$\begin{aligned}\left\| a'_\mu - \frac{1}{h}(a_{\mu+h} - a_\mu)\right\|_{L_1(0,\tau)} &\le \int_0^\tau\int_0^\tau |a(t,s)| \left|\frac{1}{h}(e^{ihs}-1) - is\right| |e^{i\mu s}|\,ds\,dt \\ &\le e^{|\mu|\tau}\int_0^\tau\int_0^\tau |a(t,s)| \left|\frac{1}{h}(e^{ihs}-1) - is\right|\,ds\,dt.\end{aligned}$$

Hence

$$\lim_{h\to 0}\left\| a'_\mu - \frac{1}{h}(a_{\mu+h} - a_\mu)\right\|_{L_1(0,\tau)} = 0.$$

Analogously it is proved that b_μ is also an entire vector function.

For every $\mu \in \mathbb{C}^1$ consider the classical resultant operator $R_0(\mu)$ acting in the space $L_1(-\tau,\tau)$ by the formula

$$(R_0(\mu)\phi)(t) = \begin{cases} a_0\phi(t) + \int_{-\tau}^\tau a_\mu(t-s)\phi(s)\,ds & (0 \le t \le \tau), \\ b_0\phi(t) + \int_{-\tau}^\tau b_\mu(t-s)\phi(s)\,ds & (-\tau \le t < 0). \end{cases} \tag{5.3}$$

The operator function $R_0(\mu)$ is entire. Indeed, put

$$(R'_0(\mu)\phi)(t) = \begin{cases} \int_{-\tau}^\tau a'_\mu(t-s)\phi(s)\,ds & (0 \le t \le \tau), \\ \int_{-\tau}^\tau b'_\mu(t-s)\phi(s)\,ds & (-\tau \le t < 0). \end{cases}$$

Then in view of estimate (4) for every $h \in \mathbb{C}^1$ we obtain

$$\begin{aligned}\left\|\frac{1}{h}(R_0(\mu+h) - R_0(\mu)) - R'_0(\mu)\right\|_{L_1(-\tau,\tau)} &\le \left\|\frac{1}{h}(a_{\mu+h} - a_\mu) - a'_\mu\right\|_{L_1(0,\tau)} \\ &\quad + \left\|\frac{1}{h}(b_{\mu+h} - b_\mu) - b'_\mu\right\|_{L_1(-\tau,0)}.\end{aligned}$$

Therefore,

$$\lim_{h \to 0} \left\| \frac{1}{h}(R_0(\mu+h) - R_0(\mu)) - R_0'(\mu) \right\|_{L_1(-\tau,\tau)} = 0 \quad (\mu \in \mathbb{C}^1).$$

Applying Theorem 0.1 to the entire functions $\mathcal{A}(\lambda,\mu)$ and $\mathcal{B}(\lambda,\mu)$ for a fixed μ, we obtain that the set of the eigenvalues of the operator function $R_0(\mu)$ coincides with the set of the points μ' for which the functions $\mathcal{A}(\lambda,\mu')$ and $\mathcal{B}(\lambda,\mu')$ have common zeros. It is not difficult to obtain that for the operator function $R_0(\mu)$ there exist points μ_0 such that the operator $R_0(\mu_0)$ is invertible. Hence the set of the eigenvalues of the operator function $R_0(\mu)$ is discrete.

Thus, system (5.2) is reduced to the family of systems of equations with one unknown

$$\begin{cases} \mathcal{A}(\lambda,\mu_j) = 0, \\ \mathcal{B}(\lambda,\mu_j) = 0, \end{cases}$$

where μ_j runs over the set of the eigenvalues of the operator function $R_0(\mu)$. In some cases, finding the eigenvalues of the operator function $R_0(\mu)$ can be reduced to finding the zeros of some entire function.

Indeed, if $a_\mu(t) \in L_2(0,\tau)$ and $b_\mu(t) \in L_2(-\tau,0)$, then the operators

$$(K(\mu)\phi)(t) = -\phi(t) + \begin{cases} a_0^{-1}R_0(\mu)\phi(t) & (0 \le t \le \tau), \\ b_0^{-1}R_0(\mu)\phi(t) & (-\tau \le t < 0) \end{cases}$$

belong to the class of the Hilbert-Schmidt operators. Therefore the set of the eigenvalues of the operator function $R_0(\mu)$ coincides with the set of the zeros of the entire function $\widetilde{\det}(I+K(\mu))$, where $\widetilde{\det}(I+K(\mu))$ denotes the regularized determinant of the operator $I+K(\mu)$ (see [2, Chap. IV, Section 2]).

Now we interchange the roles of the variables λ and μ and repeat the process described above with respect to the resultant operator

$$(R_0(\lambda)\phi)(t) \stackrel{\text{def}}{=} \begin{cases} a_0\phi(t) + \displaystyle\int_{-\tau}^{\tau} a_\lambda(t-s)\phi(s)\,ds & (0 \le t \le \tau), \\ b_0\phi(t) + \displaystyle\int_{-\tau}^{\tau} b_\lambda(t-s)\phi(s)\,ds & (-\tau \le t < 0), \end{cases} \tag{5.4}$$

where

$$a_\lambda(s) = \int_0^\tau a(t,s)e^{i\lambda t}\,dt, \quad b_\lambda(s) = \int_{-\tau}^0 b(t,s)e^{i\lambda t}\,dt.$$

We obtain that the system of equations (5.2) may be satisfied only by the eigenvalues λ of the operator $R_0(\lambda)$. By $\{\lambda_j\}$ denote the set of these eigenvalues. Finally, one concludes that all the solutions of system (5.2) are contained among the pairs (λ_j,μ_k).

5.2. Everything said above is naturally generalized to the case of matrix functions, that is, to the case of solution of the system of equations

$$\begin{cases} \mathcal{A}(\lambda,\mu)\phi = 0, \\ \mathcal{B}(\lambda,\mu)\phi = 0 \end{cases} \tag{5.5}$$

with unknown numbers λ and μ and an unknown vector $\phi \in \mathbb{C}^d$ under the assumption that

$$\mathcal{A}(\lambda,\mu) = a_0 + \int_0^\tau \int_0^\tau a(t,s)e^{i(\lambda t+\mu s)}\,dt\,ds \tag{5.6}$$

and

$$\mathcal{B}(\lambda,\mu) = b_0 + \int_{-\tau}^0 \int_{-\tau}^0 b(t,s)e^{i(\lambda t+\mu s)}\,dt\,ds, \tag{5.7}$$

where $a_0, b_0 \in L(\mathbb{C}^d)$; $a(t,s), b(-t,-s) \in L_1^{d\times d}([0,\tau]\times[0,\tau])$.

In view of Corollary 3.2, here we can restrict ourselves to applying the classical resultant operator and then reduce the problem (5.5) to solving the system of equations

$$\begin{cases} \mathcal{A}(\lambda_j,\mu_k)\phi = 0, \\ \mathcal{B}(\lambda_j,\mu_k)\phi = 0, \end{cases}$$

where μ_k runs over all eigenvalues of the operator $R_0(\mu)$ defined by equality (5.3) and λ_j runs over all eigenvalues of the operator $R_0(\lambda)$ defined by equality (5.4).

5.3. Note that one can essentially extend a number of applications of the main theorems by varying classes of matrix functions under consideration.

6. Continual analogue of the Bezoutian

Let

$$\mathcal{A}(\lambda) = 1 + \int_0^\tau a(t)e^{i\lambda t}\,dt, \quad \mathcal{B}(\lambda) = \int_{-\tau}^0 b(t)e^{i\lambda t}\,dt,$$

where $a(t) \in L_1(0,\tau)$ and $b(t) \in L_1(-\tau,0)$.

Consider the function

$$\mathcal{G}(\lambda,\mu) \overset{\text{def}}{=} \frac{i}{\lambda+\mu}\left(\mathcal{A}(\lambda)\mathcal{B}(-\mu) - e^{i\tau(\lambda+\mu)}\mathcal{A}(-\mu)\mathcal{B}(\lambda)\right) \quad (\lambda,\mu \in \mathbb{C}^1).$$

It is easy to see (see [5]) that

$$\mathcal{G}(\lambda,\mu) = \int_0^\tau \int_0^\tau \gamma(t,s)e^{i(\lambda t+\mu s)}\,dt\,ds,$$

where

$$\gamma(t,s) \overset{\text{def}}{=} a(t-s)+b(t-s)+\int_0^{\min(t,s)} \left(a(t-r)b(r-s)-b(t-r-\tau)a(r-s+\tau)\right)dr$$

and $0 \le t,s \le \tau$. Notice that $\gamma(t,s) \in L_1([0,\tau]\times[0,\tau])$.

The operator $I + \Gamma(\mathcal{A}, \mathcal{B})$ given by

$$(\Gamma(\mathcal{A}, \mathcal{B})\phi)(t) = \int_0^\tau \gamma(t, s)\phi(s)\, ds$$

acting in the space $L_1(0, \tau)$ is said to be the Bezoutian operator for the functions $\mathcal{A}(\lambda)$ and $\mathcal{B}(\lambda)$. It is a natural continual analogue of the Bezoutian operator in the discrete case (see [1, 6, 7]).

Note that $\gamma(t, 0) = a(t)$ and $\gamma(0, s) = b(-s)$.

The next theorem is a continual analogue of [1, Theorem 5.1].

Theorem 6.1. *Let*

$$\mathcal{A}(\lambda) = 1 + \int_0^\tau a(t)e^{i\lambda t}\, dt, \quad \mathcal{B}(\lambda) = 1 + \int_{-\tau}^0 b(t)e^{i\lambda t}\, dt,$$

where $a(t) \in L_1(0, \tau)$ and $b(t) \in L_1(-\tau, 0)$. Then the kernel of the Bezoutian operator $I + \Gamma(\mathcal{A}, \mathcal{B})$ of the functions $\mathcal{A}(\lambda)$ and $\mathcal{B}(\lambda)$ consists of the linear span of the system of functions

$$\phi_{jk}(t) = t^k e^{-i\lambda_j t} \quad (k = 0, 1, \dots, k_j - 1;\ j = 1, 2, \dots, l),$$

where λ_j $(j = 1, 2, \dots, l)$ are all distinct common zeros of the functions $\mathcal{A}(\lambda)$ and $\mathcal{B}(\lambda)$ and k_j is the multiplicity of the zero λ_j.

In particular,

$$\nu(\mathcal{A}, \mathcal{B}) = \dim \operatorname{Ker}(I + \Gamma(\mathcal{A}, \mathcal{B})). \tag{6.1}$$

Proof. Let us consider the following operators acting in the space $L_1(0, \tau)$:

$$(A\phi)(t) = \int_0^\tau a(t - s)\phi(s)\, ds, \qquad (B\phi)(t) = \int_0^\tau b(t - s)\phi(s)\, ds,$$

$$(\widetilde{A}\phi)(t) = \int_0^\tau a(t - s + \tau)\phi(s)\, ds, \quad (\widetilde{B}\phi)(t) = \int_0^\tau b(t - s - \tau)\phi(s)\, ds,$$

where $0 \le t \le \tau$. With the aid of these operators, the Bezoutian operator can be written in the form

$$I + \Gamma(\mathcal{A}, \mathcal{B}) = (I + A)(I + B) - \widetilde{B}\widetilde{A}.$$

We also introduce the operators

$$(H_+\phi)(t) = \phi(t - \tau) \quad (0 \le t \le \tau),$$
$$(H_-\phi)(t) = \phi(t - \tau) \quad (-\tau \le t \le 0).$$

The operator H_+ maps $L_1(-\tau, 0)$ onto $L_1(0, \tau)$ and the operator H_- is the inverse to the operator H_+. If we identify the space $L_1(-\tau, \tau)$ with the direct sum of the subspaces $L_1(-\tau, 0)$ and $L_1(0, \tau)$, then the resultant operator $R_0(\mathcal{A}, \mathcal{B})$ for the functions $\mathcal{A}(\lambda)$ and $\mathcal{B}(\lambda)$ takes the following block form:

$$R_0(\mathcal{A}, \mathcal{B}) = \begin{Vmatrix} I + H_- B H_+ & H_- \widetilde{B} \\ \widetilde{A} H_+ & I + A \end{Vmatrix}.$$

We can straightforwardly check that the equality

$$R_0(\mathcal{A},\mathcal{B}) = \begin{Vmatrix} I & H_-\widetilde{B}(I+A)^{-1} \\ 0 & I \end{Vmatrix} \begin{Vmatrix} H_-CH_+ & 0 \\ 0 & I+A \end{Vmatrix} \times \begin{Vmatrix} I & 0 \\ (I+A)^{-1}\widetilde{A}H_+ & I \end{Vmatrix} \tag{6.2}$$

holds, where $C = I + B - \widetilde{B}(I+A)^{-1}\widetilde{A}$. Since, obviously, the operators $\widetilde{B}$ and A commute, we have

$$C = (I+A)^{-1}((I+A)(I+B) - \widetilde{B}\widetilde{A}) = (I+A)^{-1}(I+\Gamma(\mathcal{A},\mathcal{B})). \tag{6.3}$$

Let $f(t) \in \operatorname{Ker} R_0(\mathcal{A},\mathcal{B})$, then in view of (6.2) and (6.3),

$$(I+\Gamma(\mathcal{A},\mathcal{B}))f_1 = 0,$$

where $f_1(t) = f(t-\tau)$ $(0 \le t \le \tau)$.

Conversely, from the equality $(I+\Gamma(\mathcal{A},\mathcal{B}))f_1(t) = 0$ it follows that $f(t)$ belongs to $\operatorname{Ker} R_0(\mathcal{A},\mathcal{B})$, where

$$f(t) = \begin{cases} f_1(t+\tau) & (-\tau \le t \le 0), \\ -(I+A)^{-1}\widetilde{A}f_1(t) & (0 \le t \le \tau). \end{cases}$$

To finish the proof, it remains to apply Theorem 0.1. □

Note also that in the case when the Bezoutian operator is invertible, its inverse is an integral operator with a kernel depending only on the difference of the arguments (see [4, 5]).

For the case $\mathcal{A}(\lambda) = \mathcal{B}(-\lambda)$ a theorem containing a continual generalization of Hermite's theorem [6] (containing, in particular, equality (6.1)) was obtained by M.G. Krein. This result was published by him only in the discrete case (see [8]).

References

[1] I.C. Gohberg and G. Heinig, *The resultant matrix and its generalizations. I. The resultant operator for matrix polynomials.* Acta Sci. Math. (Szeged) **37** (1975), 41–61 (in Russian). English translation: **this volume**. MR0380471 (52 #1371), Zbl 0298.15013.

[2] I.C. Gohberg and M.G. Krein, *Introduction to the Theory of Linear Nonselfadjoint Operators in Hilbert space.* Nauka, Moscow, 1965 (in Russian). MR0220070 (36 #3137), Zbl 0138.07803.

English translation: *Introduction to the Theory of Linear Nonselfadjoint Operators.* Amer. Math. Soc., Providence, R.I. 1969. MR0246142 (39 #7447), Zbl 0181.13504.

French translation: *Introduction à la Théorie des Opérateurs Linéaires non Auto-Adjoints Dans un Espace Hilbertien.* Dunod, Paris, 1971. MR0350445 (50 #2937).

[3] I.C. Gohberg and E.I. Sigal, *An operator generalization of the logarithmic residue theorem and the theorem of Rouché.* Matem. Sbornik, New Ser. **84(126)** (1971), 607–629 (in Russian). English translation: Math. USSR Sbornik **13** (1971), 603–625. MR0313856 (47 #2409), Zbl 0254.47046.

[4] I.C. Gohberg and A.A. Semencul, *The inversion of finite Toeplitz matrices and their continual analogues.* Matem. Issled. **7** (1972), no. 2(24), 201–223 (in Russian). MR0353038 (50 #5524), Zbl 0288.15004.

[5] I.C. Gohberg and G. Heinig, *Matrix integral operators on a finite interval with kernels depending on the difference of the arguments.* Rev. Roumaine Math. Pures Appl. **20** (1975), 55–73 (in Russian). English translation: **this volume.** MR0380495 (52 #1395), Zbl 0327.45009.

[6] M.G. Krein and M.A. Naimark, *The method of symmetric and Hermitian forms in the theory of the separation of the roots of algebraic equations.* Khar'kov, 1936 (in Russian). English translation: Linear and Multilinear Algebra **10** (1981), no. 4, 265–308. MR0638124 (84i:12016), Zbl 0584.12018.

[7] F.I. Lander, *The Bezoutian and the inversion of Hankel and Toeplitz matrices.* Matem. Issled. **9** (1974), no. 2(32), 69–87 (in Russian). MR0437559 (55 #10483), Zbl 0331.15017.

[8] M.G. Krein, *Distribution of roots of polynomials orthogonal on the unit circle with respect to a sign-alternating weight.* Teor. Funkts., Funkts. Anal. Prilozh. (Khar'kov) **2** (1966), 131–137 (in Russian). MR0201702 (34 #1584), Zbl 0257.30002.

Operator Theory:
Advances and Applications, Vol. 206, 111–125

The Spectrum of Singular Integral Operators in L_p Spaces

Israel Gohberg and Nahum Krupnik

To S. Mazur and W. Orlicz

First, we shall consider the simplest class of one-dimensional singular integral operators – the class of discrete Wiener-Hopf operators.

Let T_a be a bounded linear operator defined in the space ℓ_2 by the infinite matrix $\|a_{j-k}\|_{j,k=0}^{\infty}$, where a_j are the Fourier coefficients of some bounded function $a(\zeta)$ $(|\zeta| = 1)$.

If the function $a(\zeta)$ $(|\zeta| = 1)$ is continuous, then *the spectrum of the operator T_a consists of all points of the curve $a(\zeta)$ $(|\zeta| = 1)$ and all complex numbers λ not lying on this curve, for which*

$$\operatorname{ind}(a-\lambda) \stackrel{\text{def}}{=} \frac{1}{2\pi}[\arg(a(e^{i\theta})-\lambda)]_{\theta=0}^{\theta=2\pi} \neq 0.$$

This statement remains true (see [7]) if one replaces the space ℓ_2 by many other Banach spaces. In particular, the space ℓ_2 can be replaced by any space h_p $(1 < p < \infty)$ of the sequences of the Fourier coefficients of functions belonging to the corresponding Hardy space H_p (see [2]). The situation becomes more involved if the function $a(\zeta)$ is not continuous.

If the function $a(\zeta)$ $(|\zeta| = 1)$ is continuous from the left and has a finite number of discontinuities [of the first kind] $\zeta_1, \zeta_2, \ldots, \zeta_n$, then *the spectrum of the operator T_a in the space ℓ_2 (see [3]) consists of all points of the curve $V(a)$ resulting from the range of the function $a(\zeta)$ by adding the segments*

$$\mu a(\zeta_k) + (1-\mu)a(\zeta_k+0) \quad (0 \leq \mu \leq 1),$$

as well as of the points $\lambda \notin V(a)$, for which

$$\operatorname{ind}(a-\lambda) \stackrel{\text{def}}{=} \frac{1}{2\pi}\oint_{V(a)} d_t \arg(t-\lambda) \neq 0.$$

The paper was originally published as И.Ц. Гохберг, Н.Я. Крупник, О спектре сингулярных интегральных операторов в пространствах L_p, Studia Math. **31** (1968), 347–362. MR0236774 (38 #5068), Zbl 0179.19701.

This result does not hold for the spaces h_p with $1 < p < \infty$ if $p \neq 2$: *for a piecewise continuous function $a(\zeta)$ the spectrum of the operator T_a in h_p varies with variation of p.*

This paper is devoted to finding the spectrum of the operator T_a in the spaces h_p and solving the same problem for other singular integral operators in the spaces L_p.

Some results in this direction were obtained earlier in the paper by H. Widom [1] (see also [14]). Known methods of N.I. Mushelishvili [10] and B.V. Khvedelidze [12, 13] for solving singular integral equations with discontinuous coefficients play an important role in what follows.

The paper consists of four sections. Discrete Wiener-Hopf equations in the spaces h_p are considered in Section 1. In Section 2, the spectrum of singular integral operators with discontinuous coefficients in $L_p(\Gamma)$, where Γ consists of a finite number of closed contours, is studied. As an application, in Section 3, estimates for the norms of some singular integral operators in $L_p(\Gamma)$ are obtained. In particular, for some L_p the exact value of the norm of the Hilbert transform is calculated. In the last section the above mentioned results are generalized to some symmetric spaces.

1. The spectrum of discrete Wiener-Hopf equations in h_p spaces

1.1. Let H_p $(1 < p < \infty)$ be the Hardy space, that is, the space of all functions $f(\zeta)$ analytic in the disk $|\zeta| < 1$ with the norm

$$\|f\|_{H_p} = \lim_{\varrho \uparrow 1} \left(\int_0^{2\pi} |f(\varrho e^{i\theta})|^p \, d\theta \right)^{1/p} \quad (< \infty).$$

By h_p denote the Banach space isometric to H_p that consists of all sequences $\xi = \{\xi_j\}_0^\infty$ of the Fourier coefficients of functions in H_p.

By Λ denote the set of all piecewise continuous functions on the unit circle $(|\zeta| = 1)$ that are continuous from the left.

In this section the spectra of operators generated by the matrices of the form $\|a_{j-k}\|_{j,k=0}^\infty$, where a_j $(j = 0, \pm 1, \dots)$ are the Fourier coefficients of a function $a(\zeta) \in \Lambda$, are studied in h_p $(1 < p < \infty)$. The operator T_a generated by the above matrix is a bounded linear operator in every space h_p $(1 < p < \infty)$.

1.2. Let a and b be a pair of points in the complex plane and p be a number in the interval $(2, \infty)$. By $\nu_p(a, b)$ we denote the circular arc joining the points a, b and having the following two properties:

(α) the segment ab is seen from the interior points of the arc $\nu_p(a, b)$ under the angle $2\pi/p$;

(β) the orientation from the point a to b along the arc $\nu_p(a, b)$ is counter-clockwise.

In the case $1 < p < 2$ we set $\nu_p(a, b) = \nu_q(b, a)$ $(p^{-1} + q^{-1} = 1)$, and in the case $p = 2$ by $\nu_2(a, b)$ we denote the segment ab.

Let $a(\zeta)$ be an arbitrary function in Λ and ζ_k ($|\zeta_k| = 1$ and $k = 1, 2, \dots, n$) be all its discontinuity points. To the function $a(\zeta)$ and a number p ($1 < p < \infty$) we assign the continuous closed curve $V_p(a)$ resulting from the range of the function $a(\zeta)$ by adding the arcs $\nu_p(a(\zeta_k), a(\zeta_k + 0))$ ($k = 1, 2, \dots, n$). We orient the curve $V_p(a)$ in the natural manner. That is, on the intervals of continuity of the function $a(\zeta)$, the motion along the curve $V_p(a)$ agrees with the motion of the variable ζ counterclockwise; and along the arcs $\nu_p(a(\zeta_k)a(\zeta_k+0))$, the curve $V_p(a)$ is oriented from $a(\zeta_k)$ to $a(\zeta_k + 0)$.

We say that a function $a(\zeta) \in \Lambda$ is *p-nonsingular* if the curve $V_p(a)$ does not pass through the point $\lambda = 0$.

The winding number of the curve $V_p(a)$ about the point $\lambda = 0$, that is, the number

$$\frac{1}{2\pi} \oint_{V_p(a)} d \arg t$$

is said to be the *index* (more precisely, *p-index*) of the p-nonsingular function $a(\zeta)$ and is denoted by $\operatorname{ind}_p a$.

If a function $a(\zeta) \in \Lambda$ is not continuous, then obviously its index depends on the number p.

Note also that, in contrast to the case of continuous functions, the p-index of the product of two p-nonsingular functions may not be equal to the sum of the p-indices of those functions.

However it can be easily seen that *if the multiples f and g ($\in \Lambda$) do not have common points of discontinuity, then the p-nonsingularity of the functions f and g implies the p-nonsingularity of their product and the identity*

$$\operatorname{ind}_p(fg) = \operatorname{ind}_p f + \operatorname{ind}_p g.$$

1.3. The main result of this section is the following.

Theorem 1. *Let $a(\zeta) \in \Lambda$. The operator T_a is a Φ_+-operator or a Φ_--operator*[1] *in the space h_p if and only if the function $a(\zeta)$ is p-nonsingular.*

If the function $a(\zeta)$ is p-nonsingular, then

1. *for $\operatorname{ind}_p a > 0$ the operator T_a is left-invertible in the space h_p and*

$$\dim \operatorname{coker} T_a|h_p = \operatorname{ind}_p a;$$

2. *for $\operatorname{ind}_p a < 0$ the operator T_a is right-invertible in the space h_p and*

$$\dim \ker T_a|h_p = -\operatorname{ind}_p a;$$

3. *for $\operatorname{ind}_p a = 0$ the operator T_a is two-sided invertible in h_p.*

This theorem immediately implies the following.

[1] An operator A is said to be a Φ_+-operator (resp. Φ_--operator) if it is normally solvable and $\dim \ker A < \infty$ (resp. $\dim \operatorname{coker} A < \infty$). If A is a Φ_--operator and a Φ_+-operator simultaneously, then it is called a Φ-operator.

Theorem 2. *Suppose* $a(\zeta) \in \Lambda$. *The spectrum of the operator* T_a *in the space* h_p $(1 < p < \infty)$ *consists of all points of the curve* $V_p(a)$ *and the points* $\lambda \notin V_p(a)$, *for which* $\operatorname{ind}_p(a - \lambda) \neq 0$.

Proof of Theorem 1 will be given in Section 2. Here we illustrate Theorem 2 by considering as an example the operator T_g defined in h_p $(1 < p < \infty)$ by the matrix

$$\left\| \frac{1}{\pi i(j - k + 1/2)} \right\|_{j,k=0}^{\infty} .$$

This operator is a truncation of the discrete Hilbert transform. The corresponding function

$$g(e^{i\theta}) = \frac{1}{\pi i} \sum_{j=-\infty}^{\infty} \frac{1}{j + 1/2} e^{i\theta} = e^{-i\theta/2} \quad (0 < \theta \leq 2\pi)$$

has exactly one discontinuity on the unit circle at the point $\zeta = 1$. In view of Theorem 2 the spectrum $\sigma_p(T_g)$ of the operator T_g depends on p and is the set bounded by the half-circle $e^{i\tau}$ $(\pi \leq \tau \leq 2\pi)$ and the circular arc $\nu_p(-1, 1)$.

For $2 < p < \infty$ we have $\operatorname{ind}_p g = 0$. Therefore for these values of p the operator T_g is invertible in h_p. For $1 < p < 2$ we get $\operatorname{ind}_p g = -1$. Hence for these values of p the operator T_g is right-invertible in h_p and $\dim \ker T_g = 1$. The operator T_g is not one-sided invertible only in the space h_2 $(= \ell_2)$ (moreover, it is neither a Φ_+-operator nor a Φ_--operator).

The spectrum $\sigma_p(T_g)$ always contains interior points except for the case $p = 4$. In the latter case the spectrum $\sigma_p(T_g)$ consists of the half-circle

$$\nu_4(-1, 1) = \{e^{i\tau} : \pi \leq \tau \leq 2\pi\}.$$

2. The spectrum of singular integral operators in L_p spaces

By F^+ denote a bounded connected closed set on the complex plane with the boundary Γ consisting of a finite number of simple closed smooth oriented curves, that is, $\Gamma = \bigcup_{j=0}^{m} \Gamma_j$. Let F^- be the closure of the complement of F^+ to the whole plane. We shall assume that $0 \in F^+ \setminus \Gamma$. By F_j^- denote the connected (bounded if $j \neq 0$) part of the set F^- with the boundary Γ_j.

The set of all piecewise continuous functions that are continuous from the left on Γ is denoted by $\Lambda(\Gamma)$.

Let $t_1, t_2, \ldots, t_n$ be all discontinuity points of a function $g(t) \in \Lambda(\Gamma)$. To the function $g(t)$ we assign the curve $V_p(g)$ consisting of a finite number of closed oriented continuous curves resulting from the range of the function $g(t)$ by adding the n arcs $\nu_p(g(t_k), g(t_k + 0))$.

We say that the function $g(t)$ is *p-nonsingular* if $0 \notin V_p(g)$.

The winding number of the contour $V_p(g)$ about the point $\lambda = 0$, that is, the number

$$\frac{1}{2\pi}\oint_{V_p(g)} d\arg t$$

is said to be the *index* (*p-index*) of a p-nonsingular function $g(t)$ and is denoted by $\operatorname{ind}_p g$.

Analogously to the case of the unit circle, if two p-nonsingular functions do not have common discontinuity points, then their product is a p-nonsingular function and the p-index of the product is equal to the sum of the p-indices of the multiples.

Consider a singular integral operator $A = c(t)I + d(t)S$, where $c(t)$ and $d(t)$ belong to $\Lambda(\Gamma)$ and

$$(S\varphi)(t) = \frac{1}{\pi i}\int_\Gamma \frac{\varphi(\tau)}{\tau - t}d\tau.$$

If one introduces the notation

$$c(t) + d(t) = a(t), \quad c(t) - d(t) = b(t), \quad (I+S)/2 = P, \quad (I-S)/2 = Q,$$

then the operator A can be written in the form $A = a(t)P + b(t)Q$.

First consider the case $b(t) \equiv 1$.

Theorem 3. *The operator $A_g = gP + Q$, where $g(t) \in \Lambda(\Gamma)$, is a Φ_+-operator or a Φ_--operator in the spaces $L_p(\Gamma)$ $(1 < p < \infty)$ if and only if the function $g(t)$ is p-nonsingular.*

If the function $g(t)$ is p-nonsingular, then

1. *for $\operatorname{ind}_p g > 0$ the operator A_g is left-invertible in the space $L_p(\Gamma)$ and*

$$\dim\operatorname{coker} A_g = \operatorname{ind}_p g;$$

2. *for $\operatorname{ind}_p g < 0$ the operator A_g is right-invertible in the space $L_p(\Gamma)$ and*

$$\dim\ker A_g = -\operatorname{ind}_p g;$$

3. *for $\operatorname{ind}_p g = 0$ the operator A_g is invertible in $L_p(\Gamma)$.*

The idea of the proof of the sufficiency portion of Theorem 3 is borrowed from the theory of singular integral equations with discontinuous coefficients (see [10, 13]). According to the usual line of reasoning in this theory we first prove the sufficiency of the conditions of Theorem 3 for a special (in some sense the simplest) function $\psi(t) \in \Lambda(\Gamma)$. Then the general case will be considered with the aid of this simplest case.

Let t_k $(k = 1, 2, \ldots, n)$ be some points on the contour Γ and s_k $(s_k \notin \Gamma)$ be points chosen by the following rule: if $t_k \in \Gamma_0$, then $s_k \in F^+$; if $t_k \in \Gamma_j$ $(j = 1, 2, \ldots, m)$, then $s_k \in F_j^-$. The contour Γ_j containing the point t_k will be also denoted by $\Gamma^{(k)}$.

By $\psi_k(t)$ denote the function continuous everywhere on the contour Γ, except for possibly the point t_k, and defined by

$$\psi_k(t) = \begin{cases} (t - s_k)^{\varepsilon\gamma_k} & \text{for} \quad t \in \Gamma^{(k)}, \\ 1 & \text{for} \quad t \in \Gamma \setminus \Gamma^{(k)}, \end{cases}$$

where $\varepsilon = 1$ if $t_k \in \Gamma_0$ and $\varepsilon = -1$ if $t_k \notin \Gamma_0$ and γ_k are complex numbers satisfying

$$\frac{1-p}{p} < \operatorname{Re} \gamma_k < \frac{1}{p}.$$

It is easy to see that the function $\psi(t) = \psi_1(t) \dots \psi_n(t)$ is p-nonsingular and $\operatorname{ind}_p \psi = 0$.

Lemma 1. *The operator $A_\psi = \psi(t)P + Q$ is invertible in $L_p(\Gamma)$.*

Proof. Each function $\psi_k(t)$ can be factorized (see, e.g., [12]): $\psi_k(t) = \psi_k^-(t)\psi_k^+(t)$, where

$$\psi_k^{(\varepsilon)}(t) = \begin{cases} (t - t_k)^{\varepsilon\gamma_k} & (t \in \Gamma^{(k)}), \\ 1 & (t \in \Gamma \setminus \Gamma^{(k)}), \end{cases}$$

$$\psi_k^{(-\varepsilon)}(t) = \begin{cases} \left(\dfrac{t - s_k}{t - t_k}\right)^{\varepsilon\gamma_k} & (t \in \Gamma^{(k)}), \\ 1 & (t \in \Gamma \setminus \Gamma^{(k)}). \end{cases}$$

Here $\psi^{(\varepsilon)}$ (resp. $\psi^{(-\varepsilon)}$) denotes ψ_k^+ (resp. ψ_k^-) for $t_k \in \Gamma_0$ and ψ_k^- (resp. ψ_k^+) for $t_k \in \Gamma \setminus \Gamma_0$.

The function $\psi(t)$ admits a factorization $\psi(t) = \psi_-(t)\psi_+(t)$, where $\psi_\pm(t) = \psi_1^\pm(t) \dots \psi_n^\pm(t)$. Consider the operator $B = (\psi_+^{-1}P + \psi_-Q)\psi_-^{-1}I$. Taking into account that $P + Q = I$ and $P - Q = S$, the operator B can be represented in the form

$$B = \frac{1}{2}[(\psi^{-1} + 1)I + (\psi^{-1} - 1)\psi_- S \psi_-^{-1} I].$$

From Khvedelidze's theorem [13, p. 24] on the boundedness of the operator S in the L_p space with weight it follows that the operator $\psi_- S \psi_-^{-1} I$ is bounded in $L_p(\Gamma)$, whence the operator B is also bounded in $L_p(\Gamma)$.

It is easy to see that for Hölder continuous functions $\chi(t)$ $(t \in \Gamma)$, the equalities

$$(\psi_+^{-1}P + \psi_-Q)\psi_-^{-1}(\psi P + Q)\chi = (\psi P + Q)(\psi_+^{-1}P + \psi_-Q)\psi_-^{-1}\chi = \chi$$

hold.

Thus the operator A_ψ is invertible in $L_p(\Gamma)$ and $A_\psi^{-1} = (\psi_+^{-1}P + \psi_-Q)\psi_-^{-1}I$. The lemma is proved. □

Proof of Theorem 3. Let t_k ($\in \Gamma_k$) ($k = 1, 2, \dots, n$) be all discontinuity points of the function $g(t)$. Since $g(t)$ is a p-nonsingular function, the quotient $g(t_k)/g(t_k+0)$ can be written in the form

$$\frac{g(t_k)}{g(t_k+0)} = \left|\frac{g(t_k)}{g(t_k+0)}\right| e^{2\pi i \alpha_k}, \tag{1}$$

where $(1-p)/p < \alpha_k < 1/p$.

By γ_k denote the following numbers:

$$\gamma_k = \alpha_k + \frac{1}{2\pi i} \ln \left|\frac{g(t)}{g(t_k+0)}\right|.$$

To the points t_k and the numbers γ_k we assign the function $\psi(t)$. The quotient $g(t)/\psi(t)$ is a continuous function because

$$\frac{\psi_k(t_k)}{\psi_k(t_k+0)} = e^{2\pi i \gamma_k} = \frac{g(t_k)}{g(t_k+0)}.$$

The function $g(t)$ can be written as the product $g(t) = \psi(t) r(t)(1 + m(t))$, where $r(t)$ is a rational function that does not have poles and zeros on the contour Γ, $\operatorname{ind}_p g = \operatorname{ind} r$, and the maximum of the absolute value of the function $m(t)$ is so small that the operator $A_{\psi+m\psi} = \psi P + Q + m\psi P$ is invertible in $L_p(\Gamma)$ simultaneously with the operator A_ψ.

The function $r(t)$ can be factorized (see [4]) as follows: $r(t) = r_-(t) t^\kappa r_+(t)$, where $r_+(t)$ (resp. $r_-(t)$) is a rational function with poles and zeros in the domain $F^- \cup \{\infty\}$ (resp. F^+), and $\kappa = \operatorname{ind} r$ ($= \operatorname{ind}_p g$).

Let $\kappa \geq 0$. Then it is easy to verify the identity

$$gP + Q = r_-(\psi P + Q + \psi m P)(t^\kappa P + Q)(r_+ P + r_-^{-1} Q).$$

The operators $\psi P + Q + \psi m P$ and $r_+ P + r_-^{-1} Q$ are invertible in $L_p(\Gamma)$ and the operator $t^\kappa P + Q$ is left-invertible in $L_p(\Gamma)$ and $\dim \operatorname{coker}(t^\kappa P + Q) = \kappa$. The operator

$$(r_+^{-1} P + r_- Q)(t^{-\kappa} P + Q)(\psi P + Q + \psi m P)^{-1} r_-^{-1} I$$

is an inverse from the left to $gP + Q$. From here it follows that if $\kappa = 0$, then the operator A_g is invertible and if $\kappa > 0$, then the operator A_g is only left-invertible in $L_p(\Gamma)$ and $\dim \operatorname{coker} A_g = \operatorname{ind}_p g$.

Let $\kappa < 0$. In that case we will use the identity

$$gt^{-\kappa} P + Q = (gP + Q)(t^{-\kappa} P + Q).$$

Since $\operatorname{ind}_p(gt^{-\kappa}) = 0$, the operator $gt^{-\kappa} P + Q$ is invertible in $L_p(\Gamma)$. In addition the operator $t^{-\kappa} P + Q$ is right-invertible and $\dim \operatorname{coker}(t^{-\kappa} P + Q) = -\kappa$. From here it follows that the operator A_g is only right-invertible in $L_p(\Gamma)$ and $\dim \ker A_g = -\operatorname{ind}_p g$.

The necessity part of the theorem is proved by contradiction[2]. Assume that A_g is a Φ_+-operator or a Φ_--operator in $L_p(\Gamma)$ and $0 \in V_p(\Gamma)$. First consider the

[2]The idea of this proof is borrowed from [9].

case when there exists a neighborhood U of the origin in which $V_p(\Gamma)$ is a simple smooth line. Since A_g is a $\Phi_\pm$-operator, one can find a neighborhood $U_1 \subset U$ of the origin such that for all $\lambda \in U_1$ the operator $A_{g-\lambda}$ is a $\Phi_\pm$-operator and $\operatorname{ind} A_{g-\lambda} = \operatorname{ind} A_g$ (recall that $\operatorname{ind} A = \dim\ker A - \dim\operatorname{coker} A$). On the other hand, if one takes two points $\lambda_1, \lambda_2 \in U_1$ which lie in different domains separated by $V_p(\Gamma)$, then $\operatorname{ind}_p(g-\lambda_1) \neq \operatorname{ind}_p(g-\lambda_2)$. Then as we have shown above $\operatorname{ind} A_{g-\lambda_1} \neq \operatorname{ind} A_{g-\lambda_2}$, which is impossible.

We develop the proof in the general case by contradiction as well. Assume that A_g is a $\Phi_\pm$-operator in $L_p(\Gamma)$ and $0 \in V_p(\Gamma)$. Then one can choose a function $b(t) \in \Lambda(\Gamma)$ satisfying the three conditions:

(α) $\sup_{t\in\Gamma} |g(t)-b(t)| < \varepsilon$, where ε is so small that the operator A_b is a $\Phi_\pm$-operator[3];

(β) the curve $V_p(b)$ is a simple smooth arc in some neighborhood of the origin;

(γ) $0 \in V_p(b)$.

As we have shown, the latter condition contradicts the former two conditions. The theorem is proved. □

Theorem 1 is easily derived from the previous theorem.

Proof of Theorem 1. By C denote the isometric operator mapping each function $f(\zeta)$ in H^p to the vector $\{f_j\}_0^\infty \in h_p$ of its Fourier coefficients. It is easy to see that

$$T_a = CPA_aC^{-1},$$

where A_a is the singular integral operator $aP+Q$ for which the unit circle ($|\zeta| = 1$) plays the role of the contour Γ.

Without difficulty it is proved that the operator $PA_a|H_p$ can be one-sided invertible (it is a $\Phi_\pm$-operator) if and only if the operator A_a so is in the space $L_p(\Gamma)$. If the operator $PA_a|H_p$ is one-sided invertible, then

$$\dim\ker A_a = \dim\ker PA_a|H_p, \quad \dim\operatorname{coker} A_a = \dim\operatorname{coker} PA_a|H_p.$$

The theorem is proved. □

Theorem 4. *Let $c(t) \in \Lambda(\Gamma)$ and $d(t) \in \Lambda(\Gamma)$. The operator $A = c(t)I + d(t)S$ ($A = c(t)I + Sd(t)I$) is a Φ_+-operator or a Φ_--operator in $L_p(\Gamma)$, $1 < p < \infty$, if and only if the following two conditions are fulfilled:*

(α) $\inf_{t\in\Gamma} |c(t) - d(t)| > 0$,

(β) *the function $(c(t) + d(t)/(c(t) - d(t))$ is p-nonsingular.*

If these conditions are fulfilled and $\kappa = \operatorname{ind}_p(c+d)/(c-d)$, then

1. *for $\kappa < 0$ the operator A is right-invertible in $L_p(\Gamma)$ and* $\dim\ker A = -\kappa$;
2. *for $\kappa > 0$ the operator A is left-invertible in $L_p(\Gamma)$ and* $\dim\operatorname{coker} A = \kappa$;
3. *for $\kappa = 0$ the operator A is invertible in $L_p(\Gamma)$.*

[3]The existence of such a number ε follows from the stability theorem for $\Phi_\pm$-operators under small perturbations [5].

Proof. We develop the proof for the operator $A = c(t)I + d(t)S$. We represent it in the form $A = a(t)P + b(t)Q$, where $a(t) = c(t) + d(t)$ and $b(t) = c(t) - d(t)$.

The sufficiency of the conditions of the theorem follows from Theorem 3. To apply this theorem for the proof of the necessity, it remains to show that if A is a Φ_+-operator or a Φ_--operator, then $\inf_{t\in\Gamma} |b(t)| > 0$. We divide the proof of this fact into two steps.

1. Let us show that if the functions $a(t)$ and $b(t)$ $(\in \Lambda(\Gamma))$ are rational in each segment of their continuity and $b(t_0) = 0$, where t_0 is some point of continuity of the function $b(t)$, then the operator $aP + bQ$ is neither Φ_+-operator nor Φ_--operator.

   Indeed, since under these conditions $b_1(t) = (t^{-1} - t_0^{-1})^{-1} b(t) \in \Lambda(\Gamma)$, the operator $aP + bQ$ can be written in the form
   $$aP + bQ = (aP + b_1Q)(P + (t^{-1} + t_0^{-1})Q). \tag{2}$$
   If the operator $aP + bQ$ were a Φ_+-operator, then from identity (2) it would follow that the operator $P + (t^{-1} - t_0^{-1})Q$ is a Φ_+-operator, which is impossible due to Theorem 3. Thus, $aP + bQ$ cannot be a Φ_+-operator.

   Since the functions $a(t)$ and $b(t)$ are piecewise rational, in every neighborhood of the origin there exists a point λ such that for the operator $A - \lambda I$ conditions (α) and (β) of the theorem are fulfilled, whence the operator $A - \lambda I$ is a Φ-operator. From here it follows that if A were a Φ_--operator, then it would be a Φ-operator, but we have shown that it is not even a Φ_+-operator.
2. Let us show that if the operator A is a Φ_+-operator or a Φ_--operator, then $\inf_{t\in\Gamma} |b(t)| > 0$. Assume the contrary, i.e., assume that A is a Φ_+-operator (or a Φ_--operator) and there is a point $t_0 \in \Gamma$, at which either $b(t_0) = 0$ or $b(t_0+0) = 0$. Choose two piecewise rational functions $a_1(t)$ and $b_1(t)$ $(\in \Lambda(\Gamma))$ such that $|a(t) - a_1(t)| < \delta$ and $|b(t) - b_1(t)| < \delta$, where δ is so small that the operator $a_1P + b_1Q$ is a Φ_+-operator (or a Φ_--operator). Obviously, in this case the function $b_1(t)$ can be chosen so that the condition $b_1(t_0) = 0$ is fulfilled and t_0 is a continuity point of the function $b_1(t)$. The latter contradicts what has been proved in Step 1.

The theorem is proved. □

It is not difficult to show that conditions (α) and (β) in the formulation of Theorem 4 can be replaced by an equivalent condition: for all $\mu \in [0,1]$ and $t \in \Gamma$,
$$a(t)b(t+0)E(p,\mu) + b(t)F(p,\mu) \neq 0,$$
where
$$E(p,\mu) = \exp\left(\frac{2\pi i\mu(p-2)}{p}\right) - \exp\left(\frac{-4\pi i}{p}\right), \quad F(p,\mu) = 1 - \exp\left(\frac{2\pi i\mu(p-2)}{p}\right)$$
for $p > 2$,
$$E(p,\mu) = F(q,\mu), \quad F(p,\mu) = E(q,\mu)$$

for $1 < p < 2$ and $1/p + 1/q = 1$, and

$$E(2, \mu) = \mu, \quad F(2, \mu) = 1 - \mu.$$

The last theorem allows us to find Φ-domains[4] of singular integral operators.

Theorem 5. *Let $a(t) \in \Lambda(\Gamma)$ and $b(t) \in \Lambda(\Gamma)$. The complement to the Φ-domain of the operator $A = aP + bQ$ ($A = PaI + QbI$) consists of the ranges of the functions $a(t)$ and $b(t)$ and the set of all complex numbers λ, each of which for some $\mu \in [0, 1]$ satisfies at least one of the equations*

$$(a(t_k) - \lambda)(b(t_k + 0) - \lambda)E(p, \mu) + (a(t_k + 0) - \lambda)(b(t_k) - \lambda)F(p, \mu) = 0, \quad (3)$$

where t_k ($k = 1, 2, \dots, n$) are all discontinuity points of the functions $a(t)$ and $b(t)$.

Consider several examples of sets G_k of complex numbers λ satisfying equation (3) corresponding to the discontinuity point t_k.

Let t_1 be a discontinuity point of the function $a(t)$ only. Then

$$G_1 = \nu_p(a(t_1), a(t_1 + 0)) \cup \{b(t_1)\}.$$

Analogously, if t_2 is a discontinuity point of the function $b(t)$ only, then

$$G_2 = \nu_p(b(t_2 + 0), b(t_2)) \cup \{a(t_2)\}.$$

Let t_3 be a discontinuity point of both functions $a(t), b(t)$ and $a(t_3) = b(t_3)$. Then

$$G_3 = \nu_p(b(t_3 + 0), a(t_3 + 0)) \cup \{a(t_3)\}.$$

If t_4 is a common discontinuity point of the functions $a(t), b(t)$ and, for instance, $a(t_4) = b(t_4 + 0) = 1$, $b(t_4) = a(t_4 + 0) = -1$, then the set G_4 is the circle centered at the point $-i\cot(\pi/p)$ of radius $R = 1/\sin(\pi/p)$.

In the example $A = d(t)S$, where a function $d(t)$ ($\in \Lambda(\Gamma)$ takes only two values 0 and 1, the complement to the Φ-domain of the operator A is the set $G = \nu_p(-1, 1) \cup \nu_p(1, -1) \cup \{0\}$.

Note that all results of this section can be extended to paired and transposed to paired Wiener-Hopf equations.

3. Estimate for the norm of the singular integral operator

In this sections some estimates from below for the norms of P, Q, and S in the space $L_p(\Gamma)$ will be obtained. Moreover, for some values of p ($p = 2^n$ and $p = 2^n/(2^n - 1)$, $n = 1, 2, \dots$), the exact value for the norm of the Hilbert transform will be calculated[5].

Denote by $\mathfrak{P}_p$ the set of all compact operators in $L_p(\Gamma)$.

[4]The Φ-domain of an operator A is the set of all complex numbers λ such that the operator $A - \lambda I$ is a Φ-operator.

[5]The results of this section were obtained by the authors in [15] using a different method.

Theorem 6. *For every $p > 2$ the following estimates hold:*

$$\inf_{T\in\mathfrak{P}_p} \|P+T\|_p \geq \frac{1}{\sin \pi/p}, \quad \inf_{T\in\mathfrak{P}_p} \|Q+T\|_p \geq \frac{1}{\sin \pi/p}, \tag{4}$$

$$\inf_{T\in\mathfrak{P}_p} \|S+T\|_p \geq \cot \frac{\pi}{2p}. \tag{5}$$

For $1 < p < 2$ estimates (4) and (5) remain valid with p replaced by q (where $p^{-1}+q^{-1}=1$) on the right-hand sides.

Proof. Assume that

$$\inf_{T\in\mathfrak{P}_p} \|P+T\|_p < 1/\sin \frac{\pi}{p}$$

for some p and consider the operator $aP+Q$, where $a(t) \in \Lambda(\Gamma)$ is the function taking only two values

$$a(t) = \left(\cos \frac{\pi}{p}\right) \exp\left(\pm \frac{\pi i}{p}\right).$$

Since $|a(t)-1| = \sin \pi/p$, one has $\inf \|(a-1)P\|_p < 1$. Hence the operator

$$I + (a-1)P = aP + Q$$

is a Φ-operator, but this is impossible because the function $a(t)$ is p-singular.

For the proof of the second inequality in (4) consider the function

$$a(t) = \left(\sec \frac{\pi}{p}\right) \exp\left(\pm \frac{\pi i}{p}\right).$$

Then $|(1-a)/a| = \sin \pi/p$. The operator $aP+Q = a(I+((1-a)/a)Q)$ is not a Φ-operator in $L_p(\Gamma)$ because the function $a(t)$ is not p-nonsingular. This implies the second inequality in (4).

Inequality (5) can be proved analogously if one takes the function $a(t) = \exp(\pm i\pi/p)$ and uses the identity

$$aP + Q = \frac{a+1}{2}\left(I + \frac{a-1}{a+1}S\right). \qquad \square$$

Theorem 7. *Let $\Gamma = \{\zeta : |\zeta| = 1\}$. Then for all $n = 1, 2, \ldots,$*

$$\|S\|_p = \begin{cases} \cot \frac{\pi}{2p} & \text{if } \ p = 2^n, \\ \tan \frac{\pi}{2p} & \text{if } \ p = 2^n/(2^n-1). \end{cases} \tag{6}$$

Proof. From inequality (5) it follows that if $p > 2$, then $\|S\|_p \geq \cot \pi/2p$. Let us prove the reverse inequality. Let $\varphi(t)$ be an arbitrary function satisfying the Hölder condition on the unit circle. Then it is easy to see that

$$\varphi^2 + (S\varphi)^2 = 2[(P\varphi)^2 + (Q\varphi)^2] = 2S[(P\varphi)^2 - (Q\varphi)^2] = 2S(\varphi \cdot S\varphi),$$

that is[6], $(S\varphi)^2 = \varphi^2 + 2S(\varphi \cdot S\varphi)$. From this identity it follows that

$$\|S\varphi\|_{2p}^2 \leq 2\|S\|_p\|\varphi\|_{2p}\|S\varphi\|_{2p} + \|\varphi\|_{2p}^2.$$

Hence

$$\frac{\|S\varphi\|_{2p}}{\|\varphi\|_{2p}} \leq \|S\|_p + \sqrt{1 + \|S\|_p^2},$$

which implies that

$$\|S\|_{2p} \leq \|S\|_p + \sqrt{1 + \|S\|_p^2}.$$

Taking into account that $\|S\|_2 = 1$, from the last inequality we obtain

$$\|S\|_{2^n} \leq \cot \frac{\pi}{2^{n+1}}.$$

Thus equality (6) is proved for $p = 2^n$. For $p = 2^n/(2^n - 1)$ it follows by passing to the adjoint operator. The theorem is proved. □

4. The spectrum of singular integral operators in symmetric spaces

In this section the results obtained above are extended to more general symmetric spaces.

Let us start with definitions. Real-valued measurable functions $x(s)$ and $y(s)$ on the segment $[0, 1]$ are said to be *equimeasurable* if

$$\text{mes}\{s : x(s) > \tau\} = \text{mes}\{s : y(s) > \tau\}$$

for every τ.

A Banach space E of all complex-valued measurable functions on $[0, 1]$ is said to be *symmetric* if the following three conditions are fulfilled.

1. If $|x(s)| \leq |y(s)|$, $y(s) \in E$, and $x(s)$ is measurable on $[0, 1]$, then $x(s) \in E$ and $\|x\|_E \leq \|y\|_E$.
2. If functions $|x(s)|$ and $|y(s)|$ are equimeasurable and $y(s) \in E$, then $x(s) \in E$ and $\|x\|_E = \|y\|_E$.
3. Let E' be the set of all measurable functions $y(s)$ on $[0, 1]$ such that

$$\|y\|_{E'} \stackrel{\text{def}}{=} \sup_{\|x\|_E \leq 1} \int_0^1 |x(s)y(s)|\, ds < \infty.$$

Then

$$\|x\|_E = \sup_{\|y\|_{E'} \leq 1} \int_0^1 |x(s)y(s)|\, ds.$$

[6]This identity was used for the first time by M. Cotlar [8] in the proof of the boundedness of the Hilbert transform (see also [6], pages 120–121 of the Russian original).

By $\chi(s)$ denote the characteristic function of the segment $[0, s]$. The function $\omega(s) = \|\chi(s)\|_E$ is said to be the *fundamental function* of the space E.

Let E be a symmetric Banach space on $[0, 1]$, Γ be a contour defined in Section 2, and $t = \eta(s)$ $(0 \leq s \leq 1)$ be its (piecewise smooth) parametric equation.

A Banach space F of all complex-valued measurable functions $\varphi(t)$ on Γ such that $\varphi(\eta(s)) \in E$ and

$$\|\varphi(t)\|_F \overset{\text{def}}{=} \|\varphi(\eta(s))\|_E$$

is said to be *symmetric* with respect to the parametrization $t = \eta(s)$ of the contour Γ.

Put

$$\liminf_{s \to 0} \frac{\omega(2s)}{\omega(s)} = m(F), \quad \limsup_{s \to 0} \frac{\omega(2s)}{\omega(s)} = M(F).$$

We will need the following two theorems due to E.M. Semenov[7].

Theorem A. *Let $0 \leq \alpha_j \leq 1$ $(j = 1, 2)$ and for a symmetric space F the inequalities*

$$2^{\alpha_1} < m(F), \quad M(F) < 2^{\alpha_2}$$

be fulfilled.

If a linear operator A is bounded in the spaces $L_p(\Gamma)$ for all $p \in (1/\alpha_2, 1/\alpha_1)$, then the operator A is bounded in F.

Theorem B. *The operator S defined by the equality*

$$(S\varphi)(t) = \frac{1}{\pi i} \int_\Gamma \frac{\varphi(\tau)}{\tau - t} \, d\tau$$

is bounded in F if and only if

$$1 < m(F), \quad M(F) < 2. \tag{7}$$

If, in particular, F is an Orlicz space, then condition (7) is equivalent to the reflexivity of the space F.

Assume that $1 < m(F) = M(F) < 2$ are fulfilled in the space F. Then Theorems 3–6 of Sections 2–3 remain true if one replaces in the formulations the space L_p by the space F and the number p by the number $\varrho = 1/\log_2 M(F)$.

The proofs for the space F proceed in the same way as for the space L_p. It is necessary only to explain why the operator $A_\psi = \psi P + Q$ constructed by using the ϱ-nonsingular function $g(t)$ (see the proof of Theorem 3) is invertible in F. Indeed, since the function $g(t)$ is ϱ-nonsingular, there exists a number $\varepsilon > 0$ such that the function $g(t)$ is p-nonsingular for every $p \in (\varrho - \varepsilon, \varrho + \varepsilon)$. Then from Lemma 1 it follows that the operator A_ψ is invertible in the spaces L_p for all $p \in (\varrho - \varepsilon, \varrho + \varepsilon)$ and

$$A_\psi^{-1} = (\psi_+^{-1} P + \psi_- Q)\psi_-^{-1} I.$$

From Theorem A it follows that the operator A_ψ^{-1} is bounded in F. Hence A_ψ is invertible in F.

---

[7]These theorems were proved by Semenov [11] for symmetric spaces on $[0, 1]$. However, as it was kindly communicated to us by E.M. Semenov, they can be extended to spaces F.

In the case $m(F) \neq M(F)$ one can give only sufficient conditions for a singular integral operator to be a Φ-operator in F.

Theorem 8. *Let $c(t) \in \Lambda(\Gamma)$, $d(t) \in \Lambda(\Gamma)$ and for a symmetric space F the conditions $1 < m(F)$ and $M(F) < 2$ be fulfilled. If* $\inf |c(t) - d(t)| > 0$ $(t \in \Gamma)$ *and the function $(c(t) + d(t))/(c(t) - d(t))$ is p-nonsingular for all p satisfying*

$$(\log_2 M(F))^{-1} \leq p \leq (\log_2 m(F))^{-1}, \tag{8}$$

then for the operator $A = c(t)I + d(t)S$ $(A = c(t)I + Sd(t)I)$ in the space F the following statements hold:

1. *for*

$$\kappa = \text{ind}_p[(c+d)/(c-d)] > 0$$

the operator A is left-invertible and $\dim \operatorname{coker} A = \kappa$;

2. *for $\kappa < 0$ the operator A is right-invertible and* $\dim \ker A = -\kappa$;
3. *for $\kappa = 0$ the operator A is invertible.*

Note that the results of Section 1 can be extended to spaces more general than h_p in the same way.

References

[1] H. Widom, *Singular integral equations in L_p.* Trans. Amer. Math. Soc. **97** (1960), 131–160. MR0119064 (22 #9830), Zbl 0109.33002.

[2] K. Hoffman, *Banach Spaces of Analytic Functions.* Prentice-Hall Inc., Englewood Cliffs, N. J., 1962. Russian translation: Izdatel'stvo Inostrannoi Literatury, Moscow, 1963. MR0133008 (24 #A2844), Zbl 0117.34002.

[3] I.C. Gohberg, *Teoplitz matrices of Fourier coefficients of piecewise-continuous functions.* Funkcional. Anal. Prilozhen. **1** (1967), no. 2, 91–92 (in Russian). English translation: Funct. Anal. Appl. **1** (1967), no. 2, 166–167. MR0213909 (35 #4763), Zbl 0159.43101.

[4] I.C. Gohberg, *A factorization problem in normed rings, functions of isometric and symmetric operators and singular integral equations.* Uspehi Mat. Nauk **19** (1964), no. 1(115), 71–124 (in Russian). English translation: Russ. Math. Surv. **19** (1964) 63–114. MR0163184 (29 #487), Zbl 0124.07103.

[5] I.C. Gohberg and M.G. Krein, *The basic propositions on defect numbers, root numbers and indices of linear operators.* Uspehi Mat. Nauk (N.S.) **12** (1957), no. 2(74), 43–118 (in Russian). English translation: Amer. Math. Soc. Transl. (2) **13** (1960), 185–264. MR0096978 (20 #3459), MR0113146 (22 #3984), Zbl 0088.32101.

[6] I.C. Gohberg and M.G. Krein, *Theory and Applications of Volterra Operators in Hilbert Space.* Nauka, Moscow, 1967 (in Russian). English translation: Amer. Math. Soc., Providence, RI, 1970. MR0218923 (36 #2007), Zbl 0168.12002.

[7] I. Gohberg and I.A. Feldman, *Projection Methods for Solving Wiener-Hopf Equations.* Akad. Nauk Moldav. SSR, Kishinev, 1967 (in Russian). MR0226325 (37 #1915).

[8] M. Cotlar, *A unified theory of Hilbert transforms and ergodic theorems.* Rev. Mat. Cuyana **1** (1955), 105–167. MR0084632 (18,893d).

[9] M.G. Krein, *Integral equations on a half-line with kernel depending upon the difference of the arguments.* Uspehi Mat. Nauk **13** (1958), no. 5 (83), 3–120 (in Russian). English translation: Amer. Math. Soc. Transl. (2) **22** (1962), 163–288. MR0102721 (21 #1507), Zbl 0088.30903.

[10] N.I. Mushelishvili, *Singular Integral Equations.*

1st Russian edition, OGIZ, Moscow, Leningrad, 1946, MR0020708 (8,586b).

English translation of 1st Russian edition: Noordhoff, Groningen, 1953, MR0058845 (15,434e), Zbl 0051.33203.

Reprinted by Wolters-Noordhoff Publishing, Groningen, 1972, MR0355494 (50 #7968), by Noordhoff International Publishing, Leyden, 1977, MR0438058 (55 #10978), by Dover Publications, 1992 and 2008.

2nd Russian edition, revised, Fizmatgiz, Moscow, 1962. MR0193453 (33 #1673), Zbl 0103.07502.

German translation of 2nd Russian edition: *Singuläre Integralgleichungen.* Akademie-Verlag, Berlin, 1965. Zbl 0123.29701.

3rd Russian edition, corrected and augmented, Nauka, Moscow, 1968. MR0355495 (50 #7969), Zbl 0174.16202.

[11] E.M. Semenov, *Interpolation of linear operators and estimates of Fourier coefficients.* Dokl. Akad. Nauk SSSR **176** (1967), 1251–1254 (in Russian). English translation: Soviet Math. Dokl. **8** (1967), 1315–1319. MR0221312 (36 #4364), Zbl 0162.44601.

[12] B.V. Khvedelidze, *The Riemann-Privalov boundary-value problem with a piecewise continuous coefficient.* Trudy Gruzin. Politehn. Inst. (1962), no. 1 (81), 11–29 (in Russian). MR0206306 (34 #6125).

[13] B.V. Khvedelidze, *Linear discontinuous boundary problems in the theory of functions, singular integral equations and some of their applications.* Trudy Tbiliss. Mat. Inst. Razmadze **23** (1956), 3–158 (in Russian). MR0107148 (21 #5873).

[14] E. Shamir, *L^p solution of Riemann-Hilbert systems with piecewise continuous coefficients.* Dokl. Akad. Nauk SSSR **167** (1966) 1000–1003 (in Russian). English translation: Soviet Math. Dokl. **7** (1966), 530–533. MR0203411 (34 #3263), Zbl 0161.32301.

[15] I.C. Gohberg and N.Ya. Krupnik, *Norm of the Hilbert transformation in the L_p space.* Funkcional. Anal. Prilozhen. **2** (1968), no. 2, 91–92 (in Russian). English translation: Funct. Anal. Appl. **2** (1968), no. 2, 180–181. Zbl 0177.15503.

Operator Theory:
Advances and Applications, Vol. 206, 127–133

On an Algebra Generated by the Toeplitz Matrices in the Spaces h_p

Israel Gohberg and Nahum Krupnik

Let H_p $(1 < p < \infty)$ be the Banach Hardy space of all functions $\varphi(\zeta)$ that are analytic inside the circle $|\zeta| = 1$ with the norm

$$\|\varphi\|_{H_p} = \lim_{\varrho \uparrow 1} \left(\int_0^{2\pi} |\varphi(\varrho e^{i\theta})|^p \, d\theta \right)^{1/p}.$$

By h_p denote the Banach space isometric to H_p that consists of all numerical sequences $\xi = \{\xi_n\}_0^\infty$ of the Fourier coefficients of functions in H_p. Let Λ be the set of all piecewise continuous functions on the unit circle $|\zeta| = 1$ that are continuous from the left. To each function $a(\zeta) \in \Lambda$ we associate the operator T_a defined in h_p by the Toeplitz matrix $\|a_{j-k}\|_{j,k=1}^\infty$, where a_k are the Fourier coefficients of the function $a(\zeta)$. Obviously, for each p $(1 < p < \infty)$ the operator T_a is a bounded linear operator in h_p. By $\mathfrak{R}$ denote the algebra of all sums of products of operators of the form T_a with $a \in \Lambda$.

In the present paper the Banach algebra $\mathfrak{A}_p$ is considered. It is the closure of the algebra $\mathfrak{R}$ in the operator norm of the space h_p. The results presented below generalize results from the paper [1] obtained there for the case $p = 2$. As in the case $p = 2$, the set $\mathfrak{F}_p$ of all compact operators in h_p is the minimal two-sided ideal of the algebra $\mathfrak{A}_p$ and the quotient algebra $\mathfrak{A}_p/\mathfrak{F}_p$ is a commutative Banach algebra.

In the case of arbitrary p $(1 < p < \infty)$, the maximal ideal space of the algebra $\mathfrak{A}_p/\mathfrak{F}_p$ is the cylinder $|\zeta| = 1$, $0 \le \mu \le 1$, equipped with a special topology such that the neighborhoods of a point (ζ_0, μ_0) are defined by one of the following equalities:

$$\begin{aligned} u(\zeta_0, 0) &= \{(\zeta, \mu) : \varphi_0 - \delta < \arg\zeta < \varphi_0, 0 \le \mu \le 1\} \cup \{(\zeta_0, \mu) : 0 \le \mu < \varepsilon\}, \\ u(\zeta_0, 1) &= \{(\zeta, \mu) : \varphi_0 < \arg\zeta < \varphi_0 + \delta, 0 \le \mu \le 1\} \cup \{(\zeta_0, \mu) : \varepsilon < \mu \le 1\}, \quad (1) \\ u(\zeta_0, \mu_0) &= \{(\zeta_0, \mu) : \mu_0 - \delta_1 < \mu < \mu_0 + \delta_2\} \quad (\mu_0 \ne 0, 1), \end{aligned}$$

The paper was originally published as И.Ц. Гохберг, Н.Я. Крупник, Об алгебре, порождённой тёплицевыми матрицами в пространствах h_p, Матем. Исслед. 4 (1969), № 3, 54–62. MR0399922 (53 #3763a), Zbl 0254.47045.

where $0 < \delta_1 < \mu_0$, $0 < \delta_2 < 1 - \mu_0$, $0 < \varepsilon < 1$, $\varphi_0 = \arg \zeta_0$. It is natural to refer to the function $A_p(\zeta, \mu)$ ($|\zeta| = 1$; $0 \le \mu \le 1$) on the maximal ideal space of $\mathfrak{A}_p/\mathfrak{F}_p$ that corresponds to an operator $A \in \mathfrak{A}_p$ as the *p-symbol* of the operator A. In this case, if the operator A is given by

$$A = \sum_{j=1}^{n} \prod_{k=1}^{m} T_{a_{jk}}, \tag{2}$$

where $a_{jk} \in \Lambda$, then its p-symbol ($p \neq 2$) is defined by

$$\begin{aligned} A_p(\zeta, \mu) = \sum_{j=1}^{n} \prod_{k=1}^{m} \Bigg(& \frac{\sin(1-\mu)\theta}{\sin\theta} \exp(i\mu\theta) a_{jk}(\zeta) \\ & + \frac{\sin \mu\theta}{\sin\theta} \exp(i(\mu - 1)\theta) a_{jk}(\zeta + 0) \Bigg), \end{aligned} \tag{3}$$

where $\theta = \pi(p-2)/p$, $|\zeta| = 1$, and $0 \le \mu \le 1$. The definition of the symbol $A_p(\zeta, \mu)$ depends essentially on p and is different from the definition in the case $p = 2$:

$$A_2(\zeta, \mu) = \sum_{j=1}^{n} \prod_{k=1}^{m} \left[(1-\mu) a_{jk}(\zeta) + \mu a_{jk}(\zeta + 0)\right].$$

The latter can be obtained from (3) by passing to the limit as $p \to 2$ (or $\theta \to 0$).

It will be proved below that an operator $A \in \mathfrak{A}_p$ is a Φ-operator[1] if and only if its symbol is different from zero. The index of an operator A is also expressed via its symbol. The range of the symbol $A_p(\zeta, \mu)$ is a continuous closed curve, which can be naturally oriented. The index of this curve (that is, its winding number about the origin) taken with the opposite sign is equal to the index of the operator A.

1. Toeplitz operators in the spaces h_p

1.1. In this section auxiliary propositions on Toeplitz operators generated by the Fourier coefficients of piecewise continuous functions in h_p are obtained. All these statements are generalizations to the case of arbitrary p ($1 < p < \infty$) of theorems obtained in the authors' papers [3, 4].

Lemma 1. *Let $a(\zeta), b(\zeta) \in \Lambda$ and $p \in (1, \infty)$. The operator $K = T_a T_b - T_{ab}$ is compact in h_p if and only if the functions $a(\zeta)$ and $b(\zeta)$ do not have common points of discontinuity.*

Proof. Let P be the orthogonal projection of $L_2(|\zeta| = 1)$ onto H_2 and $\mathcal{U}$ be the isometric operator mapping each function $f(\zeta) \in H^p$ to the vector $\{f_j\} \in h_p$ of its Fourier coefficients[2]. The operator K can be represented in the form $K = \mathcal{U}T\mathcal{U}^{-1}$,

[1] For the definition of a Φ-operator, see [2].

[2] It is known that the operator P is bounded in every L_p ($1 < p < \infty$) and that it projects $L_p(|\zeta| = 1)$ onto H_p.

where $T = PaPbP - PbPaP$. The operator T is bounded in each space $L_p(|\zeta| = 1)$. In view of Krasnosel'skii's theorem [5, Theorem 1], the operator T is compact in all spaces L_p $(1 < p < \infty)$ if and only if it is compact in one of them. Because it is proved in [1] that the operator K is compact in $h_2(= \ell_2)$ if and only if the functions $a(\zeta)$ and $b(\zeta)$ do not have common points of discontinuity, this completes the proof. □

Analogously, with the aid of the M.A. Krasnosel'skii theorem and the results of the paper [1], the following statement is proved.

Lemma 2. *For every pair of functions $a(\zeta), b(\zeta) \in \Lambda$ the operator $T_aT_b - T_bT_a$ is compact in h_p $(1 < p < \infty)$.*

1.2. Let $F(\zeta) = \|a_{jk}(\zeta)\|_{j,k=1}^r$ be a matrix function with entries $a_{jk} \in \Lambda$, let h_p^r be the space of vectors $x = (x_1, \dots, x_r)$ with components $x_j \in h_p$, and let T_F be the bounded linear operator in h_p^r defined by the matrix $\|T_{a_{jk}}\|_{j,k=1}^r$. To the operator T_F and a number p $(1 < p < \infty)$ we assign the matrix function (p-symbol)

$$F_p(\zeta, \mu) = \|(T_{a_{jk}})_p(\zeta, \mu)\|_{j,k=1}^r,$$

where $(T_{a_{jk}})_p(\zeta, \mu)$ is the p-symbol of the operator $T_{a_{jk}}$ defined by equality (3). The range of the function $\det F_p(\zeta, \mu)$ is a closed curve. We orient this curve in such a way that the motion along the curve $\det F_p(\zeta, \mu)$ agrees with the motion of ζ along the circle counterclockwise at the continuity points of the matrix function $F(\zeta)$, and the motion along the complementary matrix arcs corresponds to the motion of μ from 0 to 1.

Theorem 1. *The operator T_F is a Φ-operator in h_p^r $(1 < p < \infty)$ if and only if its p-symbol is non-degenerate:*

$$\det F_p(\zeta, \mu) \neq 0 \quad (|\zeta| = 1,\ 0 \leq \mu \leq 1). \tag{4}$$

If condition (4) is fulfilled, then the index of the operator T_F is calculated by the formula

$$\operatorname{ind} T_F = -\operatorname{ind} \det F_p(\zeta, \mu).$$

In the case $r = 1$ this theorem is proved in [3]. For other r it is proved in the same way as an analogous theorem for singular integral operators in the spaces L_p (see [4]).

Theorem 2. *Let A be an operator defined in h_p by equality (2). The operator A is a Φ-operator in h_p if and only if its p-symbol*[3] *is not equal to zero:*

$$A_p(\zeta, \mu) \neq 0.$$

If this condition is fulfilled, then

$$\operatorname{ind} A = -\operatorname{ind} A_p(\zeta, \mu). \tag{5}$$

[3] Recall that the p-symbol $A_p(\zeta, \mu)$ of the operator A is defined by (3).

For the case $p = 2$ this theorem is proved in [1], for other values of p the proof is developed analogously.

Theorem 3. *Let A be an operator defined in h_p by equality (2) and $A_p(\zeta,\mu)$ be its p-symbol. Then*

$$\inf_{T\in\mathfrak{F}_p} \|A+T\|_p \geq \max_{|\zeta|=1, 0\leq\mu\leq 1} |A_p(\zeta,\mu)|. \tag{6}$$

Proof. Assume that for some operator $A \in \mathfrak{R}$ inequality (6) fails. Then there exists a point (ζ_0, μ_0), where $|\zeta_0| = 1$ and $0 \leq \mu_0 \leq 1$, and an operator $K_0 \in \mathfrak{F}_p$, such that $\|A + K_0\|_p < |A_p(\zeta_0, \mu_0)|$. Let $B = (A_p(\zeta_0, \mu_0))^{-1}A$ and $K = (A_p(\zeta_0, \mu_0))^{-1}K_0$. Then $\|B + K\|_p < 1$. Therefore $I - B$ is a Φ-operator in h_p. This contradicts Theorem 2 because $(I - B)_p(\zeta_0, \mu_0) = 0$. The theorem is proved. □

2. Algebra generated by the Toeplitz operators

2.1. Let $\mathfrak{A}_p$ be the closure of the set $\mathfrak{R}$ of the operators of the form (2) in the algebra $\mathfrak{B}_p$ of all bounded linear operators in h_p. To each operator $A \in \mathfrak{R}$ we assign its p-symbol $A_p(\zeta,\mu)$ defined by equality (3). From inequality (6) it follows that the symbol $A_p(\zeta,\mu)$ does not depend on a representation of the operator A in the form (2). Inequality (6) allows us to define the symbol $A_p(\zeta,\mu)$ for each operator $A \in \mathfrak{A}_p$ as the uniform limit of a sequence of the symbols of operators $A_n \in \mathfrak{R}$ tending to the operator A in the norm.

By $\widehat{\mathfrak{A}}_p$ denote the quotient algebra $\mathfrak{A}_p/\mathfrak{F}_p$ with the usual norm. The coset in $\widehat{\mathfrak{A}}_p$ that contains an operator A is denoted by $\widehat{A}$. From Theorem 3 it follows that the same symbol corresponds to all operators in a coset $\widehat{A} \in \widehat{\mathfrak{A}}_p$. We will denote it by $\widehat{A}_p(\zeta,\mu)$.

Theorem 4. *The maximal ideal space of the quotient algebra $\widehat{\mathfrak{A}}_p$ $(1 < p < \infty)$ is homeomorphic to the cylinder $\mathfrak{M} = \{|z| = 1, 0 \leq \mu \leq 1\}$ equipped with the topology defined by the neighborhoods (1). The symbol $\widehat{A}_p(\zeta,\mu)$ $(|\zeta| = 1, 0 \leq \mu \leq 1)$ is a function of an element $\widehat{A} \in \widehat{\mathfrak{A}}_p$ on the maximal ideal space $\mathfrak{M}$ of $\widehat{\mathfrak{A}}_p$.*

Proof. By f_{ζ_0,μ_0} denote the functional defined on the algebra $\widehat{\mathfrak{A}}_p$ by the equality

$$f_{\zeta_0,\mu_0}(\widehat{A}) = \widehat{A}(\zeta_0,\mu_0).$$

From Theorem 3 it follows that f_{ζ_0,μ_0} $(|\zeta_0| = 1,\ 0 \leq \mu \leq 1)$ is a multiplicative functional. Therefore the set

$$M_{\zeta_0,\mu_0} = \{\widehat{A} :\ \widehat{A}_p(\zeta_0,\mu_0) = 0\}$$

is a maximal ideal of the algebra $\widehat{\mathfrak{A}}_p$. Let us prove that the algebra $\widehat{A}_p$ does not have other maximal ideals. Let M_0 be some maximal ideal of the algebra $\widehat{\mathfrak{A}}_p$. First, we shall show that there exists a point ζ_0 $(|\zeta_0| = 1)$ such that $\widehat{T}_a(M_0) = a(\zeta_0)$ for all functions $a(\zeta)$ continuous on the circle $\Gamma = \{\zeta : |\zeta| = 1\}$. Assume the contrary, that is, suppose that for every point $\tau \in \Gamma$ there exists a continuous function $x_\tau(\zeta)$ such that $\widehat{T}_{x_\tau}(M_0) \neq x_\tau(\tau)$. It is obvious that 1) $\widehat{T}_{x_\tau - \alpha_\tau} \in M_0$, where $\alpha_\tau = \widehat{T}_{x_\tau}(M_0)$, and 2) $|x_\tau(\zeta) - \alpha_\tau| \geq \delta_\tau > 0$ in some neighborhood $u(\tau)$. Let $u(\tau_1), \dots, u(\tau_n)$ be a finite cover of the circle Γ and $\delta = \min\limits_{1 \leq k \leq n} \delta_{\tau_k}$. Then

$$y(\zeta) = \sum_{k=1}^{n} |x_{\tau_k}(\zeta) - \alpha_{\tau_k}|^2 \neq 0.$$

Hence $\widehat{T}_y$ is an invertible element in $\widehat{\mathfrak{A}}_p$, but this is impossible because

$$\widehat{T}_y = \sum_{k=1}^{n} \widehat{T}_{x_\tau - \alpha_\tau} \widehat{T}_{\overline{x_\tau} - \overline{\alpha_\tau}}.$$

Thus $\widehat{T}_y \in M_0$.

Let us show that for every function $x(\zeta) \in \Lambda$ continuous at the point ζ_0 the equality $\widehat{T}_x(M_0) = x(\zeta_0)$ holds. To this end we note that if a function $x(\zeta) \in \Lambda$ is continuous at the point ζ_0, then it can be represented in the form $x(\zeta) = y(\zeta) + a(\zeta)z(\zeta)$, where $a(\zeta) \in \Lambda$, $y(\zeta)$ and $z(\zeta)$ are continuous on Γ and $z(\zeta_0) = 0$. Then from the equality $\widehat{T}_x(M_0) = \widehat{T}_y(M_0) + \widehat{T}_a(M_0)\widehat{T}_z(M_0)$ it follows that

$$\widehat{T}_x(M_0) = x(\zeta_0).$$

By $\nu_p(\mu)$ $(p \neq 2)$ denote the circular arc, which is the range of the function

$$\frac{\sin \mu\theta}{\sin \theta} \exp(i(\mu - 1)\theta),$$

where $\theta = \pi(p-2)/p$ and μ varies from 0 to 1, and by $\nu_2(\mu)$ denote the segment $[0, 1]$. Consider a function $\chi(\zeta) \in \Lambda$ having the following properties: $\chi(\zeta_0) = 0$; $\chi(\zeta_0 + 0) = 1$, where ζ_0 is the point on the circle that has been found before; $\chi(\zeta)$ is continuous everywhere on Γ except for the point ζ_0, the range of $\chi(\zeta)$ coincides with the arc $\nu_p(\mu)$. From Theorem 2 it follows that the spectrum of the element $\widehat{T}_x$ in the algebra $\widehat{\mathfrak{A}}_p$ coincides with the set $\nu_p(\mu)$. Hence there exists a number $\mu_0 \in [0, 1]$ such that $\widehat{T}_x(M_0) = \nu_p(\mu_0)$.

Let us pass to the last stage of the proof. We will show that for every element $\widehat{A} \in \widehat{\mathfrak{A}}_p$ the equality $\widehat{A}(M_0) = \widehat{A}_p(\zeta_0, \mu_0)$ holds. It is easy to see that it is sufficient to prove this claim for the case $\widehat{A} = \widehat{T}_x$, where x is an arbitrary function in Λ. Fix a function $x(\zeta) \in \Lambda$ and let $y(\zeta) = x(\zeta) - C$, where C is a constant chosen so that the p-symbol of the operator T_y is nowhere zero. The function $b(\zeta) = y(\zeta)/[\chi(\zeta)y(\zeta_0 + 0) - y(\zeta_0)) + y(\zeta_0)]$ is continuous at the point ζ_0. Hence $\widehat{T}_b(M_0) = b(\zeta_0) = 1$. For the function $a(\zeta) = \chi(\zeta)[y(\zeta_0 + 0) - y(\zeta_0)] + y(\zeta_0)$ we have

$$\widehat{T}_a(M_0) = \nu_p(\mu_0)[y(\zeta_0 + 0) - y(\zeta_0)] + y(\zeta_0) = (\widehat{T}_y)_p(\zeta_0, \mu_0).$$

Therefore $\widehat{T}_x(M_0) = (\widehat{T}_x)_p(\zeta_0, \mu_0)$. The theorem is proved. □

Theorem 5. *An operator $A \in \mathfrak{A}_p$ is a Φ-operator in h_p if and only if its p-symbol is nowhere zero:*

$$\det A_p(\zeta, \mu) \neq 0 \quad (|\zeta| = 1,\ 0 \leq \mu \leq 1). \tag{7}$$

If condition (7) *is fulfilled, then*

$$\operatorname{ind} A = -\operatorname{ind} A_p(\zeta, \mu). \tag{8}$$

Proof. Let $A \in \mathfrak{A}_p$ and $A_p(\zeta, \mu) \neq 0$. According to Theorem 4, the element $\widehat{A}$ is invertible in $\widehat{\mathfrak{A}}_p$. Hence A is a Φ-operator in h_p. Since $\operatorname{ind} A$ and $A_p(\zeta_0, \mu_0)$ are continuous functionals, we see that equality (8) follows from equality (5). The sufficiency part of the theorem is proved. The proof of the necessity part is developed by contradiction. Assume that the operator A is a Φ-operator and $A_p(\zeta_0, \mu_0) = 0$. Then one can find an operator $B \in \mathfrak{R}$ such that B is a Φ-operator and $B_p(\zeta_0, \mu_0) = 0$, which is impossible. The theorem is proved. □

2.2. Let h_p^n be the Banach space of all vectors $x = (x_1, \dots, x_n)$ with entries $x_k \in h_p$ and $\mathfrak{A}^{(p)}_{n\times n}$ be the algebra of all bounded linear operators in h_p^n of the form $A = \|A_{jk}\|_{j,k=1}^n$, where $A_{jk} \in \mathfrak{A}_p$.

To each operator $A \in \mathfrak{A}^{(p)}_{n\times n}$ and the number $p \in (1, \infty)$ assign the matrix function (p-symbol)

$$A_p(\zeta, \mu) = \|(A_{jk})_p(\zeta, \mu)\|_{j,k=1}^n,$$

where $(A_{jk})_p(\zeta, \mu)$ is the p-symbol of the operator A_{jk}.

Theorem 6. *An operator $A \in \mathfrak{A}^{(p)}_{n\times n}$ is a Φ-operator in h_p if and only if its p-symbol is non-degenerate, that is,*

$$\det A_p(\zeta, \mu) \neq 0. \tag{9}$$

If condition (9) *is fulfilled, then*

$$\operatorname{ind} A = -\operatorname{ind} \det A_p(\zeta, \mu). \tag{10}$$

Proof. Let $A = \|A_{jk}\|_{j,k=1}^n$. Since the operators A_{jk} pairwise commute modulo a compact operator, we conclude (see [6, p. 108]) that the operator A is a Φ-operator in h_p^n if and only if the operator $\det A$ is a Φ-operator in h_p. From here and Theorem 5 it follows that condition (9) is fulfilled if and only if the operator A is a Φ-operator in h_p^n.

Equality (10) for the operator A satisfying condition (9) is proved by induction on n. Indeed, one can find an operator $B = \|B_{jk}\| \in \mathfrak{A}^{(p)}_{n\times n}$ (sufficiently close to the operator A) that satisfies the following conditions: B (resp. B_{11}) is a Φ-operator in h_p^n (resp. h_p) and $\operatorname{ind} B = \operatorname{ind} A$. The last condition allows us to represent the operator B in the form $B = MN + K$, where $M = \|M_{jk}\|_{j,k=1}^n \in \mathfrak{A}^{(p)}_{n\times n}$ is a lower triangular operator matrix all of whose entries on the main diagonal are the identity operators, $N = \|N_{jk}\|_{j,k=1}^n \in \mathfrak{A}^{(p)}_{n\times n}$ is an operator matrix such that all the entries of the first column are zero except for the first entry, and K is a compact operator in h_p^n. Since $\operatorname{ind} B = \operatorname{ind} N$, we have $\operatorname{ind} A = \operatorname{ind} N$.

Obviously, $\operatorname{ind} N = \operatorname{ind} N_{11} + \operatorname{ind} \widetilde{N}$, where $\widetilde{N} = \|N_{jk}\|_{j,k=2}^{n}$. According to what has been proved, $\operatorname{ind} N_{11} = -\operatorname{ind}(N_{11})_p(\zeta, \mu)$, and by the induction hypothesis, $\operatorname{ind} \widetilde{N} = -\operatorname{ind} \det \widetilde{N}_p(\zeta, \mu)$. Taking into account that

$$\det B_p(\zeta, \mu) = \det N_p(\zeta, \mu) = (N_{11})_p(\zeta, \mu) \det \widetilde{N}_p(\zeta, \mu)$$

and that the functions $\det B_p(\zeta, \mu)$ and $\det A_p(\zeta, \mu)$ are sufficiently close to each other, we get $\operatorname{ind} N = -\operatorname{ind} A_p(\zeta, \mu)$. The theorem is proved. □

References

[1] I.C. Gohberg and N.Ya. Krupnik, *On the algebra generated by Toeplitz matrices.* Funkts. Anal. Prilozh. **3** (1969), no. 2, 46–56 (in Russian). English translation: Funct. Anal. Appl. **3** (1969), 119–127. MR0250082 (40 #3323), Zbl 0199.19201.

[2] I.C. Gohberg and M.G. Krein, *The basic propositions on defect numbers, root numbers and indices of linear operators.* Uspehi Mat. Nauk (N.S.) **12** (1957), no. 2(74), 43–118 (in Russian). English translation: Amer. Math. Soc. Transl. (2) **13** (1960), 185–264. MR0096978 (20 #3459), MR0113146 (22 #3984), Zbl 0088.32101.

[3] I.C. Gohberg and N.Ya. Krupnik, *The spectrum of singular integral operators in L_p spaces.* Studia Math. **31** (1968), 347–362 (in Russian). English translation: **this volume**. MR0236774 (38 #5068), Zbl 0179.19701.

[4] I.C. Gohberg and N.Ya. Krupnik, *Systems of singular integral equations in L_p spaces with weight.* Dokl. Akad. Nauk SSSR **186** (1969) 998–1001 (in Russian). English translation: Soviet Math. Dokl. **10** (1969), 688–691. MR0248566 (40 #1818), Zbl 0188.18302.

[5] M.A. Krasnosel'skii, *On a theorem of M. Riesz.* Dokl. Akad. Nauk SSSR **131** (1960), 246–248 (in Russian). English translation: Soviet Math. Dokl. **1** (1960), 229–231. MR0119086 (22 #9852), Zbl 0097.10202.

[6] I. Gohberg and I.A. Feldman, *Projection Methods for Solving Wiener-Hopf Equations.* Akad. Nauk Moldav. SSR, Kishinev, 1967 (in Russian). MR0226325 (37 #1915).

Operator Theory:
Advances and Applications, Vol. 206, 135–144

On Singular Integral Equations with Unbounded Coefficients

Israel Gohberg and Nahum Krupnik

Algebras generated by singular integral operators with piecewise continuous coefficients are studied in the papers [1, 2, 3, 4]. The results obtained there allow us to obtain theorems on solvability and index formulas for singular integral operators of new types.

In the present paper two examples of applications of the results of [1, 2, 3, 4] are presented. The first example is related to operators of the form

$$A = a_0(t)I + \sum_{k=1}^{n} a_k(t)h_k(t)Sh_k^{-1}(t)b_k(t)I. \tag{0.1}$$

Recall that the operator of singular integration S is defined by

$$(S\varphi)(t) = \frac{1}{\pi i}\int_\Gamma \frac{\varphi(\tau)}{\tau - t}\,d\tau \quad (t \in \Gamma),$$

where Γ is some simple oriented contour consisting of a finite number of closed and open Lyapunov curves; the coefficients $a_0(t)$, $a_k(t)$ and $b_k(t)$ $(k = 1, \dots, n)$ are piecewise continuous functions; and $h_k(t)$ has the form

$$h_k(t) = \prod_{j=1}^{r} (t - t_j)^{\delta_j^{(k)}}, \tag{1}$$

where t_j are some points of the contour Γ. The operator A is considered in the weighted space L_p $(1 < p < \infty)$.

The second example is the operator of the form

$$B = c(t)I + d(t)S + g(t)R, \tag{0.2}$$

The paper was originally published as И.Ц. Гохберг, Н.Я. Крупник, О сингулярных интегральных уравнениях с неограниченными коэффициентами, Матем. Исслед. **5** (1970), № 3(17), 46–57. MR0291892 (45 #980), Zbl 0228.47037.

where $c(t)$, $d(t)$, $g(t)$ are continuous functions; the contour Γ is the segment $[0,1]$, and the operator R is defined by the equality

$$R = \frac{1}{\pi i}\left(\ln\frac{b-t}{t-a}S - S\ln\frac{b-t}{t-a}I\right).$$

A separate section is devoted to investigating the conditions of the boundedness of the operator (0.1) in the weighted spaces L_p.

1. Auxiliary propositions

By $t_1,\dots,t_N$ and $t_{N+1},\dots,t_{2N}$ denote the starting and terminating points, respectively, of all open arcs of the contour Γ. Let $t_{2N+1},\dots,t_m$ be fixed points on Γ that do not coincide with the endpoints $t_1,\dots,t_{2N}$. Let $\beta_1,\dots,\beta_m$ be some real numbers. The space L_p on the contour Γ with the weight

$$\varrho(t) = \prod_{k=1}^{m} |t-t_k|^{\beta_k}$$

is denoted by $L_p(\Gamma,\varrho)$.

If the numbers p and $\beta_1,\dots,\beta_m$ satisfy the conditions $1 < p < \infty$ and $-1 < \beta_k < p-1$ $(k = 1,\dots,m)$, then, according to Khvedelidze's theorem [5], the operator S is bounded in the space $L_p(\Gamma,\varrho)$.

By $\Lambda = \Lambda(\Gamma)$ denote the set of all piecewise continuous functions that are continuous from the left.

For the sequel it is convenient to introduce the operators $P = (I+S)/2$ and $Q = I - P$. Then the usual singular integral operator

$$A_0 = cI + dS \quad (c, d \in \Lambda)$$

in the space $L_p(\Gamma,\varrho)$ can be represented in the form $A_0 = aP + bQ$, where $a(t), b(t) \in \Lambda$. Following [1, 2, 3], to the operator A_0 assign its symbol. Let us introduce some notation for defining the symbol. Let $t \in \Gamma$ and $0 \le \mu \le 1$. By $\theta = \theta(t)$, $\ell = \ell(t,\mu)$, and $\nu = \nu(t,\mu)$ denote, respectively, the following functions:

$$\theta(t) = \begin{cases} \pi - 2\pi(1+\beta_k)/p & \text{for } \ t = t_k \quad (k = 1,\dots,m), \\ \pi - 2\pi/p & \text{for } \ t \in \Gamma, \quad t \ne t_1,\dots,t_m, \end{cases}$$

$$\ell(t,\mu) = \begin{cases} \dfrac{\sin(\theta\mu)\exp(i\theta\mu)}{\sin\theta\exp(i\theta)} & \text{if } \ \theta \ne 0, \\ \mu & \text{if } \ \theta = 0, \end{cases}$$

and

$$\nu(t,\mu) = \sqrt{\ell(t,\mu)(1-\ell(t,\mu))}.$$

The matrix function $\mathcal{A}(t,\mu)$ $(t \in \Gamma, 0 \le \mu \le 1)$ of second order defined by the equalities

$$\mathcal{A}(t,\mu) = \begin{Vmatrix} \ell(t,\mu)a(t) + (1-\ell(t,\mu))b(t) & 0 \\ 0 & b(t) \end{Vmatrix}$$

for $t = t_k$ and $k = 1, \dots, N$,

$$\mathcal{A}(t,\mu) = \begin{Vmatrix} (1-\ell(t,\mu))a(t) + \ell(t,\mu)b(t) & 0 \\ 0 & b(t) \end{Vmatrix}$$

for $t = t_k$ and $k = N+1, \dots, 2N$, and

$$\mathcal{A}(t,\mu) = \begin{Vmatrix} \ell(t,\mu)a(t+0) + (1-\ell(t,\mu))a(t) & \nu(t,\mu)(b(t+0) - b(t)) \\ \nu(t,\mu)(a(t+0) - a(t)) & \ell(t,\mu)b(t) + (1-\ell(t,\mu))b(t+0) \end{Vmatrix}$$

for $t \in \Gamma$ such that $t \neq t_1, \dots, t_{2N}$, is called the *symbol* of the operator A_0.

The *symbol* of the operator

$$A = \sum_{j=1}^{k} A_{j1}A_{j2}\dots A_{js}, \tag{1.1}$$

where $A_{jr} = a_{jr}P + b_{jr}Q$ is defined by the equality

$$\mathcal{A}(t,\mu) = \sum_{j=1}^{k} \mathcal{A}_{j1}(t,\mu)\mathcal{A}_{j2}(t,\mu)\dots\mathcal{A}_{js}(t,\mu),$$

where $\mathcal{A}_{jr}(t,\mu)$ is the symbol of the operator A_{jr}.

In [3] it is shown that the symbol $\mathcal{A}(t,\mu)$ of an operator A does not depend on a representation of the operator A in the form (1.1).

By $\mathfrak{A}$ denote the closure of the set of all operators of the form (1.1) in the algebra $\mathfrak{B}$ of all bounded linear operators acting in $L_p(\Gamma, \varrho)$.

To each operator $A \in \mathfrak{A}$ one can assign the symbol in the following manner. Let A_n be a sequence of operators of the form (1.1) that converges in the norm to the operator $A \in \mathfrak{A}$. Then [1, 2, 3] the sequence of symbols $\mathcal{A}_n(t,\mu)$ converges uniformly with respect to t and μ to some matrix $\mathcal{A}(t,\mu)$ of order two. It does not depend on the choice of a sequence A_n tending to A, and is called the symbol of the operator A in the space $L_p(\Gamma, \varrho)$. Note that this symbol depends on the operator and also on the number p and the weight $\varrho(t)$.

By $\mathfrak{A}_\sigma$ denote the algebra of all symbols corresponding to the operators in $\mathfrak{A}$ and by $\mathfrak{F}$ denote the two-sided ideal of $\mathfrak{B}$ consisting of all compact operators. In papers [1, 2, 3] it is shown that if $A \in \mathfrak{A}$, $T \in \mathfrak{F}$, and $B = A + T$, then $B \in \mathfrak{A}$ and $\mathcal{B}(t,\mu) = \mathcal{A}(t,\mu)$. The mapping $\{A+T\}_{T\in\mathfrak{F}} \to \mathcal{A}(t,\mu)$ is a homomorphism of the algebra $\mathfrak{A}/\mathfrak{F}$ onto $\mathfrak{A}_\sigma$.

The next statement (see [1, 2, 3]) plays an important role in what follows.

Theorem 1.1. *An operator $A \in \mathfrak{A}$ is a Φ_+-operator or a Φ_--operator in the space $L_p(\Gamma, \varrho)$ if and only if the condition*

$$\det \mathcal{A}(t,\mu) \neq 0 \quad (t \in \Gamma,\ 0 \leq \mu \leq 1) \tag{1.2}$$

is fulfilled.

Let condition (1.2) hold and $\mathcal{A}(t,\mu) = \|s_{jk}(t,\mu)\|_{j,k=1}^2$. Then the operator A is a Φ-operator in $L_p(\Gamma, \varrho)$, all its regularizers[1] *belong to the algebra $\mathfrak{A}$, and the index of the operator A in $L_p(\Gamma, \varrho)$ is determined by the equality*

$$\operatorname{ind} A = -\frac{1}{2\pi}\left\{\arg \frac{\det \mathcal{A}(t,\mu)}{s_{22}(t,0)s_{22}(t,1)}\right\}_{(t,\mu)\in\Gamma\times[0,1]}. \tag{1.3}$$

The number $\frac{1}{2\pi}\{\arg f(t,\mu)\}_{(t,\mu)\in\Gamma\times[0,1]}$ on the right-hand side of equality (1.3) denotes the counter-clockwise winding number of the curve $f(t,\mu)$ in the complex plane about the point $\lambda = 0$ (see [3] for more details).

2. On the boundedness of the operator of singular integration

In this section a necessary and sufficient condition for the boundedness of the operator S in the space $L_p(\Gamma, \varrho)$ is investigated. This allows us to obtain a criterion for the boundedness of the operator $h(t)Sh^{-1}(t)I$, where $h(t)$ is a function of the form (1), in the space $L_p(\Gamma, \varrho)$. The main result of this section is the following.

Theorem 2.1. *The operator S is bounded in the space $L_p(\Gamma, \varrho)$ $(1 < p < \infty)$ if and only if the numbers β_k satisfy the conditions*

$$-1 < \beta_k < p-1 \quad (k = 1, \dots, m). \tag{2.1}$$

Proof. The sufficiency of the hypotheses of this theorem was proved in the paper by B.V. Khvedelidze [5]. The unboundedness of the operator S in some cases, when condition (2.1) is violated, was proved by S.A. Freidkin [6]. We shall prove the necessity of the conditions of the theorem in the full generality[2].

Assume that the operator S is bounded in the space $L_p(\Gamma, \varrho)$. Fix some number r $(r = 1, \dots, m)$. By $\Gamma_r \subset \Gamma$ denote some one-sided neighborhood of the point t_r that does not contain all other points t_j $(j \neq r)$. It is easy to see that the operator S is bounded in the space $L_p(\Gamma_r, |t-t_r|^\delta)$, where $\delta = \beta_r$. Let us show that $-1 < \delta < p-1$. Assume that this inequality does not hold. Then there exists a number λ $(0 < \lambda \leq 1)$ such that either $\delta\lambda = -1$ or $\delta\lambda = p-1$. Since the operator S is bounded in the spaces $L_p(\Gamma_r)$ and $L_p(\Gamma_r, |t-t_r|^\delta)$, by the Stein interpolation theorem (see [7]), the operator S is bounded in the space $L_p(\Gamma_r, |t-t_r|^{\lambda\delta})$. Thus our problem is reduced to showing that the operator S is unbounded in the spaces $L_p(\Gamma_r, |t-t_r|^{-1})$ and $L_p(\Gamma_r, |t-t_r|^{p-1})$. Assume for definiteness that t_r is the starting point of the open contour Γ_r. By $\widetilde{t}$ denote the terminating point of this

[1] An operator B is said to be a regularizer of an operator B if $AB - I \in \mathfrak{F}$ and $BA - I \in \mathfrak{F}$.
[2] The used method of the proof is different from the method of [6].

contour. In the space $L_p(\Gamma_r)$ consider the operator $A = I - b(t)S$, where $b(t)$ is a continuous function on Γ_r that satisfies the conditions $1 - b^2(t) \neq 0$, $b(\tilde{t}) = 0$, and

$$b(t_r) = \begin{cases} i\tan(\pi/p) & \text{if} \quad p \neq 2, \\ 2 & \text{if} \quad p = 2. \end{cases}$$

The symbol $\mathcal{A}(t,\mu)$ of the operator A in the space $L_p(\Gamma_r)$ (see Section 1) is defined by the equality

$$\mathcal{A}(t,\mu) = \begin{cases} \left\| \begin{matrix} (1-b(t))\ell(t,\mu) + (1+b(t))(1-\ell(t,\mu)) & 0 \\ 0 & 1+b(t) \end{matrix} \right\| & \text{for } t = t_r, \\ \left\| \begin{matrix} 1-b(t) & 0 \\ 0 & 1+b(t) \end{matrix} \right\| & \text{for } t \in \Gamma_r, t \neq t_r. \end{cases}$$

Since in the case $p \neq 2$,

$$\frac{\ell(t,\mu)}{\ell(t,\mu)-1} = \frac{\sin(\theta\mu)\exp(2\pi i/p)}{\sin(\theta(1-\mu))}$$

and

$$\frac{1+b(t_r)}{1-b(t_r)} = \exp\left(\frac{2\pi i}{p}\right),$$

we see that for $\mu = 1/2$ the equality

$$\frac{\ell(t_r,1/2)}{\ell(t_r,1/2)-1} = \frac{1+b(t_r)}{1-b(t_r)}$$

holds. Therefore $\det\mathcal{A}(t_r,1/2) = 0$.

Analogously, for $p = 2$,

$$\frac{\ell(t,\mu)}{\ell(t,\mu)-1} = \mu(\mu-1), \quad \frac{1+b(t_r)}{1-b(t_r)} = -3.$$

Hence $\det\mathcal{A}(t_r,3/4) = 0$.

Thus the operator A is not a $\Phi_\pm$-operator in the space $L_p(\Gamma_r)$. It is not difficult to check that if $p_1 \neq p$, then the symbol $\mathcal{A}_1(t,\mu)$ of the operator A in the space $L_{p_1}(\Gamma_r)$ is not degenerated. Therefore the operator A is a Φ-operator in $L_{p_1}(\Gamma_r)$. Moreover, by using the results of [3], we obtain that the operator A is one-sided invertible in the space $L_{p_1}(\Gamma_r)$. Assume that the function $b(t)$ satisfies the Hölder condition on Γ_r and $p_1 < p$. Then one of the inverse operators $A^{(-1)}$ to the operator A in $L_{p_1}(\Gamma_r)$ can be obtained by the formula (see, e.g., [8])

$$A^{(-1)} = \frac{1}{1-b^2}I - \frac{b}{1-b^2}z(t)Sz^{-1}(t)I, \tag{2.2}$$

where $z(t) = g(t)|t-t_r|^{-1/p}$ and $g(t)$, $g^{-1}(t)$ are continuous functions on Γ_r.

Assume that the operator S is bounded in the space $L_p(\Gamma_r, |t-t_r|^{-1})$. Then it is easy to verify that the operator $|t-t_r|^{-1/p}S|t-t_r|^{1/p}I$ is bounded in the space $L_p(\Gamma_r)$. This implies the boundedness of the operator $A^{(-1)}$ in the space $L_p(\Gamma_r)$, and whence, this implies also the one-sided invertibility of the operator

A in $L_p(\Gamma_r)$. But the latter is impossible because the operator A is not a $\Phi_\pm$-operator in $L_p(\Gamma_r)$. Thus, it has been proved that the operator S is unbounded in $L_p(\Gamma_r, |t-t_r|^{-1})$ for every p $(1<p<\infty)$.

It remains to verify the unboundedness of the operator S in $L_p(\Gamma_r, |t-t_r|^{p-1})$. Assume that it is bounded. Then the operator $|t-t_r|^{(p-1)/p}S|t-t_r|^{(1-p)/p}I$ is bounded in $L_p(\Gamma_r)$. If one takes $p_1 > p$, then an operator inverse to A from one of the sides can be written in the form (2.2). Moreover, in this case $z(t) = g(t)|t-t_r|^{(p-1)/p}$. By analogy with the previous case we arrive at a contradiction. The theorem is proved. □

Theorem 2.2. *Let δ_k be some real numbers and*

$$h(t) = \prod_{k=1}^{m} (t-t_k)^{\delta_k}.$$

The operator $h(t)Sh^{-1}(t)I$ is bounded in the space $L_p(\Gamma, \varrho)$ $(1<p<\infty)$ with weight

$$\varrho(t) = \prod_{k=1}^{m} |t-t_k|^{\beta_k}$$

if and only if the conditions

$$-\frac{1+\beta_k}{p} < \delta_k < 1 - \frac{1+\beta_k}{p} \tag{2.3}$$

are fulfilled.

Proof. It is easy to see that the operator $hSh^{-1}I$ is bounded in the space $L_p(\Gamma, \varrho)$ if and only if the operator S is bounded in the space $L_p(\Gamma, \varrho_1)$, where

$$\varrho_1(t) = \prod_{k=1}^{m} |t-t_k|^{\lambda_k}, \quad \lambda_k = \beta_k + p\delta_k.$$

Thus condition (2.3) follows from Theorem 2.1. The theorem is proved. □

3. Operators of the form (0.1)

As above, let $t_1, \ldots, t_m$ be fixed points of the contour Γ and $L_p(\Gamma, \varrho)$ be the space with weight

$$\varrho(t) = \prod_{k=1}^{m} |t-t_k|^{\beta_k},$$

where the numbers β_k satisfy the conditions $-1 < \beta_k < p-1$ $(1<p<\infty$; $k=1,\ldots,m)$. By Ω denote the set of all functions of the form

$$h(t) = \prod_{k=1}^{m} (t-t_k)^{\delta_k}$$

with the numbers δ_k satisfying the conditions

$$-\frac{1+\beta_k}{p} < \delta_k < 1 - \frac{1+\beta_k}{p}. \tag{3.1}$$

According to Theorem 2.2, $H = h(t)Sh^{-1}(t)I$ is a bounded linear operator acting in the space $L_p(\Gamma, \varrho)$.

For simplicity we restrict ourselves in this section to the case when the contour Γ consists of one closed curve surrounding the point $z = 0$.

To the operator H in $L_p(\Gamma, \varrho)$ assign the matrix function $\mathcal{H}(t, \mu)$ ($t \in \Gamma$, $0 \le \mu \le 1$) defined by the equality

$$\mathcal{H}(t,\mu) = \begin{Vmatrix} 1 & \mathcal{U}(t,\mu) \\ 0 & -1 \end{Vmatrix}, \tag{3.2}$$

where

$$\mathcal{U}(t,\mu) = \begin{cases} \dfrac{4\nu(t_k,\mu)\cos(\pi\delta_k)\exp(\pi i\delta_k)}{2\ell(t_k,\mu)\cos(\pi\delta_k)\exp(\pi i\delta_k)+1} & \text{for} \quad t = t_k \ (k = 1, \dots, m), \\ 0 & \text{for} \quad t \in \Gamma, \ t \neq t_1, \dots, t_m. \end{cases}$$

Lemma 3.1. *Let A be an operator of the form*

$$A = x_0(t)I + \sum_{j=1}^{n} x_j(t)h_j(t)Sh_j^{-1}(t)y_j(t)I, \tag{3.3}$$

where functions x_j ($j = 0, 1, \dots, n$) and $y_j(t)$ ($j = 1, \dots, n$) belong to Λ and $h_j(t)$ are functions in Ω. Then the operator A belongs to the algebra $\mathfrak{A}$ and its symbol $\mathcal{A}(t,\mu)$ has the form

$$\mathcal{A}(t,\mu) = \mathcal{X}_0(t,\mu) + \sum_{j=1}^{n} \mathcal{X}_j(t,\mu)\mathcal{H}_j(t,\mu)\mathcal{Y}_j(t,\mu), \tag{3.4}$$

where $\mathcal{X}_j(t,\mu)$ and $\mathcal{Y}_j(t,\mu)$ are the symbols of the operators $x_j(t)I$ and $y_j(t)I$, respectively, and $\mathcal{H}_j(t,\mu)$ is the matrix function of the form (3.2) that corresponds to the operator $H_j = h_j(t)Sh_j^{-1}(t)I$.

Proof. Clearly, it is sufficient to prove the lemma for the operator

$$H = h(t)Sh^{-1}(t)I, \quad \text{where} \quad h(t) \in \Omega.$$

Let

$$h(t) = \prod_{k=1}^{m} (t - t_k)^{\delta_k},$$

where the numbers δ_k satisfy conditions (3.1). By $f_k(t)$ denote the function $f_k(t) = t^{-\delta_k}$, which is continuous on Γ except for possibly at the point t_k, and put

$$f(t) = \prod_{k=1}^{m} f_k(t).$$

It is not difficult to verify that

$$\begin{aligned}&\big(h(t)P\big(h(t)f(t)\big)^{-1}I+Q\big)(PfP+Q)\varphi\\&=(PfP+Q)\big(h(t)P\big(h(t)f(t)\big)^{-1}I+Q\big)\varphi=\varphi\end{aligned}$$

for each function $\varphi(t)$ satisfying the Hölder condition on Γ. Since the operator $h(t)Ph^{-1}(t)I$ is bounded in the space $L_p(\Gamma,\varrho)$, we have

$$hP(hf)^{-1}I+Q=(PfP+Q)^{-1}.$$

From here it follows that

$$PhPh^{-1}I=P(PfP+Q)^{-1}fI.$$

It is easy to check that $PhPh^{-1}I=hPh^{-1}I$. Hence $hPh^{-1}I=P(PfP+Q)^{-1}fI$. Thus,

$$H=2P(PfP+Q)^{-1}fI-I.$$

From Theorem 1.1 it follows that $H\in\mathfrak{A}$. It can be checked straightforwardly that the symbol of the operator H is the matrix function

$$\left\|\begin{matrix}1 & \dfrac{2\nu(t,\mu)(f(t+0)-f(t))}{\ell(t,\mu)(f(t+0)-f(t))+f(t)}\\ 0 & -1\end{matrix}\right\|.$$

Taking into account that $f(t_k+0)/f(t_k)=\exp(2\pi i\delta_k)$, it is easy to verify that the symbol of the operator H coincides with the matrix function $\mathcal{H}(t,\mu)$. The lemma is proved. □

From Theorem 1.1 and the proved lemma one can deduce various conclusions. In particular, they imply the following.

Theorem 3.1. *The operator A given by equality* (3.3) *is a Φ_+-operator or a Φ_--operator in the space $L_p(\Gamma,\varrho)$ if and only if its symbol $\mathcal{A}(t,\mu)$ given by equality* (3.4) *satisfies the condition*

$$\det\mathcal{A}(t,\mu)\neq 0\quad(t\in\Gamma,\ 0\leq\mu\leq 1).\tag{3.5}$$

If condition (3.5) *is fulfilled and $\mathcal{A}(t,\mu)=\|s_{jk}(t,\mu)\|_{j,k=1}^2$, then the operator A is a Φ-operator and*

$$\operatorname{ind}A=-\frac{1}{2\pi}\left\{\arg\frac{\det\mathcal{A}(t,\mu)}{s_{22}(t,0)s_{22}(t,1)}\right\}_{(t,\mu)\in\Gamma\times[0,1]}.$$

4. Operators of the form (0.2)

In this section, for simplicity we assume that $\Gamma=[a,b]$. We suppose that all operators act in the space $L_2(a,b)$.

Under these conditions, a formula for the symbol for the operator

$$A_0=c(t)I+d(t)S\quad(t\in[a,b])$$

with continuous coefficients, as well as for more general operators $A = \sum_{k=0}^{n} a_k(t)S^k$, is essentially simplified.

It turns out (see [4]) that if the functions $a_k(t)$ $(k = 0, 1, \dots, n)$ are continuous on $[a, b]$, then the symbol $\mathcal{A}$ of the operator

$$A = \sum_{k=0}^{n} a_k(t)S^k \tag{4.1}$$

is defined by

$$\mathcal{A}(t, z) = \sum_{k=0}^{n} a_k(t)z^k \quad (t \in [a, b],\ z \in [-1, 1]). \tag{4.2}$$

Let $\mathcal{L}$ be the boundary of the rectangle $\{a \le t \le b; -1 \le z \le 1\}$. In [4] it is shown that the operator A defined by equality (4.1) is a Φ_+-operator or a Φ_--operator in $L_2(a, b)$ if and only if the condition

$$\mathcal{A}(t, z) \neq 0 \quad ((t, z) \in \mathcal{L}) \tag{4.3}$$

is fulfilled. If it holds, then the operator A is a Φ-operator and its index in the space $L_2(a, b)$ is calculated by the formula

$$\operatorname{ind} A = \operatorname{ind} \mathcal{A}(t, z),$$

where $\operatorname{ind} \mathcal{A}(t, z)$ is the winding number of the continuous curve $\mathcal{A}(t, z)$ about the point $\lambda = 0$ when the points (t, z) run through the contour $\mathcal{L}$ in the positive direction.

Let $c(t)$, $d(t)$, and $g(t)$ be continuous functions on $[a, b]$ and

$$R = \frac{1}{\pi i}\left(\ln \frac{b-t}{t-a} S - S \ln \frac{b-t}{t-a} I\right).$$

The operator

$$B = c(t)I + d(t)S + g(t)R$$

is an operator of the form (4.1) and its symbol is given by the equality

$$\mathcal{B}(t, z) = c(t) - g(t) + d(t)z + g(t)z^2. \tag{4.4}$$

Indeed, with the aid of the Poincaré-Bertrand formula [8] it is easy to derive that $R = S^2 - I$. Hence the operator B can be represented in the form (4.1) and formula (4.4) is a corollary of formula (4.2). In particular, this implies the following.

Theorem 4.1. *The operator $B = c(t)I + d(t)S + g(t)R$ is a Φ_+-operator or a Φ_--operator in $L_2(a, b)$ if and only if the function $c(t) - g(t) + d(t)z + g(t)z^2$ is not equal to zero on $\mathcal{L}$. If this condition is satisfied then the operator B is a Φ-operator and*

$$\operatorname{ind} A = \operatorname{ind}(c(t) - g(t) + d(t)z + g(t)z^2).$$

References

[1] I.C. Gohberg and N.Ya. Krupnik, *Algebra generated by one-dimensional singular integral operators with piecewise continuous coefficients.* Funkcional. Anal. Prilozhen. **4** (1970), no. 3, 26–36 (in Russian). English translation: Funct. Anal. Appl. **4** (1970), no. 3, 193–201. MR0270164 (42 #5057), Zbl 0225.45005.

[2] I.C. Gohberg and N.Ya. Krupnik, *On symbols of one-dimensional singular integral operators on an open contour.* Dokl. Akad. Nauk SSSR **191** (1970), 12–15 (in Russian). English translation: Soviet Math. Dokl. **11** (1970), 299–303. MR0264466 (41 #9060), Zbl 0205.40402.

[3] I.C. Gohberg and N.Ya. Krupnik, *Singular integral operators with piecewise continuous coefficients and their symbols.* Izv. Akad. Nauk SSSR Ser. Mat. **35** (1971), 940–964 (in Russian). English translation: Math. USSR Izvestiya **5** (1971), no. 4, 955–979. MR0291893 (45 #981), Zbl 0235.47025.

[4] I.C. Gohberg and N.Ya. Krupnik, *Singular integral equations with continuous coefficients on a composed contour.* Matem. Issled. **5** (1970), no. 2(16), 89–103 (in Russian). English translation: **this volume**. MR0447996 (56 #6306), Zbl 0223.45005.

[5] B.V. Khvedelidze, *Linear discontinuous boundary problems in the theory of functions, singular integral equations and some of their applications.* Trudy Tbiliss. Mat. Inst. Razmadze **23** (1956), 3–158 (in Russian). MR0107148 (21 #5873).

[6] S.A. Freidkin, *On the unboundedness of the singular operator.* Kishinev University, Sect. Estesstv. Eksperim. Nauk, 1970, 84–86 (in Russian).

[7] E.M. Stein, *Interpolation of linear operators.* Trans. Amer. Math. Soc. **83** (1956), 482–492. MR0082586 (18,575d), Zbl 0072.32402.

[8] N.I. Mushelishvili, *Singular Integral Equations.*

1st Russian edition, OGIZ, Moscow, Leningrad, 1946, MR0020708 (8,586b).

English translation of 1st Russian edition: Noordhoff, Groningen, 1953, MR0058845 (15,434e), Zbl 0051.33203.

Reprinted by Wolters-Noordhoff Publishing, Groningen, 1972, MR0355494 (50 #7968), by Noordhoff International Publishing, Leyden, 1977, MR0438058 (55 #10978), by Dover Publications, 1992 and 2008.

2nd Russian edition, revised, Fizmatgiz, Moscow, 1962. MR0193453 (33 #1673), Zbl 0103.07502.

German translation of 2nd Russian edition: *Singuläre Integralgleichungen.* Akademie-Verlag, Berlin, 1965. Zbl 0123.29701.

3rd Russian edition, corrected and augmented, Nauka, Moscow, 1968. MR0355495 (50 #7969), Zbl 0174.16202.

Operator Theory:
Advances and Applications, Vol. 206, 145–156

Singular Integral Equations with Continuous Coefficients on a Composed Contour

Israel Gohberg and Nahum Krupnik

In this paper the algebra generated by singular integral operators A of the form

$$(A\varphi)(t) = a(t)\varphi(t) + \frac{b(t)}{\pi i}\int_\Gamma \frac{\varphi(\tau)}{\tau - t}\,d\tau \quad (t \in \Gamma), \tag{0.1}$$

where Γ is an oriented contour in the complex plane that consists of a finite number of closed and open simple Lyapunov curves, $a(t)$ and $b(t)$ are continuous functions on Γ, is studied. The operators of the form (0.1) will be considered in the space $L_p(\Gamma, \varrho)$ $(1 < p < \infty)$ with weight

$$\varrho(t) = \prod_{k=1}^{2N} |t - \alpha_k|^{\beta_k},$$

where α_k $(k = 1, \dots, N)$ are the starting points and α_k $(k = N+1, \dots, 2N)$ are the terminating points of the corresponding open arcs of the contour Γ, and the numbers β_k satisfy the conditions $-1 < \beta_k < p - 1$. In what follows we will denote the space $L_p(\Gamma, \varrho)$ by $\mathcal{L}_\nu$, where the vector ν is defined by $\nu = (p, \beta_1, \dots, \beta_{2N})$.

By $\mathfrak{A}_\nu$ denote the Banach algebra, which is the closure in the operator norm of the set of the operators of the form

$$B = \sum_{j=1}^{s} \prod_{k=1}^{r} A_{jk}, \tag{0.2}$$

where A_{jk} are operators of the form (0.1), which act in the space $\mathcal{L}_\nu$, and r, s are arbitrary natural numbers.

A space curve denoted by $\widetilde{\Gamma}$ appears in the formulations of the main results. It consists of two copies of the curve Γ, which lie in the planes $z = 1$ and $z = -1$, as well as the straight segments parallel to the z-axis, which connect the starting points and the terminating points, respectively, of the open arcs of the contour Γ.

The paper was originally published as И.Ц. Гохберг, Н.Я. Крупник, Сингулярные интегральные уравнения с непрерывными коэффициентами на составном контуре, Матем. Исслед. **5** (1970), № 2(16), 89–103. MR0447996 (56 #6306), Zbl 0223.45005.

In other words, the curve $\widetilde{\Gamma}$ consists of all points (x, y, z) such that the conditions

$$x + iy \in \Gamma, \quad -1 \le z \le 1, \quad (1 - z^2) \prod_{k=1}^{2N} (x + iy - \alpha_k) = 0$$

hold. We orient the contour $\widetilde{\Gamma}$ in such a way that the direction along $\widetilde{\Gamma}$ in the plane $z = 1$ coincides with the direction along Γ and is opposite to its direction in the plane $z = -1$.

The main results of the paper are the following statements.

The algebra $\mathfrak{A}_\nu$ contains the two-sided ideal γ_ν of all compact operators acting in the space $\mathcal{L}_\nu$. The quotient algebra $\mathfrak{A}_\nu/\gamma_\nu$ is a commutative Banach algebra and its maximal ideal space is homeomorphic to the curve $\widetilde{\Gamma}$.

Let an operator A belong to $\mathfrak{A}_\nu$ and $\widehat{A}$ be the element of $\mathfrak{A}_\nu/\gamma_\nu$ that contains the operator A. By $A_\nu(t, z)$ $((t, z) \in \widetilde{\Gamma})$ denote the function of an element $\widehat{A}$ on the maximal ideal space $\widetilde{\Gamma}$ of the algebra $\mathfrak{A}_\nu/\gamma_\nu$. It is natural to refer to the function $A_\nu(t, z)$ as the symbol of all operators $A \in \widehat{A}$. If the operator A is defined by equality (0.1), then its symbol is given by the formula

$$A_\nu(t, z) = a(t) + b(t)\Omega_\nu(t, z),$$

where

$$\Omega_\nu(t, z) = \begin{cases} \dfrac{z(1 + a_k^2) - i(1 - z^2)a_k}{1 + z^2 a_k^2} & \text{for} \quad t = \alpha_k \quad (k \le N), \\ z & \text{for} \quad t \in \widetilde{\Gamma}, \quad t \ne \alpha_k, \\ \dfrac{z(1 - a_k^2) + i(1 - z^2)a_k}{1 + z^2 a_k^2} & \text{for} \quad t = \alpha_k \quad (k > N) \end{cases}$$

and $a_k = \cot(\pi(1 + \beta_k)/p)$.

In other words, the function $A_\nu(t, z)$ is defined in the planes $z = 1$ and $z = -1$ by the equalities

$$A_\nu(t, 1) = a(t) + b(t), \quad A_\nu(t, -1) = a(t) - b(t),$$

and on each straight segment $t = \alpha_k$ $(k = 1, \ldots, 2N)$ the range of the function $A_\nu(\alpha_k, z)$ circumscribes some circular arc (or the straight segment) connecting the points $z_k = a(\alpha_k) + b(\alpha_k)$ and $\zeta_k = a(\alpha_k) - b(\alpha_k)$ on the plane. The segment $z_k\zeta_k$ is seen from the points of the arc $A_\nu(\alpha_k, z)$ at the angle

$$\psi = \min(2\pi(1 + \beta_k)/p, 2\pi - 2\pi(1 + \beta_k)/p).$$

An operator $A \in \mathfrak{A}_\nu$ is a $\Phi_\pm$-operator in the space $\mathcal{L}_\nu$ if and only if its symbol $A_\nu(t, z)$ differs from zero on the contour $\widetilde{\Gamma}$. The increment of the function $(\arg A_\nu(t, z))/(2\pi)$ along the contour $\widetilde{\Gamma}$ is equal to the index of the operator A in the space $\mathcal{L}_\nu$.

The above-mentioned results are obtained in this paper in a more general case when the algebra $\mathfrak{A}_\nu$ is generated by the operators of the form (0.1) with matrix coefficients $a(t)$ and $b(t)$.

Note that in the case $\nu = (2, 0, \ldots, 0)$, that is, $\mathcal{L}_\nu = L_2(\Gamma)$, the algebra $\mathfrak{A}_\nu/\gamma_\nu$ is isometric to the algebra $C(\widetilde{\Gamma})$ of all continuous functions on the contour $\widetilde{\Gamma}$. This statement follows from more general results by the authors [1] for the space $L_2(\Gamma)$.

A description of the spectrum of the operators of the form (0.1) with differentiable coefficients $a(t)$ and $b(t)$ when $\nu = (2, 0, \ldots, 0)$ and $\Gamma = [0, 1]$ and a description of the maximal ideals of the algebra $\mathfrak{A}_\nu/\gamma_\nu$ in this case were obtained earlier by J. Schwartz [2].

1. Auxiliary propositions and theorems on solvability

Let us introduce the following notation: let $C_n(\Gamma)$ be the algebra of all continuous matrix functions on Γ of order n; let $\Lambda_n(\Gamma)$ be the set of all piecewise continuous matrix functions on Γ of order n; let $\mathcal{L}_\nu^n$ be the Banach space of the vector functions $\varphi = (\varphi_1, \ldots, \varphi_n)$ with components $\varphi_j \in \mathcal{L}_\nu$; and let S be the operator defined in $\mathcal{L}_\nu^n$ by the equality $S\{\varphi_j\}_{j=1}^n = \{S\varphi_j\}_{j=1}^n$, where

$$(S\varphi_j)(t) = \frac{1}{\pi i}\int_\Gamma \frac{\varphi_j(\tau)}{\tau - t}\,d\tau \quad (t \in \Gamma,\ \varphi_j \in \mathcal{L}_\nu).$$

Assume that the contour Γ_0 consists of only closed nonintersecting curves, $M(t) \in \Lambda_n(\Gamma_0)$, and $t_1, \ldots, t_m$ are all discontinuity points of the matrix function $M(t)$. To the matrix function $M(t)$ and the vector $\nu = (p, \beta_1, \ldots, \beta_m)$ assign the continuous matrix curve $V_{M,\nu}(t, \mu)$ obtained by adding the m matrix arcs $W_k(M, \mu)$ $(0 \le \mu \le 1)$ to the range of $M(t)$, where

$$W_k(M,\mu) = \begin{cases} \dfrac{\exp(i\mu\theta_k)\sin(1-\mu)\theta_k}{\sin\theta_k} M(t_k - 0) & \\ \quad + \dfrac{\exp(i(\mu-1)\theta_k)\sin\mu\theta_k}{\sin\theta_k} M(t_k + 0) & \text{if} \quad \theta_k \neq 0, \\ (1-\mu)M(t_k - 0) + \mu M(t_k + 0) & \text{if} \quad \theta_k = 0, \end{cases}$$

and $\theta_k = \pi - 2\pi(1 + \beta_k)/p$ (see [3]).

We say that a matrix function $M(t) \in \Lambda_n(\Gamma_0)$ is ν-nonsingular if

$$\det V_{M,\nu}(t, \mu) \neq 0 \quad (t_0 \in \Gamma_0,\ 0 \le \mu \le 1).$$

We orient the range of the function $\det V_{M,\nu}(t, \mu)$ so that the motion along the curve $\det V_{M,\nu}(t, \mu)$ agrees with the variation of t along Γ_0 in the positive direction at the continuity points of the matrix function $M(t)$ and with the variation of μ from 0 to 1 along the complementary arcs. If $\det V_{M,\nu}(t, \mu) \neq 0$, then by $\operatorname{ind}\det V_{M,\nu}(t, \mu)$ denote the counterclockwise winding number of the curve $\det V_{M,\nu}(t, \mu)$ about the point $\lambda = 0$.

Theorem 1.1. *Let $F_0(t), G_0(t) \in \Lambda_n(\Gamma_0)$ and $t_1, \ldots, t_m$ be all discontinuity points of the matrix functions $F_0(t)$ and $G_0(t)$. The operator $B = F_0 I + G_0 S$ is a Φ-operator in the space $L_p^n(\Gamma_0, \prod |t - t_k|^{\beta_k})$ $(1 < p < \infty, -1 < \beta_k < p - 1)$ if and only if the following two conditions are fulfilled:*

1) $\det(F_0(t+0) - G_0(t+0)) \neq 0$, $\det(F_0(t-0) - G_0(t-0)) \neq 0$;
2) *the matrix function* $M(t) = (F_0(t) - G_0(t))^{-1}(F_0(t) + G_0(t))$ *is* ν*-nonsingular.*

If these conditions are fulfilled, then the index $\kappa(B)$ *of the operator* B *is calculated by the formula*

$$\kappa(B) = -\operatorname{ind}\det V_{M,\nu}(t,\mu). \tag{1.1}$$

A proof of this theorem is given in [3].

Theorem 1.2. *Let a contour* Γ *consist of a finite number of closed and open arcs and* $F(t), G(t) \in C_n(\Gamma)$*. The operator* $A = F(t)I + G(t)S$ *is a* Φ*-operator in* $\mathcal{L}_\nu^n$ *if and only if the condition*

$$\det(F(t) + \Omega_\nu(t,z)G(t)) \neq 0 \quad ((t,z) \in \widetilde{\Gamma}) \tag{1.2}$$

holds. If (1.2) *is fulfilled, then the index* $\kappa(A, \mathcal{L}_\nu^n)$ *of the operator* A *in the space* $\mathcal{L}_\nu^n$ *is calculated by the equality*

$$\kappa(A, \mathcal{L}_\nu^n) = \operatorname{ind}\det(F(t) + \Omega_\nu(t,z)G(t)). \tag{1.3}$$

Proof. Let condition (1.2) be fulfilled. From this condition (for $z = -1$) it follows that $\det(F(t) - G(t)) \neq 0$ $(t \in \Gamma)$. Let $\Gamma_0 (\supset \Gamma)$ be some contour consisting of a finite number of closed Lyapunov curves. Define the functions $F_0(t)$ and $G_0(t)$ on Γ_0 by the equalities

$$F_0(t) = \begin{cases} F(t) & \text{if} \quad t \in \Gamma, \\ E & \text{if} \quad t \in \Gamma_0 \setminus \Gamma, \end{cases} \qquad G_0(t) = \begin{cases} G(t) & \text{if} \quad t \in \Gamma, \\ 0 & \text{if} \quad t \in \Gamma_0 \setminus \Gamma, \end{cases}$$

where E is the identity matrix of order n.

It is easy to see that

$$\det(F_0(t+0) - G_0(t+0)) \neq 0, \quad \det(F_0(t-0) - G_0(t-0)) \neq 0.$$

By $M(t)$ denote the matrix function $M(t) = (F_0(t) - G_0(t))^{-1}(F_0(t) + G_0(t))$. This matrix function is continuous everywhere on Γ_0 except possibly at the points $\alpha_1, \dots, \alpha_{2N}$. The equations of the matrix arcs $W_k(M,\mu)$ corresponding to the matrix function $M(t)$ at the points α_k $(k \leq N)$ can be written in the form

$$W_k(M,\mu) = \begin{cases} (F(\alpha_k) - G(\alpha_k))^{-1} & \\ \quad \times \left[F(\alpha_k) + \dfrac{i(\cos\theta_k - \exp(i(2\mu-1)\theta_k))}{\sin\theta_k} G(\alpha_k)\right] & \text{if} \quad \theta_k \neq 0, \\ (F(\alpha_k) - G(\alpha_k))^{-1}[F(\alpha_k) + (2\mu - 1)G(\alpha_k)] & \text{if} \quad \theta_k = 0. \end{cases}$$

Introduce the following notation:

$$z = \begin{cases} \dfrac{1}{a_k} \tan \dfrac{(2\mu-1)\theta_k}{2} & \text{if} \quad \theta_k \neq 0, \\ 2\mu - 1 & \text{if} \quad \theta_k = 0. \end{cases}$$

If μ ranges over the segment $[0,1]$, then z ranges over the segment $[-1,1]$. It is easy to check that

$$i\left(\cot\theta_k - \frac{\exp(i(2\mu-1)\theta_k)}{\sin\theta_k}\right) = \frac{z(1+a_k^2)-i(1-z^2)a_k}{1+a_k^2z^2}.$$

From here it follows that

$$W_k(M,\mu) = (F(\alpha_k)-G(\alpha_k))^{-1}(F(\alpha_k)+\Omega_\nu(\alpha_k,z)G(\alpha_k)) \quad (k \le N). \tag{1.4}$$

Analogously, setting

$$z = \begin{cases} \dfrac{1}{a_k}\tan\dfrac{(1-2\mu)\theta_k}{2} & \text{if} \quad \theta_k \neq 0, \\ 1-2\mu & \text{if} \quad \theta_k = 0, \end{cases}$$

we get equality (1.4) for the points $\alpha_{N+1},\dots,\alpha_{2N}$.

Thus,

$$F(t)+\Omega_\nu(t,z)G(t) = \begin{cases} F(t)+G(t) & \text{if} \quad z=1,\ t\in\Gamma, \\ (F(\alpha_k)-G(\alpha_k))W_k(M,\mu) & \text{if} \quad t=\alpha_k, \\ F(t)-G(t) & \text{if} \quad z=-1,\ t\in\Gamma. \end{cases} \tag{1.5}$$

From here it follows that the matrix function $M(t)$ is ν-nonsingular. Hence the operator $B = F_0I + G_0S$ is a Φ-operator in the space $L_p^n(\Gamma_0,\varrho)$. From Theorem 1.1 and equality (1.5) it follows also that if the operator B is a Φ-operator, then $\det(F(t)+\Omega_\nu(t,z)G(t)) \neq 0$ $((t,z)\in\widetilde{\Gamma})$. Let $\det(F(t)+\Omega_\nu(t,z)G(t)) \neq 0$. Then from formula (1.1) with the aid of equality (1.5) it is easy to derive that

$$\kappa(B, L_p^n(\Gamma_0,\varrho)) = \operatorname{ind}\det(F(t)+\Omega_\nu(t,z)G(t)).$$

Let us show that all these statements are valid also for the operator A. To this end, we embed the space $L_p^n(\Gamma,\varrho)$ into the space $L_p^n(\Gamma_0,\varrho)$ assuming that all the functions in $L_p(\Gamma,\varrho)$ are equal to zero on the complementary contour $\Gamma_0\setminus\Gamma$ (cf. [5]). It is easy to see that the subspace $L_p^n(\Gamma,\varrho)$ is an invariant subspace of the operator B and its restriction to $L_p^n(\Gamma,\varrho)$ coincides with A. Moreover, the equality $\widetilde{P}B\widetilde{P} = \widetilde{P}$ holds, where $\widetilde{P}$ is the projection of the space $L_p^n(\Gamma_0,\varrho)$ onto $L_p^n(\Gamma_0\setminus\Gamma,\varrho)$ parallel to $L_p^n(\Gamma,\varrho)$. From here it follows that the operator A is a Φ-operator in the space $L_p^n(\Gamma,\varrho)$ if and only if the operator B is a Φ-operator in $L_p^n(\Gamma_0,\varrho)$ and $\kappa(A, L_p^n(\Gamma,\varrho)) = \kappa(B, L_N^p(\Gamma_0,\varrho))$. The theorem is proved. □

Theorem 1.3. *Let $a_{jk}, b_{jk} \in C(\Gamma)$ and $A_{jk} = a_{jk}I + b_{jk}S$. The operator*

$$A = \sum_{j=1}^r \prod_{k=1}^s A_{jk}$$

is a Φ-operator in the space $\mathcal{L}_\nu$ if and only if the condition

$$A_\nu(t,z) = \sum_{j=1}^r \prod_{k=1}^s (a_{jk}(t)+\Omega_\nu(t,z)b_{jk}(t)) \neq 0$$

holds. If this condition is fulfilled, then

$$\kappa(A, \mathcal{L}_\nu) = \operatorname{ind} A_\nu(t, z). \tag{1.6}$$

Recall that the counterclockwise winding number of the continuous plane curve $A_\nu(t, z)$ about the point $\lambda = 0$ when (t, z) ranges over the oriented curve $\widetilde{\Gamma}$ is denoted by $\operatorname{ind} A_\nu(t, z)$.

The proof of the above theorem is based on the following auxiliary statement.

Lemma 1.1. *Let x_{jk} $(j = 1, \dots, r;\ k = 1, \dots, s)$ be elements of some, in general, noncommutative algebra with unit e and $x = \sum_{j=1}^{r} \prod_{k=1}^{s} x_{jk}$. Then the identity*

$$\begin{pmatrix} X & Z \\ Y & 0 \end{pmatrix} = \begin{pmatrix} E & 0 \\ T & -e \end{pmatrix} \begin{pmatrix} E & 0 \\ 0 & x \end{pmatrix} \begin{pmatrix} X & Z \\ 0 & e \end{pmatrix} \tag{1.7}$$

holds, where X is the square matrix of order $r(s+1)$ given by

$$X = \begin{pmatrix}
\overbrace{e & & -x_{11}}^{r+1} & & & & & & & & \\
& \ddots & & \ddots & & & & & & & \\
& & \ddots & & -x_{r1} & & & & & & \\
& & & \ddots & & -x_{12} & & & & & \\
& & & & \ddots & & \ddots & & & & \\
& & & & & \ddots & & -x_{r2} & & & \\
& & & & & & \ddots & & \ddots & & \\
& & & & & & & \ddots & & -x_{1s} & \\
& & & & & & & & \ddots & & \ddots \\
& & & & & & & & & \ddots & -x_{rs} \\
& & & & & & & & & & \ddots \\
& & & & & & & & & & e
\end{pmatrix}$$

and whose entries are equal to zero except for the entries of the two diagonals, E is the identity matrix of order $r(s+1)$, Z is the one-column matrix whose lower r entries are equal to e and all other entries are equal to zero,

$$Y = (\overbrace{e, \dots, e}^{r}, \overbrace{0, \dots, 0}^{rs})$$

and

$$T = (\overbrace{e, \dots, e}^{r}, x_{11}, \dots, x_{r1}, x_{11}x_{12}, \dots, x_{r1}x_{r2}, \dots, x_{11}\cdots x_{1s}, \dots, x_{r1}\cdots x_{rs}).$$

Equality (1.7) is checked straightforwardly. By $\Theta(x_{jk})$ denote the matrix on the left-hand side of equality (1.7).

Proof of Theorem 1.3. Substitute $x_{jk} \mapsto A_{jk}$ and $e \mapsto I$ in the formulas for the blocks of (1.7). From the resulting equality it follows that the operator A is (is not) a Φ-operator in the space $\mathcal{L}_\nu$ if and only if the operator $\Theta(A_{jk})$ is (resp. is not) a Φ-operator in the space $\mathcal{L}_\nu^n$ ($n = rs + r + 1$) and the indices of these operators in the corresponding spaces coincide:

$$\kappa(A, \mathcal{L}_\nu) = \kappa(\Theta(A_{jk}), \mathcal{L}_\nu^n).$$

The operator $\Theta(A_{jk})$ is the singular integral operator with matrix coefficients

$$\Theta(A_{jk}) = F(t)I + G(t)S.$$

The straightforward verification shows that

$$F(t) + \Omega_\nu(t,z)G(t) = \Theta(a_{jk}(t) + \Omega_\nu(t,z)b_{jk}(t)), \tag{1.8}$$

where the matrix on the right-hand side of the latter equality is constructed from elements of the algebra $C(\widetilde{\Gamma})$. From equalities (1.7) and (1.8) it follows that

$$\det(F(t) + \Omega_\nu(t,z)G(t)) = A_\nu(t,z).$$

It remains to apply Theorem 1.2, which implies all the assertions of the theorem. □

Corollary 1.1. *The complement to the set of the Φ-points*[1] *of the operator*

$$A = \sum \prod (a_{jk}I + b_{jk}S) \quad (a_{jk}(t), b_{jk}(t) \in C(\Gamma))$$

coincides with the range of the function

$$A_\nu(t,z) = \sum \prod (a_{jk}(t) + \Omega_\nu(t,z)b_{jk}(t)) \quad ((t,z) \in \widetilde{\Gamma}).$$

Theorem 1.4. *Let*

$$A = \sum_{j=1}^{r} \prod_{k=1}^{s} (a_{jk}I + b_{jk}S) \quad (a_{jk}(t), b_{jk}(t) \in C(\Gamma))$$

and

$$A_\nu(t,z) = \sum_{j=1}^{r} \prod_{k=1}^{s} (a_{jk}(t) + \Omega_\nu(t,z)b_{jk}(t)) \quad ((t,z) \in \widetilde{\Gamma}).$$

Then the inequality

$$\inf_{T \in \gamma_\nu} \|A + T\| \geq \max_{(t,z) \in \widetilde{\Gamma}} |A_\nu(t,z)| \tag{1.9}$$

holds.

Proof. Since for every point $(t_0, z_0) \in \widetilde{\Gamma}$ the number $A_\nu(t_0, z_0)$ is not a Φ-point of the operator A, we have $\inf \|A + T\| \geq |A_\nu(t_0, z_0)|$. This implies inequality (1.9). The theorem is proved. □

[1]A complex number λ is said to be a Φ-point of an operator A if the operator $A - \lambda I$ is a Φ-operator (see [4]).

2. Quotient algebra

In this section the structure of the maximal ideals of the quotient algebra $\mathfrak{A}_\nu/\gamma_\nu$ is studied.

Lemma 2.1. *The set γ_ν is a minimal two-sided ideal of the algebra $\mathfrak{A}_\nu$ and the quotient algebra $\mathfrak{A}_\nu/\gamma_\nu$ is commutative.*

Proof. Let $a(t), b(t) \in C(\Gamma)$. Then the operator $T = a(t)(tS - St)b(t)I$ belongs to the algebra $\mathfrak{A}_\nu$. The operator T has rank one:

$$(T\varphi)(t) = \frac{1}{\pi i}\int_\Gamma a(t)b(\tau)\varphi(\tau)\,d\tau. \tag{2.1}$$

The set of the operators of the form (2.1) is dense in the set of all rank one operators. This yields the first statement of the lemma.

Let $a(t)$ be an arbitrary function in $C(\Gamma)$. Then $a(t)$ is the uniform limit of a sequence of rational functions $r_n(t) \in C(\Gamma)$. It is easy to check that the operator $r_n(t)S - Sr_n(t)I$ is of finite rank. Hence $a(t)S - Sa(t)I \in \gamma_\nu$. From here it follows that the commutant of the algebra $\mathfrak{A}_\nu$ is contained in γ_ν.

The lemma is proved. □

In view of what has been proved, the quotient algebra $\mathfrak{A}_\nu/\gamma_\nu$ is commutative. By $\mathfrak{R}$ denote the set of all operators of the form

$$A = \sum_{j=1}^{r}\prod_{k=1}^{s}(a_{jk}I + b_{jk}S) \quad (a_{jk}(t), b_{jk}(t) \in C(\Gamma)).$$

Since the operators $a(t)S - Sa(t)I$ are compact, we see that each operator $A \in \mathfrak{R}$ can be represented in the form

$$A = \sum_{k=0}^{m} a_k(t)S^k + T, \tag{2.2}$$

where $a_k(t) \in C(\Gamma)$, $T \in \gamma_\nu$. To each operator A of the form (2.2) we assign the symbol (more precisely, the ν-symbol) defined by the equality

$$A_\nu(t,z) = \sum_{k=0}^{m} a_k(t)(\Omega_\nu(t,z))^k \quad ((t,z) \in \widetilde{\Gamma}).$$

From inequality (1.9) it follows that the symbol of the operator A does not depend on a manner of representation of the operator A in the form (2.2). Inequality (1.9) allows us to define the ν-symbol of an operator $A \in \mathfrak{A}_\nu$ as the uniform limit of a sequence of the symbols of operators $A_n \in \mathfrak{R}$ converging to the operator A in the norm of the algebra $\mathfrak{A}_\nu$. From inequality (1.9) it also follows that the same symbol corresponds to all operators in a coset $\widehat{A} \in \mathfrak{A}_\nu/\gamma_\nu$. Let us agree to denote it by $A_\nu(t,z)$ or $\widehat{A}_\nu(t,z)$.

Theorem 2.1. *The set* M_{t_0,z_0} $((t_0, z_0) \in \widetilde{\Gamma})$ *of all elements of* $\widehat{A} \in \mathfrak{A}_\nu/\gamma_\nu$ *such that* $\widehat{A}_\nu(t_0, z_0) = 0$ *is a maximal ideal of the algebra* $\mathfrak{A}_\nu/\gamma_\nu$.

All maximal ideals of the algebra $\mathfrak{A}_\nu/\gamma_\nu$ *are of the form* M_{t_0,z_0}. *The symbol* $\widehat{A}_\nu(t, z)$ *is a function of an element* $\widehat{A} \in \mathfrak{A}_\nu/\gamma_\nu$ *on the maximal ideal space of the algebra* $\mathfrak{A}_\nu/\gamma_\nu$:

$$\widehat{A}(M_{t_0,z_0}) = \widehat{A}_\nu(t_0, z_0).$$

Proof. Since the functional $\psi_{t_0,z_0}(\widehat{A}) = \widehat{A}_\nu(t_0, z_0)$ $((t_0, z_0) \in \widetilde{\Gamma})$ is a multiplicative functional on the algebra $\mathfrak{A}_\nu/\gamma_\nu$, we deduce that (see [6]) the set M_{t_0,z_0} of the zeros of this functional is a maximal ideal of the algebra $\mathfrak{A}_\nu/\gamma_\nu$.

Let M be an arbitrary maximal ideal. By t_0 denote the number $(\widehat{tI})(M)$. Since the spectrum of the element $\widehat{tI}$ coincides with the contour Γ, we get $t_0 \in \Gamma$. It is not difficult to show that for each function $a(t)$ continuous on the contour Γ, the number $(\widehat{a(t)I})(M)$ coincides with $a(t_0)$. Let us show that there exists a number z_0 $(-1 \le z_0 \le 1)$ such that $\widehat{S}(M) = \Omega_\nu(t_0, z_0)$. Consider separately two cases.

Case I: the point t_0 is not an endpoint of an open arc of the contour Γ. By B denote the operator defined in $\mathcal{L}_\nu$ by the equality

$$B = (I - S^2) \prod_{j=1}^{2N} (t - \alpha_j)I.$$

It is easy to see that $B_\nu(t, z) \equiv 0$. From Theorem 1.3 it follows that each point $\lambda \neq 0$ is a Φ-point of the operator B. The latter fact means that the spectrum of the element $\widehat{B}$ in the algebra $\mathfrak{B}/\gamma_\nu$, where $\mathfrak{B}_\nu$ is the Banach algebra of all bounded linear operators in $\mathcal{L}_\nu$, consists of one point $\lambda = 0$.

In this case it is easy to see that, when passing to the subalgebra $\mathfrak{A}_\nu/\gamma_\nu$, the spectrum of the element $\widehat{B}$ does not change. From here it follows that $\widehat{B}(M) = 0$. However, $t_0 \neq \alpha_j$, whence $1 - (\widehat{S}(M))^2 = 0$. Let us denote the number $\widehat{S}(M)$ by z_0. Since $z_0^2 = 1$, we have $\Omega_\nu(t_0, z_0) = z_0$. Thus $\Omega_\nu(t_0, z_0) = \widehat{S}(M)$.

Case II: the point t_0 coincides with one of the endpoints of open arcs. In this case we prove the existence of the point z_0 by contradiction. Assume that $\widehat{S}(M) \neq \Omega_\nu(t_0, z)$ $(-1 \le z \le 1)$. Then it turns out that one can choose a function $a(t)$ satisfying the following two conditions: 1) $a(t_0) = -\widehat{S}(M)$ and 2) the point $\lambda = 0$ belongs to an unbounded regularity component of the operator $C = a(t)I + S$ in the algebra $\mathfrak{B}_\nu$ (below we will give an example of such a function).

From condition 2) it follows that the point $\lambda = 0$ belongs to an unbounded connected regularity component of the element $\widehat{C}$ in the algebra $\mathfrak{B}_\nu/\gamma_\nu$. Therefore (see [7]), this component still consists of regular points of the element $\widehat{C}$ when passing to the algebra $\mathfrak{A}_\nu/\gamma_\nu$. Hence $\widehat{C}$ is invertible in the algebra $\mathfrak{A}_\nu/\gamma_\nu$. The latter is impossible because $\widehat{C}(M) = a(t_0) + \widehat{S}(M) = 0$.

Let us give an example of a function $a(t)$ satisfying conditions 1) and 2). First, let $\operatorname{Im} \widehat{S}(M) \neq 0$. By ℓ denote the straight line in the plane passing through the points $t = 0$ and $t = \widehat{S}(M)$. This straight line crosses the curve $t = \Omega_\nu(t_0, z) - \widehat{S}(M)$ $(-1 \leq z \leq 1)$ at a unique point $\widetilde{t}$. Note that $\widetilde{t} \neq 0$ because $\widehat{S}(M) \neq \Omega_\nu(t_0, z)$. We choose a point $\widetilde{z}$ on the straight line ℓ sufficiently far and such that the functions $\Omega_\nu(\alpha_j, z) + \widetilde{z}$ $(\alpha_j \neq t_0, -1 \leq z \leq 1)$ do not take real values and the point $t = 0$ does not belong to the segment joining the points $\widetilde{t}$ and $\widetilde{z}$. We choose $a(t)$ as a continuous function on Γ whose range fills in the segment joining the points $-\widehat{S}(M)$ and $\widetilde{z}$, and moreover, $a(t_0) = -\widehat{S}(M)$ and $a(\alpha_j) = \widetilde{z}$ $(\alpha_j \neq t_0)$. If $\operatorname{Im} \widehat{S}(M) = 0$, then we choose the straight line ℓ passing through the point $-\widehat{S}(M)$ perpendicularly to the real axis. The rest of the construction is developed analogously. From Theorem 1.3 it follows that the complement to the set of the Φ-points of the operator C in $\mathcal{L}_\nu$ consists of the two segments $\lambda = a(t) + 1$ and $\lambda = a(t) - 1$; the $2N - 1$ circular arcs (or segments) $\lambda = \Omega_\nu(\alpha_j, z) + \widetilde{z}$ $(\alpha_j \neq t_0)$, and the circular arc (or segment) $\lambda = \Omega_\nu(t_0, z) - \widehat{S}(M)$. It is not difficult to observe that the point $\lambda = 0$ belongs to an unbounded regularity component of the operator C in the algebra $\mathfrak{B}_\nu$.

Thus we have shown the existence of a point $(t_0, z_0) \in \widetilde{\Gamma}$ such that $\widehat{S}(M) = \Omega_\nu(t_0, z_0)$ and $(\widehat{f(t)I})(M) = f(t_0)$ for each function $f(t) \in C(\Gamma)$. From here it already follows that $\widehat{A}(M) = \widehat{A}_\nu(t_0, z_0)$ for all $\widehat{A} \in \mathfrak{A}_\nu/\gamma_\nu$. Hence $M = M_{t_0, z_0}$. The theorem is proved. □

3. Normal solvability and index of operators in the algebra $\mathfrak{A}_\nu^{(n)}$

Theorem 3.1. *Let $A \in \mathfrak{A}_\nu$. The operator A is a Φ_+-operator or a Φ_--operator in the space $\mathcal{L}_\nu$ if and only if the condition $A_\nu(t, z) \neq 0$ $((t, z) \in \widetilde{\Gamma})$ holds. If this condition is fulfilled, then the index $\kappa(A, \mathcal{L}_\nu)$ of the operator A in the space $\mathcal{L}_\nu$ is calculated by the formula*

$$\kappa(A, \mathcal{L}_\nu) = \operatorname{ind} A_\nu(t, z). \tag{3.1}$$

Proof. Let $A_\nu(t, z) \neq 0$. Then from Theorem 2.1 it follows that the coset $\widehat{A}$ containing the operator A is invertible in the quotient algebra $\mathfrak{A}_\nu/\gamma_\nu$. Hence it is invertible in the algebra $\mathfrak{B}_\nu/\gamma_\nu$. From here it follows that (see, e.g., [8]) the operator A is a Φ-operator in $\mathcal{L}_\nu$. Since the functionals $\kappa(A, \mathcal{L}_\nu)$ and $\operatorname{ind} A_\nu(t, z)$ are continuous on the set of Φ-operators acting in $\mathcal{L}_\nu$, we see that formula (3.1) is a corollary of formula (1.6).

Let us prove the necessity of the hypotheses of the theorem. First we show that if the operator A is a Φ-operator, then $A_\nu(t, z) \neq 0$ $((t, z) \in \widetilde{\Gamma})$. Assume the contrary, that is, that the operator A is a Φ-operator and $A_\nu(t_0, z_0) = 0$. Then there exists a Φ-operator $B \in \mathfrak{R}$ such that $B_\nu(t_0, z_0) = 0$, but this contradicts Theorem 1.3.

Let the operator A be a $\Phi_\pm$-operator. Then it is not difficult to find $C \in \mathfrak{R}$ sufficiently close to A in the norm and such that $C_\nu(t, z) \neq 0$. In view of what has just been proved, the operator C is a Φ-operator. From here it follows that the operator A is also a Φ-operator. The theorem is proved. □

The proved theorem can be generalized to the algebra generated by singular integral operators with matrix coefficients.

Let $\mathfrak{A}_\nu^{(n)}$ be the algebra of bounded linear operators in $\mathcal{L}_\nu^n$ of the form $A = \|A_{jk}\|_{j,k=1}^n$, where $A_{jk} \in \mathfrak{A}_\nu$. We refer to the matrix function

$$A_\nu(t,z) = \|(A_{jk})_\nu(t,z)\|_{j,k=1}^n,$$

where $(A_{jk})_\nu(t,z)$ is the ν-symbol of the operator A_{jk}, as the symbol (more precisely, the ν-symbol) of the operator A.

Theorem 3.2. *Let $A \in \mathfrak{A}_\nu^{(n)}$. The operator A is a Φ_+-operator or a Φ_--operator in $\mathcal{L}_\nu^n$ if and only if the condition* $\det A_\nu(t,z) \neq 0$ $((t,z) \in \widetilde{\Gamma})$ *holds. If this condition is fulfilled, then*

$$\kappa(A, \mathcal{L}_\nu^n) = \operatorname{ind} \det A_\nu(t,z). \tag{3.2}$$

Proof. Since $A_{jk}A_{sr} - A_{sr}A_{jk} \in \gamma_\nu$ $(j,k,s,r = 1,\dots,n)$, in view of [9, Theorem 2] we see that A is a Φ-operator in $\mathcal{L}_\nu^n$ if and only if the operator $B = \det \|A_{jk}\|_{j,k=1}^n$ (the formally constructed determinant of the matrix $\|A_{jk}\|$) is a Φ-operator in $\mathcal{L}_\nu$. From the equality $\det A_\nu(t,z) = B_\nu(t,z)$ and Theorem 3.1 it follows that the condition $\det A_\nu(t,z) \neq 0$ $((t,z) \in \widetilde{\Gamma})$ is necessary and sufficient for the operator A to be a Φ-operator in $\mathcal{L}_\nu^n$.

Let the operator $A = \|A_{jk}\|_{j,k=1}^n \in \mathfrak{A}_\nu^{(n)}$ be a $\Phi_\pm$-operator. Then there exists an operator $A'_{11} \in \mathfrak{A}_\nu$ sufficiently close to the operator A_{11} and such that $(A'_{11})_\nu(t,z) \neq 0$ $((t,z) \in \widetilde{\Gamma})$. From Theorem 3.1 it follows that the operator A'_{11} is a Φ-operator. Let R be one of its regularizers. From the results of the previous section it follows that $R \in \mathfrak{A}_\nu$. It is easy to verify that

$$\begin{aligned}
&\begin{pmatrix} A'_{11} & A_{12} & \dots & A_{1n} \\ A_{21} & A_{22} & \dots & A_{2n} \\ \dots & \dots & \dots & \dots \\ A_{n1} & A_{n2} & \dots & A_{nn} \end{pmatrix} \\
&= \begin{pmatrix} I & 0 & \dots & 0 \\ A_{21}R & B_{22} & \dots & B_{2n} \\ \dots & \dots & \dots & \dots \\ A_{n1}R & B_{n2} & \dots & B_{nn} \end{pmatrix} \begin{pmatrix} A'_{11} & A_{12} & \dots & A_{1n} \\ 0 & I & \dots & 0 \\ \dots & \dots & \dots & \dots \\ 0 & 0 & \dots & I \end{pmatrix} + T,
\end{aligned} \tag{3.3}$$

where $B_{jk} = A_{jk} - RA_{j1}A_{1k} \in \mathfrak{A}_\nu$ and T is a compact operator in $\mathcal{L}_\nu^n$.

To finish the proof of the theorem, it remains to prove formula (3.2) and to show that if an operator $A \in \mathfrak{A}_\nu^{(n)}$ is a $\Phi_\pm$-operator, then it is a Φ-operator. For $n = 1$ both statements follow from the previous theorem. By using equality (3.3), it is not difficult to prove these statements by induction on n (cf. [10]). The theorem is proved. □

References

[1] I.C. Gohberg and N.Ya. Krupnik, *On symbols of one-dimensional singular integral operators on an open contour.* Dokl. Akad. Nauk SSSR **191** (1970), 12–15 (in Russian). English translation: Soviet Math. Dokl. **11** (1970), 299–303. MR0264466 (41 #9060), Zbl 0205.40402.

[2] J. Schwartz, *Some results on the spectra and spectral resolutions of a class of singular integral operators.* Commun. Pure Appl. Math. **15** (1962), 75–90. MR0163177 (29 #480), Zbl 0111.30202.

[3] I.C. Gohberg and N.Ya. Krupnik, *Systems of singular integral equations in L_p spaces with weight.* Dokl. Akad. Nauk SSSR **186** (1969) 998–1001 (in Russian). English translation: Soviet Math. Dokl. **10** (1969), 688–691. MR0248566 (40 #1818), Zbl 0188.18302.

[4] I.C. Gohberg and M.G. Krein, *The basic propositions on defect numbers, root numbers and indices of linear operators.* Uspehi Mat. Nauk (N.S.) **12** (1957), no. 2(74), 43–118 (in Russian). English translation: Amer. Math. Soc. Transl. (2) **13** (1960), 185–264. MR0096978 (20 #3459), MR0113146 (22 #3984), Zbl 0088.32101.

[5] I.C. Gohberg and N.Ya. Krupnik, *On the spectrum of singular integral operators in L_p spaces with weight.* Dokl. Akad. Nauk SSSR **185** (1969), 745–748 (in Russian). English translation: Soviet Math. Dokl. **10** (1969), 406–410. MR0248565 (40 #1817), Zbl 0188.18301.

[6] I.M. Gelfand, D.A. Raikov, and G.E. Shilov, *Commutative Normed Rings.* Gosudarstv. Izdat. Fiz.-Mat. Lit., Moscow, 1960 (in Russian). MR0123921 (23 #A1242).

English translation with a supplementary chapter: Chelsea Publishing Co., New York, 1964. MR0205105 (34 #4940).

German translation: *Kommutative Normierte Algebren.* VEB Deutscher Verlag der Wissenschaften, Berlin, 1964. MR0169072 (29 #6327), Zbl 0134.32103.

French translation with appendices by J.P. Kahane and P. Malliavin: *Les Anneaux Normes Commutatifs.* Gauthier-Villars Editeur, Paris, 1964. Zbl 0134.32104.

[7] I.C. Gohberg, *Normal solvability and the index of a function of an operator.* Izv. Akad. Nauk Mold. SSR 1963, no. 11, Ser. Estestv. Tekh. Nauk (1964), 11–25 (in Russian). MR0223918 (36 #6965), Zbl 0152.33601.

[8] I.C. Gohberg, A.S. Markus, and I.A. Feldman, *Normally solvable operators and ideals associated with them.* Izv. Akad. Nauk Mold. SSR (1960), no. 10(76), 51–70 (in Russian). English translation: Amer. Math. Soc. Transl. (2) **61** (1967), 63–84. MR0218920 (36 #2004), Zbl 0181.40601.

[9] N.Ya. Krupnik, *On the question of normal solvability and the index of singular integral equations.* Kishinev. Gos. Univ. Uchen. Zap. **82** (1965) 3–7 (in Russian). MR0205115 (34 #4950).

[10] I.C. Gohberg and N.Ya. Krupnik, *On an algebra generated by the Toeplitz matrices in the spaces h_p.* Matem. Issled. **4** (1969), no. 3, 54–62 (in Russian). English translation: **this volume**. MR0399922 (53 #3763a), Zbl 0254.47045.

Operator Theory:
Advances and Applications, Vol. 206, 157–184

On a Local Principle and Algebras Generated by Toeplitz Matrices

Israel Gohberg and Nahum Krupnik

The main topic of the present paper is the study of some Banach algebras of bounded linear operators acting in the spaces ℓ_p $(1 < p < \infty)$. Generators of these algebras are defined by Toeplitz matrices constructed from the Fourier coefficients of functions having finite limits from the left and from the right at each point.

These algebras were studied for the case of the space ℓ_2 in the paper [1] and for the case of the space h_p $(1 < p < \infty)$ [1] in the paper [2].

First important results on the above-mentioned algebras in the spaces ℓ_p $(1 < p < \infty, p \neq 2)$ were obtained very recently in the papers by R.V. Duduchava [3, 4]. Results of R.V. Duduchava are extended in this paper with the aid of a local principle. In the spaces L_p this local principle is a simplification of I.B. Simonenko's local principle [5, 6]. However, it is applicable to a much larger class of Banach spaces. A general scheme developed in [20] plays also an essential role in the present paper.

The paper consists of seven sections. In Section 1, the local principle is presented. In Section 2, first applications of the local principle are given. They reproduce or generalize some results of I.B. Simonenko [8]. In Section 3, main properties of bounded operators generated by Toeplitz matrices in ℓ_p $(1 \le p \le \infty)$ are contained. In Section 4, applications of the local principle to the study of bounded operators generated by Toeplitz matrices constructed from the Fourier coefficients of continuous functions are outlined. Theorems on inverting operators generated by Toeplitz matrices in ℓ_p constructed from the Fourier coefficients of piecewise continuous functions are contained in Section 5. An investigation of algebras generated by such operators and a symbol theory are presented in Section 6. In Section 7, the results are extended to paired equations and their transposed.

The authors wish to express their sincere gratitude to R.V. Duduchava for useful discussions.

The paper was originally published as И.Ц. Гохберг, Н.Я. Крупник, Об одном локальном принципе и алгебрах, порождённых тёплицевыми матрицами, An. Şti. Univ. "Al. I. Cuza" Iaşi Secţ. I a Mat. (N.S.) **19** (1973), 43–71. MR0399923 (53 #3763b), Zbl 0437.47019.

[1] The space of the sequences of the Fourier coefficients of the functions in the Hardy space H_p is denoted by h_p.

1. Localizing classes

1.1. Let $\mathfrak{A}$ be a Banach algebra with unit e. A set M of elements of the algebra $\mathfrak{A}$ is said to be a *localizing class* if it does not contain zero and for every pair of its elements a_1, a_2 there exists a third element $a \in M$ such that

$$a_1 a = a_2 a = a a_2 = a a_1 = a.$$

Elements x and y in $\mathfrak{A}$ are called *M-equivalent from the left* if

$$\inf_{a \in M} \|(x-y)a\| = 0.$$

The M-equivalency from the right is defined analogously. If elements x and y in $\mathfrak{A}$ are M-equivalent from the left and from the right, then we say that they are M-equivalent.

An element x of the algebra $\mathfrak{A}$ is called *M-invertible from the left (right)* if there exist elements $z \in \mathfrak{A}$ and $a \in M$ such that

$$zxa = a \quad (axz = a).$$

Lemma 1.1. *Let M be a localizing class and elements $x, y \in \mathfrak{A}$ be M-equivalent from the left (resp. right). If the element x is M-invertible from the left (resp. right), then the element y is also M-invertible from the left (resp. right).*

Proof. Let x be M-invertible from the left. Then there exist elements $z \in \mathfrak{A}$ and $a_1 \in M$ such that $zxa_1 = a_1$. Since the elements x and y are M-equivalent from the left, there is an element $a_2 \in M$ such that $\|(x-y)a_2\| < 1/\|z\|$. We choose an element $a \in M$ so that the equalities $a_1 a = a_2 a = a$ hold. Then $zya = zxa - ua$, where $u = z(x-y)a_2$. Taking into account that $zxa = a$, we obtain $zya = (e-u)a$. Since $\|u\| < 1$, we see that the element $e - u$ is invertible. Thus $z_1 ya = a$, where $z_1 = (e-u)^{-1}z$, whence the element y is M-invertible from the left.

The lemma is proved. □

1.2. A system $\{M_\gamma\}_{\gamma \in \Gamma}$ of localizing classes M_γ is said to be a *covering system* if from every set $\{a_\gamma\}_{\gamma \in \Gamma}$ of elements $a_\gamma \in M_\gamma$ one can select a finite number of elements whose sum is an invertible element.

Lemma 1.2. *Let $\{M_\gamma\}_{\gamma \in \Gamma}$ be a covering system of localizing classes. An element x in $\mathfrak{A}$ that commutes with all elements in $\bigcup_{\gamma \in \Gamma} M_\gamma$ is invertible from the left (resp. right) in the algebra $\mathfrak{A}$ if and only if x is M_γ-invertible from the left (resp. right) for each $\gamma \in \Gamma$.*

Proof. The necessity of the hypotheses of the lemma is obvious. Let us prove their sufficiency. Let an element x be M_γ-invertible from the left for each $\gamma \in \Gamma$. Then there exist elements $z_\gamma \in \mathfrak{A}$ and $a_\gamma \in M_\gamma$ such that $z_\gamma x a_\gamma = a_\gamma$. Since the system $\{M_\gamma\}$ is a covering system, from the set $\{a_\gamma\}$ one can extract a finite number of elements $a_{\gamma_1}, \ldots, a_{\gamma_N}$ such that their sum is an invertible element. Put

$$u = \sum_{j=1}^{N} z_{\gamma_j} a_{\gamma_j}.$$

Then

$$ux = \sum_{j=1}^{N} z_{\gamma_j} a_{\gamma_j} x = \sum_{j=1}^{N} z_{\gamma_j} x a_{\gamma_j} = \sum_{j=1}^{N} a_{\gamma_j}.$$

Thus the element x is invertible and

$$x^{-1} = \left(\sum_{j=1}^{N} a_{\gamma_j} \right)^{-1} u.$$

The lemma is proved analogously in the case of the invertibility from the right.

The lemma is proved. □

The proved lemmas immediately imply the following.

Theorem 1.1. *Let $\{M_\gamma\}_{\gamma \in \Gamma}$ be a covering system of localizing classes and an element x be M_γ-equivalent from the left (resp. right) to an element $y_\gamma \in \mathfrak{A}$ for each $\gamma \in \Gamma$. If the element x commutes with all elements in $\bigcup_{\gamma \in \Gamma} M_\gamma$, then it is invertible from the left (resp. right) if and only if the element y_γ is M_γ-invertible from the left (resp. right) for every $\gamma \in \Gamma$.*

1.3. The presented statements can also be interpreted from the point of view of the theory of ideals. We restrict ourselves to the case when all elements in the union $\mathfrak{M} = \bigcup_{\gamma \in \Gamma} M_\gamma$ commute with each other.

By $\mathfrak{A}_0$ denote the commutant of the set $\mathfrak{M}$. Obviously, $\mathfrak{A}_0$ is a subalgebra of the algebra $\mathfrak{A}$. The set J_γ of the elements M_γ-equivalent to zero forms a closed two-sided ideal of the algebra $\mathfrak{A}_0$. Indeed, it is necessary to check only that the unit does not belong to J_γ. Assume that $e \in J_\gamma$. Then in the class M_γ there exist elements c_n $(n = 1, 2, \dots)$ tending to zero. For each c_n there exists an element $a_n \in M_\gamma$ such that $c_n a_n = a_n$. From here it follows that $\|c_n\| \geq 1$.

By $\mathfrak{A}_\gamma$ denote the quotient algebra $\mathfrak{A}_0/J_\gamma$. By X_γ denote the coset in $\mathfrak{A}_\gamma$ that contains an element $x \in \mathfrak{A}_0$.

The coset X_γ is invertible in $\mathfrak{A}_\gamma$ if and only if the element x is M_γ-invertible in $\mathfrak{A}_0$. Indeed, if X_γ is invertible in $\mathfrak{A}_\gamma$ and $Z = X_\gamma^{-1}$, then the element $zx - e$ is M_γ-equivalent to zero. The latter means that x is M_γ-invertible in $\mathfrak{A}_\gamma$. Conversely, if x is an M_γ-invertible element in $\mathfrak{A}_0$, that is, $zxa = a$, where z is some element in $\mathfrak{A}_0$ and a is some element in M_γ, then $A_\gamma = E_\gamma$. Therefore $Z_\gamma X_\gamma = E_\gamma$.

Now Theorem 1.1 can be reformulated as follows.

Theorem 1.2. *Let $x \in \mathfrak{A}_0$. An element x is invertible in the algebra $\mathfrak{A}$ if and only if the element X_γ is invertible in $\mathfrak{A}_\gamma$ for each $\gamma \in \Gamma$.*

This theorem follows immediately from Theorem 1.1, the above remarks, and the fact that the invertibility of an element $x \in \mathfrak{A}_0$ in the algebra $\mathfrak{A}$ implies that x^{-1} belongs to the subalgebra $\mathfrak{A}_0$.

2. First example

2.1. Let Γ be a contour in the complex plane that consists of a finite number of nonintersecting simple closed Lyapunov curves.

Consider the operator of singular integration S defined by the equality

$$(S\varphi)(t) = \frac{1}{\pi i}\int_\Gamma \frac{\varphi(\tau)}{\tau - t}\,d\tau \quad (t \in \Gamma).$$

It is known that this operator is bounded in all spaces $L_p(\Gamma)$ $(1 < p < \infty)$. From the known interpolation theorem by E.M. Semenov [9] it follows that the operator S is bounded in a bulk of separable symmetric spaces including among them the reflexive Orlicz spaces, the uniformly convex Lorentz spaces, and others.

In what follows by $E(\Gamma)$ denote one of such spaces. Let $a(t) = \|a_{jk}(t)\|_{j,k=1}^n$ and $b(t) = \|b_{jk}(t)\|_{j,k=1}^n$ be matrix functions whose entries are complex-valued measurable essentially bounded functions on Γ. Matrix functions $a(t)$ and $b(t)$ are said to be equivalent at a point $t_0 \in \Gamma$ if for every $\varepsilon > 0$ there exists an open arc l_{t_0} containing the point t_0, on which

$$\operatorname*{vrai\,sup}_{t \in l_{t_0}} |a_{jk}(t) - b_{jk}(t)| < \varepsilon.$$

Let F be some set of matrix functions of order n on the contour Γ. Following I.B. Simonenko [8], a matrix $f(t)$ is said to belong to the local closure of the set F if it is equivalent to some matrix function in F at each point of the contour Γ.

A set F is said to be locally closed if the local closure of the set F is contained in F. Let P be the projection defined in $E(\Gamma)$ by the equality $P = (I + S)/2$. By $E_+(\Gamma)$ denote the range of the operator P and by $E_+^n(\Gamma)$ denote the direct sum of n copies of the space $E_+(\Gamma)$. We extend the operator P to the whole space $E^n(\Gamma)$ putting $P\{\varphi_j\}_{j=1}^n = \{P\varphi_j\}_{j=1}^n$. For each matrix function $a(t) = \|a_{jk}(t)\|$ of order n with entries in $L_\infty(\Gamma)$ by T_a denote the matrix operator defined in $E_+^n(\Gamma)$ by the equality[2]

$$T_a = P(a\varphi).$$

By $L_\infty^\Phi(\Gamma)$ denote the set of all matrix functions $a(t)$ with entries in $L_\infty(\Gamma)$ such that the operator T_a is a Φ-operator[3] in the space $E_+^n(\Gamma)$.

Theorem 2.1. *The set $L_\infty^\Phi(\Gamma)$ is locally closed.*

Proof. Let us show that this theorem is a corollary of Theorem 1.1. The role of the algebra $\mathfrak{A}$ is played by the quotient algebra $\mathfrak{A} = L(E_+^n(\Gamma))/\mathcal{J}(E_n^+(\Gamma))$, where $L(E_n^+(\Gamma))$ is the algebra of all bounded linear operators in $E_n^+(\Gamma)$ and $\mathcal{J}(E_+^n(\Gamma))$

---

[2]This operator is sometimes called a generalized matrix Wiener-Hopf operator. Note that in the case when Γ is the unit circle, $n = 1$, and the system t^k $(k = 0, \pm 1, \dots)$ forms a basis of the space $E(\Gamma)$, the Toeplitz matrix constructed from the Fourier coefficients of the function $a(t)$ corresponds to the operator T_a in this basis.

[3]An operator A is said to be a Φ-operator if its range is closed and the numbers $\dim\ker A$ and $\dim\ker A^*$ are finite.

is the two-sided ideal of $L(E_+^n(\Gamma))$ consisting of all compact operators. The coset in $\mathfrak{A}$ that contains an operator $A \in L(E_+^n(\Gamma))$ is denoted by $\widehat{A}$.

We introduce a system $\{M_\zeta\}_{\zeta\in\Gamma}$ of localizing classes in the algebra $\mathfrak{A}$. Let $\zeta_0 \in \Gamma$ and N_{ζ_0} be the set of all continuous functions on Γ, each of which is equal to 1 in some neighborhood (depending on the function) of the point ζ_0. By M_{ζ_0} denote the set of all elements $\widehat{T}_g$ of the algebra $\mathfrak{A}$ generated by the matrices $g(t) = g_0(t)E_n$, where E_n is the identity matrix of order n and $g_0(\zeta) \in N_{\zeta_0}$. One can check straightforwardly that if $a \in C(\Gamma)$, then $a(t)S - Sa(t)I \in \mathcal{J}(E(\Gamma))$. From here it follows that if $a \in C(\Gamma)$ and $b \in L_\infty(\Gamma)$, then

$$T_aT_b - T_{ab} \in \mathcal{J}(E_+(\Gamma)). \tag{2.1}$$

With the aid of this property it is easy to check that for every matrix function $f(t) = \|f_{jk}(t)\|_{j,k=1}^n$ the coset $\widehat{T}_f$ belongs to the commutant of the set $\mathfrak{M} = \bigcup_{t\in\Gamma} M_t$. Property (2.1) also implies that if matrix functions $f(t)$ and $h(t)$ are equivalent at the point ζ_0, then the cosets $\widehat{T}_f$ and $\widehat{T}_h$ are M_{ζ_0}-equivalent.

It easy to verify that the set $\{M_t\}_{t\in\Gamma}$ forms a covering system of localizing classes.

Let $f(t)$ belong to the local closure of the set $L_\infty^\Phi(\Gamma)$. It is known that the operator T_a is a Φ-operator if and only if the element $\widehat{T}_a$ is invertible in the algebra $\mathfrak{A}$. Therefore for every point $\zeta_0 \in \Gamma$ the element $\widehat{T}_f$ is M_{ζ_0}-equivalent to some element $\widehat{T}_{f_{\zeta_0}}$, which is invertible in $\mathfrak{A}$. In view of Theorem 1.1, the element $\widehat{T}_f$ is also invertible in $\mathfrak{A}$. Thus $f \in L_\infty^\Phi(\Gamma)$.

The theorem is proved. □

It can be shown analogously that if a matrix function $f(t)$ is equivalent to a matrix function $f_\zeta(t)$ at each point $\zeta \in \Gamma$ and all operators T_{f_ζ} $(\zeta \in \Gamma)$ admit a left (resp. right) regularization, then the operator T_f admits a left (resp. right) regularization.

Theorem 2.1 can be formulated also for one-dimensional matrix singular integral operators. In that form Theorem 2.1 was proved by I.B. Simonenko [8] for the spaces $L_p(\Gamma)$ $(1 < p < \infty)$. In the paper [8] this theorem was deduced from a local principle (see [5, 6]) more complicated than the one presented in Section 1.

2.2. The statement presented below follows from Theorem 2.1. It was obtained for the first time by I.B. Simonenko [10, 8].

Theorem 2.2. *Let $a(t)$ $(t \in \Gamma)$ be a matrix function with entries in $L_\infty(\Gamma)$. Suppose that for every point $t_0 \in \Gamma$ there exists a neighborhood $l_{t_0} \subset \Gamma$ of this point such that the range of the form $(a(t)\eta, \eta)$ $(\eta \in \mathbb{C}^n, \|\eta\| = 1)$ for $t \in l_{t_0}$ is located in some closed half-plane Π_{t_0} not containing the origin. Then the operator T_a is a Φ-operator in the space $L_2^n(\Gamma)_+$.*

Proof. Let ζ be an arbitrary point on Γ. By a_ζ denote a matrix function that coincides with $a(t)$ on the arc l_ζ and is defined on the complement $\Gamma \setminus l_\zeta$ so that the range of the form $(a_\zeta(t)\eta, \eta)$ $(t \in \Gamma)$ is located in the half-plane Π_ζ. It is not difficult to find a complex number γ so that the norm of the operator $\gamma a_\zeta - E_n$ acting in the space $L_2^n(\Gamma)$ is less than 1. By $|A|$ denote the quotient norm

$$|A| = \inf_{T \in \mathcal{J}(L_2^n(\Gamma))} \|A + T\|.$$

It is known (see [21]) that $|P| = 1$, whence $|P(\gamma a_\zeta - E_n)| < 1$. Since

$$Pa_\zeta = \frac{1}{\gamma}[P + P(\gamma a_\zeta - E_n)],$$

from the last inequality it follows that the operator T_{a_ζ} is a Φ-operator in $L_2^n(\Gamma)_+$. Therefore $a_\zeta \in L_\infty^\Phi(\Gamma)$. In view of Theorem 2.1, $a \in L_\infty^\Phi(\Gamma)$.

The theorem is proved. □

For $n = 1$, I.B. Simonenko [10, 8] obtained a generalization of Theorem 2.2 to the case of an arbitrary space $L_p(\Gamma)$, $(1 < p < \infty)$. That theorem can also be obtained from Theorem 2.1 with the aid of the results [11] (see also [12, 13]).

3. Some properties of operators generated by Toeplitz matrices in ℓ_p spaces

3.1. In the following sections operators generated by Toeplitz matrices in spaces ℓ_p are studied with the aid of the local principle of Section 1. Main properties of these operators are presented in this section.

By $\widetilde{\ell}_p$ $(1 \le p \le \infty)$ denote the Banach space of the sequences $\{\xi_j\}_{j=-\infty}^{\infty}$, $\xi_j \in \mathbb{C}^1$, with the norm

$$\|\xi\| = \left(\sum_{j=-\infty}^{\infty} |\xi_j|^p \right)^{1/p}.$$

By ℓ_p denote the space of the one-sided sequences $\{\xi_j\}_{j=0}^{\infty}$ with the norm

$$\|\xi\| = \left(\sum_{j=0}^{\infty} |\xi_j|^p \right)^{1/p}.$$

Let Γ_0 be the unit circle. To each function $a(\zeta) \in L_\infty(\Gamma_0)$ assign the Toeplitz matrices $\|a_{j-k}\|_{j,k=0}^{\infty}$ and $\|a_{j-k}\|_{j,k=-\infty}^{\infty}$ consisting of the Fourier coefficients a_j $(j = 0, \pm 1, \dots)$ of this function. By T_a denote the linear operator generated in the space ℓ_p $(1 \le p \le \infty)$ by the matrix $\|a_{j-k}\|_{j,k=0}^{\infty}$ and by $\widetilde{T}_a$ denote the operator generated by the matrix $\|a_{j-k}\|_{j,k=-\infty}^{\infty}$ in the Banach space $\widetilde{\ell}_p$.

Let us agree to denote by $L(\mathcal{L})$ the Banach algebra of all bounded linear operators acting in a Banach space $\mathcal{L}$.

By $\mathcal{R}_p$ $(1 \le p \le \infty)$ denote the set of all functions $a(\zeta) \in L_\infty(\Gamma_0)$ such that $T_a \in L(\ell_p)$. It is known that the sets $\mathcal{R}_1$ and $\mathcal{R}_\infty$ coincide, that they consist of all functions $a(\zeta)$ $(|\zeta| = 1)$ that may be expanded into absolutely convergent Fourier series, and that $\mathcal{R}_2 = L_\infty(\Gamma_0)$. It is easy to see that if for a function $a \in L_\infty(\Gamma_0)$

the operator $\widetilde{T}_a$ belongs to $L(\widetilde{\ell}_p)$, then $a \in \mathcal{R}_p$ and $\|T_a\| \le \|\widetilde{T}_a\|$. The converse statement is also true.

Proposition 3.1. *If $a \in \mathcal{R}_p$ $(1 \le p \le \infty)$, then $\widetilde{T}_a \in L(\widetilde{\ell}_p)$ and $\|\widetilde{T}_a\| = \|T_a\|$.*

Proof. Indeed, let $T_a \in L(\ell_p)$ and P_n $(n = 1, 2, \dots)$ be the projections defined in $\widetilde{\ell}_p$ by the equality $P_n\{\xi_j\}_{-\infty}^{\infty} = (\dots, 0, \xi_{-n}, \dots, \xi_0, \xi_1, \dots)$. It is not difficult to see that for each finitely supported sequence $\xi = (\dots, 0, \xi_k, \dots, \xi_r, 0, \dots)$ the equality

$$\lim_{n\to\infty} \|P_n \widetilde{T}_a P_n \xi - \widetilde{T}_a \xi\| = 0$$

holds and, moreover, $\|P_n \widetilde{T}_a P_n \xi\| \le \|T_a\|\,\|\xi\|$. Hence the operator $\widetilde{T}_a$ can be extended to the whole space $\widetilde{\ell}_p$ and $\|\widetilde{T}_a\| \le \|T_a\|$. □

Proposition 3.2. *If $a, b \in \mathcal{R}_p$, then $ab \in \mathcal{R}_p$ and*

$$\|T_{ab}\| \le \|T_a\|\,\|T_b\|. \tag{3.1}$$

Proof. Let $a, b \in \mathcal{R}_p$. Then in view of Proposition 3.1, $\widetilde{T}_a, \widetilde{T}_b \in L(\widetilde{\ell}_p)$. It can be checked straightforwardly that $\widetilde{T}_{ab}\xi = \widetilde{T}_a \widetilde{T}_b \xi$ for all finitely supported sequences $\xi \in \widetilde{\ell}_p$. Hence $\widetilde{T}_{ab} = \widetilde{T}_a \widetilde{T}_b$ $(\in L(\widetilde{\ell}_p))$. From here it follows that $ab \in \mathcal{R}_p$ and inequality (3.1) holds. □

Proposition 3.3. *If $a(\zeta) \in \mathcal{R}_p$, then $\overline{a(\zeta)} \in \mathcal{R}_p$ and $\|T_a\|_p = \|T_{\overline{a}}\|_p$.*

Proof. This statement follows easily from Proposition 3.1 and the equality $\|\widetilde{T}_{\overline{a}}\xi\| = \|\widetilde{T}_a \overline{\xi}\|$, where $\overline{\xi} = \{\overline{\xi_{-j}}\}_{-\infty}^{\infty}$. □

3.2. From the proved properties it is easy to deduce the following statement.

Proposition 3.4. *The set $\mathcal{R}_p$ $(1 \le p \le \infty)$ is a Banach algebra with the norm*

$$\|a\| = \|T_a\|_p$$

and, moreover,

$$\|a\| \ge \operatorname*{vrai\,sup}_{|\zeta|=1} |a(\zeta)|.$$

Proof. In view of Proposition 3.2 it is necessary only to verify the completeness of the space $\mathcal{R}_p$.

From Proposition 3.3 it follows that if $a \in \mathcal{R}_p$, then $a \in \mathcal{R}_q$ $(p^{-1} + q^{-1} = 1)$ and $\|T_a\|_p = \|T_a\|_q$. By the M. Riesz interpolation theorem,

$$\|T_a\|_2 \le \|T_a\|_p.$$

Suppose a sequence $\{a_n\}$ is a Cauchy sequence in $\mathcal{R}_p$. Then it is a Cauchy sequence in $\mathcal{R}_2$ $(= L_\infty(\Gamma_0))$. Hence there exist a function $a \in L_\infty(\Gamma_0)$ and an operator $A \in L(\ell_p)$ such that $\|T_{a_n} - T_a\|_2 \to 0$ and $\|T_{a_n} - A\|_p \to 0$ as $n \to \infty$. From here it follows that the operator T_a is bounded on a dense set in ℓ_p, whence $a \in \mathcal{R}_p$. It is easy to check that $\|a - a_n\| \to 0$. □

Let us agree on the following two notations:

$$\mathcal{R}_{[p_1,p_2]} = \bigcap_{p_1 \le p \le p_2} \mathcal{R}_p, \quad \mathcal{R}_{(p_1,p_2)} = \bigcap_{p_1 < p < p_2} \mathcal{R}_p.$$

Proposition 3.5. *If $a \in \mathcal{R}_p$, then $a \in \mathcal{R}_{[p,q]}$ for $p < 2$ and $a \in \mathcal{R}_{[q,p]}$ for $p > 2$, where $p^{-1} + q^{-1} = 1$.*

Proof. Indeed, from Proposition 3.3 it follows that if $a \in \mathcal{R}_p$, then $a \in \mathcal{R}_q$. Thus Proposition 3.5 follows from the M. Riesz interpolation theorem. □

Proposition 3.6. *Suppose $a \in \mathcal{R}_p$ $(1 \le p \le \infty)$ and r satisfies the inequality $2 < r < p$ for $p < 2$ and the inequality $p < r < 2$ for $p < 2$. Then*

$$\|T_a\|_r \le \|T_a\|_p^t \operatorname*{vrai\,sup}_{|\zeta|=1} |a(\zeta)|^{1-t},$$

where

$$t = \frac{p(2-r)}{r(2-p)}.$$

Proof. This statement is also an immediate corollary of the M. Riesz interpolation theorem. □

3.3. By V denote the algebra of functions $f : \Gamma \to \mathbb{C}^1$ of bounded variation. It is known (see [14]) that $V \subset \mathcal{R}_{(1,\infty)}$. Moreover, for an arbitrary function $a \in V$,

$$\|T_a\|_p \le k_p\big(\operatorname{vrai\,sup} |a(\zeta)| + \operatorname{var} a\big), \tag{3.2}$$

where $\operatorname{var} a$ denotes the total variation of the function $a(\zeta)$ on Γ_0, and k_p is a constant depending only on p.

A more general statement was obtained in the paper [15].

Let V_β $(\beta \ge 1)$ be the set of the functions $a(\zeta)$ such that

$$\sup \sum_{k=0}^{n-1} |a(e^{i\theta_{k+1}}) - a(e^{i\theta_k})|^\beta < \infty,$$

$$0 = \theta_0 < \theta_1 < \cdots < \theta_n = 2\pi.$$

Then $V_\beta \subset \mathcal{R}_{(2\beta/(\beta+1), 2\beta/(\beta-1))}$.

In [15] the following propositions were also obtained.

Let H_α be the collection of all functions $a(\zeta)$ such that $a(e^{i\theta})$ satisfies the Hölder condition with exponent α on some segment $[\theta_0, \theta_0 + 2\pi]$. Then

$$H_\alpha \subset \mathcal{R}_{(2(1+2\alpha)^{-1}, 2(1-2\alpha)^{-1})} \quad (0 < \alpha < 1/2).$$

Moreover, if $\beta \ge 2$ and $\delta > 0$, then

$$H_\delta \cap V_\beta \subset \mathcal{R}_{(2\beta(\beta+2)^{-1}, 2\beta(\beta-2)^{-1})}.$$

If $a(\zeta) \in H_\alpha$ $(0 < \alpha \le 1/2)$ and

$$\sum_{k=2^{-n}}^{2^n} |a_k| \le 2^{n\beta},$$

then $a \in \mathcal{R}_{(p_1,p_2)}$, where

$$p_1 = (2\alpha + 2\beta)/\beta + 2\alpha, \quad p_2 = (2\alpha + 2\beta)/\beta \quad (0 < \beta < 1/2 - \alpha).$$

With the aid of Proposition 3.4 one can essentially extend the class of examples of functions belonging to the set $\mathcal{R}_{(p_1,p_2)}$. For instance, if $x_j(\zeta) \in \mathcal{R}_p$ $(j = 1, 2, \ldots, n)$ and $y_j(\zeta) \in V$, then

$$\sum x_j(\zeta) y_j(\zeta) \in \mathcal{R}_p \quad (1 < p < \infty).$$

In particular, if the functions $x_j(\zeta)$ are expanded in absolutely convergent Fourier series and $\chi_j(\zeta)$ are characteristic functions of some arcs on the unit circle, then

$$\sum_{j=1}^{n} x_j(\zeta) \chi_j(\zeta) \in \mathcal{R}_{(1,\infty)}.$$

This class of functions was considered by R.V. Duduchava [3, 4].

3.4. In conclusion we present two more properties important for further considerations.

Proposition 3.7. *Let functions $a(\zeta)$ and $b(\zeta)$ in $\mathcal{R}_{(p-\varepsilon,p+\varepsilon)}$ $(1 < p < \infty, \varepsilon > 0)$ have finite limits from the left and from the right at each point ζ. Then the operator $T_a T_b - T_b T_a$ is compact in the space ℓ_p. Moreover, if the functions $a(\zeta)$ and $b(\zeta)$ do not have common points of discontinuity, then the operator $T_a T_b - T_{ab}$ also is compact in ℓ_p.*

Proof. Indeed, under the presented assumptions, the operators $T_a T_b - T_b T_a$ and $T_a T_b - T_{ab}$ are compact in ℓ_2 (see [1]) and are bounded in $\ell_{p\pm\varepsilon}$ and $\ell_{q\pm\varepsilon}$. Therefore, Proposition 3.7 follows straightforwardly from the Krasnosel'skii interpolation theorem [16, Theorem 1]. □

Proposition 3.8. *Let $a(\zeta) \in \mathcal{R}_p$. If the function $a(\zeta)$ is not identically zero, then the equation $T_a x = 0$ in the space ℓ_p $(1 \le p \le \infty)$ or the equation $T_{\overline{a}} y = 0$ in the space ℓ_q $(p^{-1} + q^{-1} = 1)$ has only a trivial solution.*

Proof. In the case $p = 2$ this statement is proved in [22]. Without loss of generality one can assume that $1 \le p < 2$.

We prove Proposition 3.8 by contradiction. Assume that there exist nonzero vectors $x_+ = \{x_j^+\}_0^\infty \in \ell_p$ and $y_+ = \{y_j^+\}_0^\infty \in \ell_q$ such that

$$T_a x_+ = 0, \quad T_{\overline{a}} y_+ = 0. \tag{3.3}$$

By $\widetilde{x}_+$ and $\widetilde{y}_+$ denote the vectors $\{x_j^+\}_{j=-\infty}^\infty \in \widetilde{\ell}_p$ and $\{y_j^+\}_{j=-\infty}^\infty \in \widetilde{\ell}_q$, respectively, such that $x_j^+ = y_j^+ = 0$ for $j < 0$.

It is easy to see that equalities (3.3) can be rewritten as follows:

$$\widetilde{T}_a \widetilde{x}_+ = \widetilde{x}_-, \quad \widetilde{T}_{\overline{a}} \widetilde{y}_+ = \widetilde{y}_-, \tag{3.4}$$

where the vectors $\widetilde{x}_- = \{x_j^-\}_{j=-\infty}^\infty \in \widetilde{\ell}_p$ and $\widetilde{y}_- = \{y_j^-\}_{j=-\infty}^\infty \in \widetilde{\ell}_q$ have the property $x_j^- = y_j^- = 0$ for $j \ge 0$.

We construct infinite Toeplitz matrices

$$A = \|a_{j-k}\|_{j,k=-\infty}^{\infty}, \quad X_{\pm} = \|x_{j-k}^{\pm}\|_{j,k=-\infty}^{\infty}, \quad Y_{\pm} = \|y_{j-k}^{\pm}\|_{j,k=-\infty}^{\infty}.$$

Equalities (3.4) imply the following equalities

$$AX_{+} = X_{-}, \quad Y_{+}^{*}A = Y_{-}^{*}. \tag{3.5}$$

Multiplying both sides of the second equality of (3.5) from the right by X_{+} and taking into account the first of these equalities, we obtain

$$Y_{+}^{*}X_{-} = Y_{-}^{*}X_{+}. \tag{3.6}$$

Since the matrix $Y_{+}^{*}X_{-}$ is upper triangular and $Y_{-}^{*}X_{+}$ is lower triangular, and all entries of the main diagonals of these matrices are equal to zero, we see that $Y_{+}^{*}X_{-} = 0$ and $Y_{-}^{*}X_{+} = 0$. It can be checked straightforwardly that if $Y_{+}^{*}X_{-} = 0$, then one of the matrices X_{-} or Y_{+} is equal to zero. If $Y_{+} \neq 0$, then $X_{-} = 0$. In view of (3.5), $AX_{+} = 0$. From here it follows that $a(\zeta)x_{+}(\zeta) = 0$, where

$$x_{+}(\zeta) = \sum_{j=0}^{\infty} x_{j}^{+}\zeta^{j}.$$

Since $\{x_{j}^{+}\} \in \ell_p$ and $p < 2$, we see that the function $x_{+}(\zeta)$ belongs to the Hardy space H_2. By the hypotheses of the proposition, $a(\zeta)$ is different from zero on a set of positive measure. Therefore $x_{+}(\zeta) = 0$ on this set. Thus $x_{+}(\zeta) \equiv 0$ and this contradicts the assumption. □

4. Second example and its applications

4.1. We will present an example illustrating results of Section 1. By ΠC denote the set of all functions $a(\zeta)$ ($|\zeta| = 1$) having finite limits $a(\zeta_0+0)$ and $a(\zeta_0-0)$ as ζ tends to ζ_0 clockwise and counter-clockwise, respectively, at each point $\zeta_0 \in \Gamma_0$. By ΠC_p ($\Pi C_{\langle p \rangle}$) denote the intersection of ΠC with $\mathcal{R}_p$ (respectively, $\cup_{\epsilon>0}\mathcal{R}_{(p-\epsilon,p+\epsilon)}$).

Theorem 4.1. *If a function $a(\zeta)$ belongs to $\Pi C_{\langle p \rangle}$ and $a(\zeta \pm 0) \neq 0$ for all $\zeta \in \Gamma_0$, then the function $1/a(\zeta)$ also belongs to $\Pi C_{\langle p \rangle}$.*

Proof. Obviously it is sufficient to show that $1/a(\zeta) \in \mathcal{R}_p$. We will show that the operator $\widetilde{T}_a$ is invertible in $\widetilde{\ell}_p$. Since for all $x \in \widetilde{\ell}_2$ the equality $\widetilde{T}_{a^{-1}}\widetilde{T}_a x = x$ holds, this will prove the boundedness of the operator $\widetilde{T}_{a^{-1}}$ in $\widetilde{\ell}_p$, and hence will finish the proof of the theorem.

We will develop the proof with the aid of Theorem 1.1. Let $\mathfrak{A} = L(\widetilde{\ell}_p)$. For each point $\zeta_0 \in \Gamma_0$ we introduce the localizing class $M_{\zeta_0} \subset L(\widetilde{\ell}_p)$ consisting of all operators of the form $\widetilde{T}_x$, where $x(\zeta)$ is the characteristic function of some neighborhood of the point ζ_0. It is easy to verify that the set $\{M_\zeta\}_{\zeta\in\Gamma_0}$ forms a covering system.

Moreover, if $a(\zeta) \in \mathcal{R}_p$, then the operator $\widetilde{T}_a$ belongs to the commutant of the set $\bigcup_{\zeta\in\Gamma_0} M_\zeta$. Let $a(\zeta) \in \Pi C_{\langle p\rangle}$ and let τ be some point of the unit circle Γ_0. If τ is a continuity point of the function $a(\zeta)$, then we put

$$a_\tau(\zeta) \stackrel{\text{def}}{=} a(\tau) \quad (\zeta \in \Gamma_0).$$

If $\tau = e^{i\theta}$ is a discontinuity point of the function $a(\zeta)$, then we put

$$a_\tau(e^{i\varphi}) = \begin{cases} a(\tau+0) & \text{for} \quad \theta < \varphi < \varphi + \pi, \\ a(\tau-0) & \text{for} \quad \pi + \theta < \varphi < \theta + 2\pi. \end{cases}$$

We choose a neighborhood $l(\zeta_0)$ of the point ζ_0 so that $\sup|a(\zeta) - a_\tau(\zeta)|$ ($\zeta \in l(\zeta_0)$, $\zeta \neq \zeta_0$) is sufficiently small. Let $\chi(\zeta)$ be the characteristic function of the arc $l(\zeta_0)$. From Proposition 3.7 and relation (3.2) it follows that

$$\|(\widetilde{T}_a - \widetilde{T}_{a_\tau})\widetilde{T}_\chi\| \le \|\widetilde{T}_a - \widetilde{T}_{a_\tau}\|_s^t k_s^t 2^t \sup_{\zeta\in l(\zeta_0),\zeta\neq\zeta_0} |a(\zeta) - a_\tau(\zeta)|^{1-t},$$

where the numbers s and t do not depend on $l(\zeta_0)$. Hence the operators $\widetilde{T}_a$ and $\widetilde{T}_{a_\tau}$ are M_τ-equivalent. Since $a(\zeta \pm 0) \neq 0$ ($\zeta \in \Gamma_0$), each of the operators $\widetilde{T}_{a_\tau}$ ($\tau \in \Gamma_0$) is invertible in the algebra $L(\widetilde{l}_p)$. In view of Theorem 1.1, the operator $\widetilde{T}_a$ is invertible in $L(\widetilde{\ell}_p)$. The theorem is proved. □

4.2. In this subsection, a Fredholm criterion for operators generated by Toeplitz matrices with continuous coefficients is obtained with the aid of Theorem 1.1. These results will be used in forthcoming sections.

By $C_{\langle p\rangle}$ denote the intersection of the sets $C(\Gamma)$ and $\bigcup_{\varepsilon>0} \mathcal{R}_{(p-\varepsilon,p+\varepsilon)}$ and by $C^{n\times n}_{\langle p\rangle}$ denote the set of all matrix functions of order n with entries in $C_{\langle p\rangle}$. If $a = \|a_{jk}(\zeta)\|_{j,k=1}^n$, then by T_a denote the operator $\|T_{a_{jk}}\|_{j,k=1}^n$.

Theorem 4.2. *Let $a \in C^{n\times n}_{\langle p\rangle}$. The operator T_a is a Φ_+-operator or a Φ_-operator in ℓ_p^n if and only if the condition*

$$\det a(\zeta) \neq 0 \quad (|\zeta| = 1) \tag{4.1}$$

holds. If condition (4.1) *is fulfilled, then the operator T_a is a Φ-operator.*

We will use the next statement in the proof of this theorem.

Lemma 4.1. *Suppose a sequence of functions $b_m(\zeta)$ ($|\zeta| = 1$) converges to zero in measure on Γ_0 and, for every $r \in (1,\infty)$, $\|T_{b_m}\|_r \le \alpha_r$, where the constant α_r does not depend on m. Then the sequence T_{b_m} converges to zero strongly in each space ℓ_p ($1 < p < \infty$).*

Proof. Since the sequence of the norms $\|T_{b_m}\|_p$ is bounded, it is sufficient to verify that $\|T_{b_m}x\|_p \to 0$ on some dense set. For $p = 2$ this can be easily checked. For $p > 2$ this follows from the inequality $\|T_{b_m}x\|_p \le \|T_{b_m}x\|_2$, and for $1 < p < 2$ one can apply, for instance, the relations

$$\|T_{b_m}x\|_p^p \le \|T_{b_m}x\|_2^{p-1}\|T_{b_m}x\|_{2/(3-p)} \le \|T_{b_m}x\|_2^{p-1}\|x\|\alpha_{2(3-p)}.$$

The lemma is proved. □

Proof of Theorem 4.2. Assume that condition (4.1) is fulfilled. From Theorem 4.1 it follows that $a^{-1}(\zeta) \in C^{n\times n}_{\langle p\rangle}$. By Proposition 3.7,

$$T_a T_{a^{-1}} - I \in \mathcal{J}(\ell^n_p), \quad T_{a^{-1}} T_a - I \in \mathcal{J}(\ell^n_p).$$

Hence the operator T_a is a Φ-operator. The sufficiency part of the theorem is proved.

Let us show the necessity part. Assume that $\det a(\zeta_0) = 0$, where ζ_0 is some point on Γ_0 and the operator T_a is a Φ_+-operator (if T_a is a Φ_--operator, then we can pass to the adjoint operator in the dual space). There exists a finite rank operator K and a constant $C > 0$ such that

$$\|(T_a + K)x\|_p \geq C\|x\|_p$$

for all $x \in \ell^n_p$. From here it follows that

$$\|(T_a + K)T_h\|_p \geq C\|T_h\|_p \tag{4.2}$$

for every matrix function $h(\zeta) \in \mathcal{R}^{n\times n}_p$.

Below we will construct a matrix $h(\zeta)$, for which condition (4.2) is not fulfilled. This will lead to a contradiction.

By $u = (u_1, \dots, u_n) \in \mathbb{C}^n$ denote a nonzero solution of the equation $a(\zeta_0)u = 0$ and put

$$f_k(\zeta) = \sum_{j=1}^{n} a_{jk}(\zeta)u_j \quad (k = 1, \dots, n),$$

where $a_{jk}(\zeta)$ are the entries of the matrix $a(\zeta)$. By $b(\zeta)$ denote some function that can be expanded into the absolutely convergent Fourier series

$$b(\zeta) = \sum_{k=-\infty}^{\infty} \beta_k \zeta^k \quad \left(|\zeta| = 1, \ \sum_{k=-\infty}^{\infty} |\beta_k| < \infty \right)$$

and that has the following properties.

1. The support of the function $b(\zeta)$ is contained in a neighborhood $v(\zeta_0)$ of the point ζ_0 such that
$$\sup_{\zeta \in v(\zeta_0)} |f_k(\zeta)| < \delta \quad (k = 1, 2, \dots, n), \tag{4.3}$$
where $\delta > 0$ is a given arbitrarily small number.
2. $\max |b(\zeta)| = 1$ and $\operatorname{var} b(\zeta) = 2$.

We choose a natural number N so that for the trigonometric polynomial

$$Q(\zeta) = \sum_{k=-N}^{N} \beta_k \zeta^k$$

the inequalities

$$\max |Q(\zeta)| \geq \frac{1}{2}, \quad \max |f_k(\zeta)Q(\zeta)| < 2\delta \quad (k = 1, 2, \dots, n),$$

and $\|T_Q\|_r \le 2\|T_b\|_r$ for all $r \ge 1$ hold. Note that the latter inequality is satisfied if N is chosen so that $\|T_b - T_Q\|_1 < \|T_b\|_2$. In that case $\|T_b - T_Q\|_r < \|T_b\|_2$, whence $\|T_Q\|_r \le \|T_b\|_r + \|T_b\|_2 \le 2\|T_b\|_r$.

Note that in view of (3.2), $\|T_Q\|_r \le 6k_r$. Put

$$h(\zeta) = \zeta^N Q(\zeta) = \left\| \begin{matrix} u_1 & 0 & \dots & 0 \\ u_2 & 0 & \dots & 0 \\ \dots & \dots & \dots & \dots \\ u_n & 0 & \dots & 0 \end{matrix} \right\|.$$

Then

$$T_a T_h - T_{ah} = \left\| \begin{matrix} T_{g_1} & 0 & \dots & 0 \\ T_{g_2} & 0 & \dots & 0 \\ \dots & \dots & \dots & \dots \\ T_{g_n} & 0 & \dots & 0 \end{matrix} \right\|,$$

where $g_k(\zeta) = f_k(\zeta)Q(\zeta)\zeta^N$. Since

$$\|T_{g_k}\|_p \le \|T_{f_k}\|_r^{1/2}\|T_Q\|_r^{1/2} \max |f_k(\zeta)Q(\zeta)|^{1/2} \le (12k_r\|T_{f_k}\|\delta)^{1/2},$$

where $r = 4p/(2+p)$, choosing the neighborhood $v(\zeta_0)$ one can make the norm $\|T_a T_h\|_p$ arbitrarily small.

Let $v_m(\zeta_0)$ be the sequence of the neighborhoods of the point ζ_0 that collapse to the point ζ_0. Then the corresponding sequence of polynomials $Q_m(\zeta)$ can be chosen so that it tends to zero in measure. By Lemma 4.1, the sequence TQ_n tends to zero strongly. Hence (see [17, Chapter II, Section 3]), $\|KTQ_n\|_p \to 0$. From here it follows that the norm $\|KT_h\|_p$ can also be made arbitrarily small.

Since $\max |Q(\zeta)| > 1/2$, we see that the norms $\|T_h\|_p$ are bounded from below by a constant independent of $v(\zeta_0)$.

The obtained claims contradict relation (4.2). The theorem is proved. □

5. Inversion of Toeplitz matrices

5.1. Let $a \in \Pi C_p$. We define the symbol of the operator T_a acting in ℓ_p. To this end, by $\xi_p(\mu)$ denote the function defined on the segment $[0,1]$ by the equality

$$\xi_p(\mu) = \begin{cases} \dfrac{\sin\theta\mu \exp(i\theta\mu)}{\sin\theta \exp(i\theta)} & \text{if} \quad \theta \ne 0, \\ \mu & \text{if} \quad \theta = 0, \end{cases} \tag{5.1}$$

where $\theta = \pi(2-p)/p$.

If μ runs over the segment $[0,1]$, then $\xi_p(\mu)$ runs over the circular arc (or the segment if $p = 2$) joining the points 0 and 1. For $p > 2$ (resp. $p < 2$) this arc is located in the upper (resp. lower) half-plane and from the points of this arc the segment $[0,1]$ is seen at the angle $2\pi/p$ (resp. $2\pi(p-1)/p$).

The function $a(\zeta, \mu; p)$ defined on the cylinder $X = \Gamma_0 \times [0,1]$ by the equality

$$a(\zeta, \mu; p) = \xi_p(\mu)a(\zeta + 0) + (1 - \xi_p(\mu))a(\zeta - 0)$$

is said to be the symbol of the operator T_a ($a \in \Pi C_p$) acting in ℓ_p.

Let $\zeta_1, \zeta_2, \ldots$ be all discontinuity points of the function $a(\zeta)$. Then the range of the function $a(\zeta, \mu; p)$ is the union of the range of the function $a(\zeta)$ ($|\zeta| = 1$) and a finite or countable number of circular arcs (or segments)

$$S_k = a(\zeta_k - 0) + (a(\zeta_k + 0) - a(\zeta_k - 0))\xi_p(\mu),$$

joining the points $a(\zeta_k + 0)$ and $a(\zeta_k - 0)$.

If the function $a(\zeta)$ has at least one discontinuity point, then the symbol $a(\zeta, \mu; p)$ depends on p.

Note that the symbol $a(\zeta, \mu; p)$ of the operator T_a acting in the space ℓ_p differs from the symbol $\mathcal{A}(\zeta, \mu; p)$ of the same operator acting in the space h_p (see [2]). The range of the function $\mathcal{A}(\zeta, \mu; p)$ can be obtained from the range of the function $a(\zeta, \mu; p)$ by replacing each of the arcs S_k by the arc symmetric to S_k with respect to the segment joining the points $a(\zeta_k + 0)$ and $a(\zeta_k - 0)$.

Let us define the index of the function $a(\zeta, \mu; p)$. First of all, we orient the curve $a(\zeta, \mu; p)$ so that the motion along the curve $a(\zeta, \mu; p)$ agrees with the motion of ζ along the circle counterclockwise at the continuity points of the function $a(\zeta)$; and the complementary arcs are oriented from $a(\zeta_k - 0)$ to $a(\zeta_k + 0)$.

If the function $a(\zeta) \in \Pi C_p$ has a finite number of discontinuity points and $a(\zeta, \mu; p) \neq 0$ ($|\zeta| = 1$, $0 \leq \mu \leq 1$), then the winding number of the curve $a(\zeta, \mu; p)$ about the point $\lambda = 0$ is said to be the index of the curve $a(\zeta, \mu; p)$. This index is denoted by $\operatorname{ind} a(\zeta, \mu; p)$. In the general case, if $a \in \Pi C_p$ and $a(\zeta, \mu; p) \neq 0$, then the function $a(\zeta)$ can be uniformly approximated by functions $a_n(\zeta) \in \Pi C_p$ having finite numbers of discontinuity points. It is easy to see that the sequence $\operatorname{ind} a_n(\zeta, \mu; p)$ stabilizes starting with some n. The index of the function $a(\zeta, \mu; p)$ is defined by the equality

$$\operatorname{ind} a(\zeta, \mu; p) = \lim_{n \to \infty} \operatorname{ind} a_n(\zeta, \mu; p).$$

In this section the following theorem is proved.

Theorem 5.1. *Let $a(\zeta) \in \Pi C_{\langle p \rangle}$, where $1 < p < \infty$. The operator T_a is normally solvable in the space ℓ_p if and only if the condition*

$$a(\zeta, \mu; p) \neq 0 \quad (|\zeta| = 1,\ 0 \leq \mu \leq 1) \tag{5.2}$$

holds. If condition (5.2) *is fulfilled, then the operator T_a is invertible (invertible only from the left; invertible only from the right) whenever the number*

$$\kappa = \operatorname{ind} a(\zeta, \mu; p)$$

is equal to zero (resp. positive; negative).

Under condition (5.2), *the equalities*

$$\dim \ker T_a = -\operatorname{ind} a(\zeta, \mu; p) \quad \text{for} \quad \kappa < 0,$$
$$\dim \operatorname{coker} T_a = \operatorname{ind} a(\zeta, \mu; p) \quad \text{for} \quad \kappa > 0,$$

hold.

This theorem for a certain class of functions $a(\zeta)$ was proved by R.V. Duduchava [3, 4]. Namely, in those papers it is supposed that the function $a(\zeta)$ has the form

$$a(\zeta) = \sum_{j=1}^{n} a_j(\zeta)\zeta^{\alpha_j}, \tag{5.3}$$

where the functions $a_j(\zeta)$ $(j = 1, \dots, n)$ have absolutely convergent Fourier series. In the proof of Theorem 5.1 we make a substantial use of the result by R.V. Duduchava [3, 4] concerning the function $a(\zeta) = \zeta^\alpha$. This result is formulated as follows.

Proposition 5.1. *If the condition*

$$-\frac{1}{p} < \operatorname{Re}\alpha < 1 - \frac{1}{p} \quad (1 < p < \infty)$$

holds, then the operator T_{ζ^α} is invertible in the space ℓ_p.

5.2. *Proof of Theorem* 5.1. By $\mathfrak{A}$ denote the quotient algebra $L(\ell_p)/\mathcal{J}(\ell_p)$. If $b \in \mathcal{R}_p$, then by $\widehat{T}_b$ denote the coset in $\mathfrak{A}$ containing the operator T_b. For each point ζ_0 of the unit circle Γ_0 define the set $M_{\zeta_0} \subset \mathfrak{A}$ consisting of all cosets $\widehat{T}_x$ such that functions $x(\zeta)$ are continuous on Γ_0, have finite total variation, and have the following properties: 1) $x(\zeta)$ takes the value 1 in some neighborhood of the point ζ_0; 2) $0 \le x(\zeta) \le 1$ and $\operatorname{var} x(\zeta) = 2$.

From Proposition 3.7 and Theorem 4.3 it follows immediately that the set $\{M_\zeta\}_{\zeta\in\Gamma_0}$ forms a covering system of localizing classes. As in Theorem 4.1 one can show that if $a, b \in \Pi C_{\langle p\rangle}$ and $a(\zeta_0 + 0) = b(\zeta_0 + 0)$, $a(\zeta_0 - 0) = b(\zeta_0 - 0)$ at some point $\zeta_0 \in \Gamma_0$, then $\widehat{T}_a$ and $\widehat{T}_b$ are M_{ζ_0}-equivalent.

Let condition (5.2) be fulfilled. The coset $\widehat{T}_a$ is M_τ-equivalent to the coset $a(\tau)\widehat{I}$ at each continuity point $\tau \in \Gamma_0$ of the function $a(\zeta)$. The coset $a(\tau)\widehat{I}$ is invertible in the algebra $\mathfrak{A}$ because $a(\tau) \neq 0$.

Let ζ_0 be a discontinuity point of the function $a(\zeta)$. From condition (5.2) it follows that $a(\zeta_0 \pm 0) \neq 0$. We find a function $b(\zeta) = \beta\zeta^\alpha$ $(\alpha, \beta \in \mathbb{C}^1, -1/p \le \operatorname{Re}\alpha < 1 - 1/p)$, which is continuous at each point $\zeta \neq \zeta_0$ and such that

$$a(\zeta_0 + 0) = b(\zeta_0 + 0), \quad a(\zeta_0 - 0) = b(\zeta_0 - 0), \quad a(\zeta_0) = b(\zeta_0).$$

From condition (5.2) it follows that $b(\zeta_0, \mu; p) \neq 0$ $(0 \le \mu \le 1)$. In view of the definition of the symbol, this implies that $\beta \neq 0$ and

$$e^{-2\pi i\alpha} \neq \frac{\sin\theta(\mu - 1)}{\sin\theta\mu} e^{2\pi i/p} \quad (0 \le \mu \le 1).$$

From this relation it follows that $\operatorname{Re}\alpha \neq -1/p$, whence $-1/p < \operatorname{Re}\alpha < 1 - 1/p$. Therefore, by Proposition 5.1, the operator T_b is invertible.

Since the function $a(\zeta) - b(\zeta)$ is continuous at the point ζ_0 and is equal to zero at this point, we see that the cosets $\widehat{T}_a$ and $\widehat{T}_b$ are M_{ζ_0}-equivalent. Thus, for each point $\zeta \in \Gamma_0$ the coset $\widehat{T}_a$ is M_ζ-equivalent to some invertible element.

Since, moreover, $\widehat{T}_a$ commutes with each element in M_ζ, we conclude that $\widehat{T}_a$ is invertible in the algebra $L(\ell_p)/\mathcal{J}(\ell_p)$ by Theorem 1.1. Therefore the operator $\widehat{T}_a$ is a Φ-operator in ℓ_p.

Let us show that the index of the operator T_a is calculated by the formula

$$\operatorname{Ind} T_a = -\operatorname{ind} a(\zeta, \mu; p). \tag{5.4}$$

We will use the following statement in the proof of this formula.

Lemma 5.1. *Let $a(\zeta)$ and $a_n(\zeta)$ be functions in $\Pi C_{\langle p\rangle}$ that satisfy the conditions*

$$a(\zeta, \mu; p) \neq 0 \quad (|\zeta| = 1,\ 0 \leq \mu \leq 1)$$

and $\sup |a(\zeta) - a_n(\zeta)| \to 0$. *Then, starting with some n, the equalities*

$$\operatorname{ind} a_n(\zeta, \mu; p) = \operatorname{ind} a(\zeta, \mu; p), \quad \operatorname{Ind} T_{a_n} = \operatorname{Ind} T_a \tag{5.5}$$

hold.

Proof. Indeed, since the sequence $a_n(\zeta)$ converges uniformly to $a(\zeta)$, one can easily check that $a_n(\zeta, \mu; p)$ converges uniformly to $a(\zeta, \mu; p)$ on the cylinder $\Gamma_0 \times [0, 1]$. This implies the first equality in (5.5). Since, starting with some n, the inequalities

$$|a(\zeta, \mu; p) - a_n(\zeta, \mu; p)| < \frac{1}{2} \inf |a(\zeta, \mu; p)|$$

hold, we deduce that for all λ in the segment $[0, 1]$,

$$a(\zeta, \mu; p) + \lambda\big(a_n(\zeta, \mu; p) - a(\zeta, \mu; p)\big) \neq 0 \quad (|\zeta| = 1,\ 0 \leq \mu \leq 1).$$

In view of what has been proved above, the operator $W_\lambda = T_a + \lambda T_{a_n - a}$ $(0 \leq \lambda \leq 1)$ is a Φ-operator. This implies that $\operatorname{Ind} T_a = \operatorname{Ind} T_{a_n}$. □

From the proved statement it follows that it is sufficient to prove formula (5.4) for the case when the function $a(\zeta)$ has a finite number of discontinuity points. Let $\zeta_1, \zeta_2, \ldots, \zeta_n$ be the discontinuity points of the function $a(\zeta)$. From condition (5.2) it follows that $a(\zeta_k \pm 0) \neq 0$. This allows us to choose the functions $\psi_k(\zeta) = \zeta^{\alpha_k}$ $(-1/p \leq \operatorname{Re} \alpha_k < 1 - 1/p)$, each of which is continuous everywhere except for one point $\zeta = \zeta_k$, and such that the condition

$$\frac{a(\zeta_k + 0)}{a(\zeta_k - 0)} = \frac{\psi_k(\zeta_k + 0)}{\psi_k(\zeta_k - 0)}$$

holds. As above, one can show that $\operatorname{Re} \alpha_k \neq -1/p$. Let $\psi(\zeta) = \psi_1(\zeta) \ldots \psi_n(\zeta)$. Then the function $b(\zeta) = a(\zeta)/\psi(\zeta)$ is continuous on Γ_0 and belongs to $\mathcal{R}_{\langle p\rangle}$. It is checked straightforwardly that $\operatorname{ind} \psi_k(\zeta, \mu; p) = 0$. Since the functions $\psi_k(\zeta)$ do not have common discontinuity points, we conclude that $\operatorname{ind} \psi(\zeta, \mu; p) = 0$. Moreover,

$$\operatorname{ind} a(\zeta, \mu; p) = \operatorname{ind} b(\zeta) + \operatorname{ind} \psi(\zeta, \mu; p).$$

Therefore $\operatorname{ind} a(\zeta, \mu; p) = \operatorname{ind} b(\zeta)$. In turn, the operator T_a can be represented in the form $T_a = T_b T_{\psi_1} \ldots T_{\psi_n} + T$, where $T \in \mathcal{J}(\ell_p)$. In view of R.V. Duduchava's Proposition 5.1, the operators T_{ψ_k} are invertible, whence $\operatorname{Ind} T_a = \operatorname{Ind} T_b$.

Thus, it is sufficient to prove formula (5.4) for the case when the function $a(\zeta) \in \mathcal{R}_{\langle p \rangle}$ is continuous on Γ_0. In view of what has been proved above, the continuous function can be replaced by a polynomial. Formula (5.4) for polynomials is well known (see, e.g., [17, Chapter I, Section 7]). To finish the proof of the sufficiency of the hypotheses of the theorem it remains to apply Lemma 5.1.

The proof of the necessity consists of four parts.

I. First we show that if a function $a(\zeta) \in \Pi C_{\langle p \rangle}$ is continuous on some arc $\zeta'\zeta''$, is equal to zero at some interior point ζ_0 of this arc, and is different from zero at its endpoints (that is, $a(\zeta') \neq 0$ and $a(\zeta'') \neq 0$), then the operator T_a is neither Φ_+-operator nor Φ_--operator. Let $c(\zeta)$ be an arbitrary continuous function in $\mathcal{R}_{\langle p \rangle}$ such that it is different from zero on Γ_0 and coincides with $b(\zeta)$ at the points ζ' and ζ''. By $\chi(\zeta)$ $(\zeta \in \Gamma_0)$ denote the characteristic function of the arc $\zeta'\zeta''$ and put

$$g(\zeta) = c(\zeta)(1 - \chi(\zeta)) + b(\zeta)\chi(\zeta), \quad h(\zeta) = c^{-1}(\zeta)b(\zeta)(1 - \chi(\zeta)) + \chi(\zeta).$$

From Theorem 4.1 it follows that $g, h \in \mathcal{R}_{\langle p \rangle}$. Proposition 3.7 implies that

$$T_b = T_h T_g + K' = T_g T_h + K'',$$

where $K', K'' \in \mathcal{J}(\ell_p)$. From these equalities it follows that if the operator T_b is a Φ_+-operator or a Φ_--operator, then so is the operator T_g. Since $g \in C_{\langle p \rangle}$ and $g(\zeta_0) = 0$, the latter is impossible in view of Theorem 4.2.

II. Let the operator T_a be a $\Phi_\pm$-operator. Let us show that $a(\zeta \pm 0) \neq 0$ for all $\zeta \in \Gamma_0$. If the function $a(\zeta)$ or its one-sided limit $a(\zeta \pm 0)$ is equal to zero at some point $\zeta_0 \in \Gamma_0$, then by using Proposition 3.6 and estimates (3.1)–(3.2) it is not difficult to find a $\delta > 0$ and a segment Δ containing the point ζ_0 such that the function

$$b(\zeta) = a(\zeta)(1 - \chi_\Delta)) + \delta(\zeta - \zeta_0),$$

where $\chi_\Delta(\zeta)$ is the characteristic function of the segment Δ, would satisfy the conditions of the previous paragraph and the operator T_b would be a $\Phi_\pm$-operator. But this situation is impossible as we have shown.

III. Let us prove that if the operator T_a $(a \in \Pi C_{\langle p \rangle})$ is normally solvable, then it is a Φ-operator. Let the operator T_a be normally solvable. In view of Proposition 3.8 it is a Φ_+-operator or a Φ_--operator. In this case $a(\zeta \pm 0) \neq 0$. Hence the range of the function $a(\zeta, \mu; p)$ does not fill completely any neighborhood of the origin. Let λ_n be a sequence of complex numbers tending to zero and not belonging to the range of the function $a(\zeta, \mu; p)$. From what has been proved above it follows that the operators $T_{a-\lambda_n}$ are Φ-operators. Since $\|T_{a-\lambda_n} - T_a\| \to 0$ as $n \to \infty$, we deduce that T_a is a Φ-operator in view of the stability of the index of $\Phi_\pm$-operators.

IV. It remains to show that if the operator T_a is a Φ-operator and ζ_0 is a discontinuity point of the function $a(\zeta)$, then $a(\zeta_0, \mu; p) \neq 0$ $(0 \leq \mu \leq 1)$. Let T_a be a Φ-operator. We have already shown that $a(\zeta_0 \pm 0) \neq 0$. Therefore one can choose a function $b(\zeta) = \beta\zeta^\alpha$ $(\alpha, \beta \in \mathbb{C}^1,\ -1/p \leq \operatorname{Re}\alpha < 1 - 1/p)$, which is continuous at each point $\zeta \neq \zeta_0$ and such that $b(\zeta_0+0) = a(\zeta_0+0)$ and $b(\zeta_0-0) = a(\zeta_0-0)$. The cosets $\widehat{T}_a$ and $\widehat{T}_b$ are M_{ζ_0}-equivalent and $\widehat{T}_a$ is invertible in the algebra $\mathfrak{A} = L(\ell_p)/\mathcal{J}(\ell_p)$, whence $\widehat{T}_b$ is M_{ζ_0}-invertible. Since, at the points $\eta \neq \zeta_0$, the coset $\widehat{T}_b$ is M_η-equivalent to the scalar element $b(\eta)\widehat{I}$ and $b(\eta) \neq 0$, we see that $\widehat{T}_b$ is M_η-invertible. From Lemma 1.2 it follows that the coset $\widehat{T}_b$ is invertible in the algebra $\mathfrak{A}$, whence the operator T_b is a Φ-operator.

We will prove the inequality $a(\zeta_0, \mu; p) \neq 0$ by contradiction. Assume that $a(\zeta_0, \mu_0; p) = 0$ for some μ_0 $(0 < \mu_0 < 1)$. Then

$$\frac{a(\zeta_0+0)}{a(\zeta_0-0)} = \frac{\sin\theta(\mu_0-1)}{\sin\theta\mu_0} e^{2\pi i/p}.$$

Since

$$\frac{a(\zeta_0+0)}{a(\zeta_0-0)} = \frac{b(\zeta_0+0)}{b(\zeta_0-0)} = e^{-2\pi i\alpha},$$

we see that $\operatorname{Re}\alpha = -1/p$.

Let us show that under this condition the operator T_b cannot be a Φ-operator. To this end, consider the operator function $A(\varepsilon) = T_{\zeta^{\alpha+\varepsilon}}$. With the aid of Proposition 3.6 it is not difficult to verify that $A(\varepsilon)$ is a continuous function in ε. If $0 < \varepsilon < 1$, then in view of Proposition 5.1 the operator $A(\varepsilon)$ is invertible in ℓ_p. On the other hand, if $-1 < \varepsilon < 0$, then the operator $A(\varepsilon+1)$ is invertible. Hence $\operatorname{Ind} A(\varepsilon) = \operatorname{Ind} A(\varepsilon+1) + \operatorname{Ind} T_{\zeta^{-1}} = 1$. From the theorem on the stability of the index of a Φ-operator it follows that $T_{\zeta^\alpha} = A(0)$ is not a Φ-operator.

The theorem is proved. □

5.3. By $\Pi C_p^{n\times n}$ and $\Pi C_{\langle p\rangle}^{n\times n}$ denote the sets of matrix functions $a(\zeta) = \|a_{jk}(\zeta)\|_1^n$ with entries in ΠC_p and $\Pi C_{\langle p\rangle}$, respectively, and by T_a $(a \in \Pi C_p^{n\times n})$ denote the operator defined in the space ℓ_p^n by the matrix

$$T_a = \|T_{a_{jk}}\|_{j,k=1}^n.$$

The matrix function

$$a(\zeta, \mu; p) = \|a_{jk}(\zeta, \mu; p)\|_{j,k=1}^n,$$

where $a_{jk}(\zeta, \mu; p)$ is the symbol of the operator $T_{a_{jk}}$ in the space ℓ_p, is said to be the symbol of the operator T_a.

Theorem 5.2. *Let $a \in \Pi C_{\langle p\rangle}^{n\times n}$. The operator T_a is a Φ_+-operator or a Φ_--operator in ℓ_p^n if and only if the condition*

$$\det a(\zeta, \mu; p) \neq 0 \quad (|\zeta| = 1,\ 0 \leq \mu \leq 1) \tag{5.6}$$

holds. If condition (5.6) *is fulfilled, then*

$$\operatorname{Ind} T_a = -\operatorname{ind}\det a(\zeta, \mu; p). \tag{5.7}$$

We preface the proof of the theorem with the following lemma.

Lemma 5.2. *Each matrix function $a(\zeta) \in \Pi C^{n\times n}_{\langle p\rangle}$ that has a finite number of discontinuity points $\zeta_1, \zeta_2, \dots, \zeta_n$ and satisfies the conditions $\det a(\zeta_k \pm 0) \neq 0$ can be represented in the form $a = bxc$, where $b(\zeta)$ and $c(\zeta)$ are non-degenerate matrix functions with entries in $C_{\langle p\rangle}$ and $x(\zeta)$ is a triangular matrix function in $\Pi C^{n\times n}_{\langle p\rangle}$.*

The proof of this lemma is analogous to the proof of [1, Lemma 3.1].

Proof of Theorem 5.2. If the matrix function $a(\zeta)$ has a finite number of discontinuity points, then the proof of the theorem can be developed as in the case $p = 2$ (see [1, Theorem 3.1]). For this purpose Theorem 4.2 and Lemma 5.2 presented above are essentially used. In the general case we will develop the proof with the aid of localizing classes.

Let $\{M_\zeta\}_{\zeta\in\Gamma_0}$ be a covering system of localizing classes defined in the proof of Theorem 5.1, and let E_n be the identity matrix of order n. It is easy to check that the set $\{M^n_\zeta\}_{\zeta\in\Gamma_0}$ of the elements of the form $M^n_\zeta = M_\zeta E_n$ forms a covering system of localizing classes of the algebra $L(\ell^n_p)/\mathcal{J}(\ell^n_p)$ and if $a \in \Pi C^{n\times n}_{\langle p\rangle}$, then the coset $\widehat{T}_a$ belongs to the commutant of the set $\bigcup_{\zeta\in\Gamma_0} M^n_\zeta$.

By $a_0(\zeta)$ denote a matrix function satisfying the following conditions: 1) $a_0(\zeta) \in \Pi C^{n\times n}_{\langle p\rangle}$; 2) $a_0(\zeta_0 + 0) = a(\zeta_0 + 0)$, $a_0(\zeta_0 - 0) = a(\zeta_0 - 0)$; 3) the matrix function $a_{\zeta_0}(\zeta)$ is continuous at every point $\zeta \neq \zeta_0$ and $\det a_0(\zeta) \neq 0$. From the latter property of the matrix function a_0 it follows that the coset $\widehat{T}_{a_0}$ is invertible in $L(\ell^n_p)/\mathcal{J}(\ell^n_p)$ if and only if it is $M^n_{\zeta_0}$-invertible. Since the matrix function $a_0(\zeta)$ has at most one discontinuity point, we see (as it has been noticed above) that the invertibility of the element $\widehat{T}_{a_0}$ is equivalent to the non-degeneracy of the symbol $a_0(\zeta, \mu; p)$. Taking into account that the elements $\widehat{T}_a$ and $\widehat{T}_{a_0}$ are $M^n_{\zeta_0}$-equivalent and that $a(\zeta_0, \mu; p) = a_0(\zeta_0, \mu; p)$, we get that the coset $\widehat{T}_a$ is $M^n_{\zeta_0}$-invertible if and only if $\det a(\zeta_0, \mu; p) \neq 0$. To finish the proof it remains to apply Lemma 1.2. The theorem is proved. □

6. Algebra of operators with Fredholm symbol

Let $\mathcal{L}$ be a Banach space and L_0 be some Banach subalgebra of the algebra $L(\mathcal{L})$. An algebra L_0 is said to be an algebra with Fredholm symbol (see [7])[4] if 1) one can construct a homomorphism $\pi : A \to \mathcal{A}(x)$ of the algebra L_0 onto some algebra of matrix functions $\mathcal{A}(x)$, where x runs over some set $X \subset \mathbb{R}^m$; 2) for an operator A to be a Fredholm operator in $\mathcal{L}$ it is necessary and sufficient that its symbol be non-degenerate on X:

$$\inf_{x\in X} |\det \mathcal{A}(x)| \neq 0 \quad (\mathcal{A}(x) = \pi(A)).$$

By $\mathcal{G}_p$ denote the set of all operators T_a, where $a(\zeta) \in \Pi C_{\langle p\rangle}$, by WH_p denote the smallest Banach subalgebra of the algebra $L(\ell_p)$ that contains $\mathcal{G}_p$, and by $\mathcal{K}_p$

[4] In [7], such an algebra is simply called an algebra with symbol.

denote the set of all operators of the form

$$A = \sum_{j=1}^{r} A_{j1} A_{j2} \dots A_{jk}, \tag{6.1}$$

where $A_{jl} \in \mathcal{G}_p$. Note that $\mathcal{K}_p$ is dense in WH_p.

Let $X = \Gamma_0 \times [0,1]$ and $x = (\zeta, \mu) \in X$. In the previous section, we defined the symbol $\mathcal{A}(x)$ of every operator $A \in \mathcal{G}_p$ by equality (5.1). From the definition of symbol and Theorem 5.2 it follows that

a) if $A = cI$ ($c =$ const), then $\mathcal{A}(x) = c$;
b) let $A = (A_{jl})_1^n$ and $\mathcal{A}(x) = (\mathcal{A}_{jl}(x))_1^n$; for the operator A to be a Φ-operator in ℓ_p^n it is necessary and sufficient that the condition $\det \mathcal{A}(x) \neq 0$ ($x \in X$) holds.

The symbol of the operator (6.1) is defined by the formula

$$\mathcal{A}(x) = \sum_{j=1}^{r} \prod_{l=1}^{k} \mathcal{A}_{jl}(x). \tag{6.2}$$

In [7, Theorem 1.1] it was shown that properties a) and b) imply the following statements:

1) the function $\mathcal{A}(x)$ does not depend on the manner of representation of the operator A in the form (6.1);
2) the mapping $\pi : A \to \mathcal{A}(x)$ is a homomorphism of the (nonclosed) algebra $\mathcal{K}_p$ onto the set $\pi(\mathcal{K}_p)$;
3) the operator $A \in \mathcal{K}_p$ is a Φ-operator in ℓ_p if and only if $\mathcal{A}(x) \neq 0$ ($x \in X$);
4) for every operator $A \in \mathcal{K}_p$ the inequality

$$\max_{x \in X} |\mathcal{A}(x)| \leq \|\widehat{A}\|. \tag{6.3}$$

holds.

Inequality (6.3) allows us to extend the homomorphism π from the algebra $\mathcal{K}_p$ to the whole algebra WH_p. In the same way as in [18, Lemma 4.1], one can show that $\mathcal{J}(\ell_p) \subset WH_p$. From Proposition 3.7 it follows that the quotient algebra $\widehat{WH}_p = WH_p/\mathcal{J}(\ell_p)$ is commutative. From inequality (6.3) it follows that the symbols of all operators belonging to the same coset $\widehat{A}$ coincide. We denote this common symbol by $\widehat{\mathcal{A}}(x)$. Since the functional $\varphi_x(\widehat{\mathcal{A}}) = \widehat{\mathcal{A}}(x)$ is multiplicative for every $x \in X$, we see that the set $J_x = \ker \varphi_x$ is a maximal ideal in the algebra $\widehat{WH}_p$. Let us show that all maximal ideals of the algebra $\widehat{WH}_p$ are exhausted by such ideals.

Lemma 6.1. *All maximal ideals of the algebra $\widehat{WH}_p$ are exhausted by the ideals of the form $J_x = \ker \varphi_x$.*

Proof. Let J be an arbitrary maximal ideal of the algebra $\widehat{WH}_p$. First let us show that there exists a point $\zeta_0 \in \Gamma_0$ such that for every function $a(\zeta) \in C_{\langle p \rangle}$ the Gelfand transform of an element $\widehat{T}_a$ has the form $\widehat{T}_a = a(\zeta_0)$. Assume the contrary, that is, for every point $\tau \in \Gamma_0$ there exists a function $x_\tau(\zeta) \in C_{\langle p\rangle}$ such that $\widehat{T}_{x_\tau}(J) \neq x_\tau(\tau)$. Obviously, 1) $\widehat{T}_{x_\tau-\alpha_\tau} \in J$, where $\alpha_\tau = \widehat{T}_{x_\tau}(J)$, and 2) $|x_\tau(\zeta) - \alpha_\tau| \geq \delta_\tau > 0$ in some neighborhood u_τ. Let $u(\tau_1), \dots, u(\tau_n)$ be a finite cover of the circle Γ_0 and δ be the smallest of the numbers $\delta_{\tau_1}, \dots, \delta_{\tau_n}$. Then

$$y(\zeta) = \sum_{k=1}^{n} |x_{\tau_k}(\zeta) - \alpha_{\tau_k}|^2 \neq 0.$$

In view of Propositions 3.3, 3.4 and Theorem 4.1, the element $\widehat{T}_y$ is invertible in the quotient algebra $L(\ell_p)/\mathcal{J}(\ell_p)$. Since its spectrum in this algebra is real, we see that $\widehat{T}_y$ is invertible in the algebra $\widehat{WH}_p$. But this is impossible because $\widehat{T}_y \in J$.

Let us show that for every function $a(\zeta) \in \Pi C_{\langle p\rangle}$, which is continuous at the point ζ_0, the Gelfand transform $\widehat{T}_a(J)$ also has the form $\widehat{T}_a(J) = a(\zeta_0)$. If the function $a(\zeta)$ is equal to $a(\zeta_0)$ in some neighborhood of the point ζ_0, then it can be written in the form $a(\zeta) = b(\zeta)c(\zeta) + a(\zeta_0)$, where $b(\zeta) \in \Pi C\langle p\rangle$, $c(\zeta) \in C_{\langle p\rangle}$, and $c(\zeta_0) = 0$. In view of what has been proved above, $\widehat{T}_c(J) = 0$, whence $\widehat{T}_a(J) = a(\zeta_0)$.

Let $a(\zeta) \in \Pi C_{\langle p\rangle}$ be a continuous function at the point ζ_0 and χ_δ be the characteristic function of the δ-neighborhood of the point ζ_0. Proposition 3.6 implies the equality

$$\lim_{\delta\to 0} \|T_a - T_{a(1-\chi_\delta)+a(\zeta_0)\chi_\delta}\| = \lim_{\delta\to 0} \|T_{a(\zeta)-a(\zeta_0)\chi_\delta}\| = 0.$$

Therefore, in view of what has been proved above, $\widehat{T}_a(J) = a(\zeta_0)$.

Let s be the arc of the circle, which is the range of the function $\xi_p(\mu)$ defined by equality (5.1) when μ runs from 0 to 1. It is easy to find a function $b(\zeta) \in \mathcal{R}_{\langle p\rangle}$ having the following properties: $b(\zeta_0 - 0) = 0$; $b(\zeta_0 + 0) = 1$ (where ζ_0 is the point of the circle found before); $b(\zeta)$ is continuous on Γ_0 everywhere except for the point ζ_0; and the range of the function $b(\zeta)$ coincides with the arc s.

From Theorem 5.1 it follows that the spectrum of the element $\widehat{T}_b$ in the algebra $L(\ell_p)/\mathcal{J}(\ell_p)$ coincides with the arc s. Since the complement to the spectrum of this element is connected, we deduce that (see [19]) the spectrum of the element $\widehat{T}_b$ in the algebra $\widehat{WH}_p$ also coincides with the arc s. This implies the existence of a number $\mu_0 \in [0, 1]$ such that $\widehat{T}_b(J) = \xi_p(\mu_0)$.

We move to the last stage of the proof. Let us show that for every element $\widehat{A} \in \widehat{WH}_p$ the equality $\widehat{A}(J) = \mathcal{A}(\zeta_0, \mu_0)$ holds. It is easy to see that it is sufficient to prove this statement for an operator T_a, where a is an arbitrary function in $\Pi C_{\langle p\rangle}$. Since the function

$$c(\zeta) = a(\zeta) - a(\zeta_0 + 0)b(\zeta) - a(\zeta_0 - 0)(1 - b)$$

is continuous at the point ζ_0 and is equal to zero at this point, we see that $\widehat{T}_c(J) = 0$, whence $\widehat{T}_a(J) = a(\zeta_0, \mu_0; p)$. This implies that $J = J_{x_0}$, where $x_0 = (\zeta_0, \mu_0)$. The lemma is proved. □

In the paper [7] (see [7, Theorem 4.1]) it is shown that if for a homomorphism π satisfying conditions a) and b) Lemma 6.1 is true, then the following theorem holds.

Theorem 6.1. *An operator $A \in WH_p$ is a Φ-operator in ℓ_p if and only if its symbol is different from zero on the cylinder X.*

Theorem 6.1 can be strengthened. We will do this in a more general situation.

By $WH_p^{n\times n}$ denote the algebra of all operators defined in the space ℓ_p^n by the equality $A = \|A_{jk}\|_{j,k=1}^n$, where $A_{jk} \in WH_p$. The matrix function $\mathcal{A}(x) = \|\mathcal{A}_{jk}(x)\|_1^n$ is called the symbol of the operator A, where $\mathcal{A}_{jk}(x)$ is the symbol of the operator A_{jk}.

If all operators A_{jk} belong to $\mathcal{K}_p$ and $\det \|\mathcal{A}_{jk}(x)\| \neq 0$, then the index $\operatorname{ind}\det \mathcal{A}(x)$ $(x = (\zeta, \mu))$ is defined in the same way as in the previous section. In the general case, when $A \in WH_p^{n\times n}$, the number $\operatorname{ind}\det \mathcal{A}(x)$ is defined by taking the limit.

Theorem 6.2. *An operator $A \in WH_p^{n\times n}$ is a Φ_+-operator or a Φ_--operator in ℓ_p^n if and only if its symbol is not degenerate on X:*

$$\det \mathcal{A}(x) \neq 0 \quad (x \in \Gamma_0 \times [0,1]). \tag{6.5}$$

If condition (6.5) *is fulfilled, then the operator A is a Φ-operator and*

$$\operatorname{Ind} A = -\operatorname{ind}\det \mathcal{A}(x). \tag{6.6}$$

Proof. Since the algebra WH_p is commutative, we see that condition (6.5) is necessary and sufficient for the operator A to be a Φ-operator in ℓ_p^n. This follows from Theorem 6.1 and a general statement saying that an operator $A = \|A_{jk}\|$ is a Φ-operator in ℓ_p^n if and only if the operator $\det \|A_{jk}\|$ is a Φ-operator in ℓ_p (see [17, Lemma 4.1]).

Since both sides of equality (6.6) are continuous functionals on the set of all Φ-operators in $WH_p^{n\times n}$, we see that it is sufficient to derive formula (6.6) for an operator $A \in \mathcal{K}_p^{n\times n}$. Such an operator can be represented in the form

$$A = \sum_{j=1}^{r} T_{a_{j1}} T_{a_{j2}} \dots T_{a_{js}},$$

where $a_{jl} \in \Pi C_{\langle p \rangle}$. For this operator formula (6.6) is obtained by passing to the linear dilation $T_f = \Theta(T_{a_{jl}})$ of the operator A (see [20, Section 3]). Here formula (5.7) of the present paper is used.

By the same method one can show that if the operator A is a Φ_+-operator or a Φ_--operator, then it is a Φ-operator. The theorem is proved. □

7. Algebras generated by paired operators

7.1. Let a_j $(j = 0, \pm 1, \dots)$ be the Fourier coefficients of a function $a(\zeta)$ in $\mathcal{R}_p$. Recall that $\widetilde{T}_a$ denotes the bounded linear operator defined in the space $\widetilde{\ell}_p$ by the matrix $\|a_{j-k}\|_{j,k=-\infty}^{\infty}$. By P denote the projection defined in the space $\widetilde{\ell}_p$ by the equality

$$P\{\xi_j\} = \{\dots, 0, \dots, 0, \xi_0, \xi_1, \dots\}$$

and by Q denote the complementary projection $I - P$.

To each pair of functions $a(\zeta)$ and $b(\zeta)$ in $\mathcal{R}_p$ assign the bounded linear operator $P\widetilde{T}_a + Q\widetilde{T}_b$ acting in $\widetilde{\ell}_p$. This operator is called a paired operator. An operator of the form $\widetilde{T}_a P + \widetilde{T}_b Q$ is called the transposed operator to the paired operator.

The operators $P\widetilde{T}_a + Q\widetilde{T}_b$ and $\widetilde{T}_a P + \widetilde{T}_b Q$ are defined analogously in the case when $a(\zeta)$ and $b(\zeta)$ are matrix functions in $\mathcal{R}_p^{n\times n}$. In this case the projection P is extended to the whole space ℓ_p^n by the equality $P\{x_k\}_1^n = \{Px_k\}_1^n$ $(x_k \in \ell_p)$ and the projection Q is equal to $I - P$ as before.

Let $a(\zeta), b(\zeta) \in C_p^{n\times n}$. The matrix function $\mathcal{A}(\zeta, \mu; p)$ $(|\zeta| = 1,\ 0 \le \mu \le 1)$ of order $2n$ defined by

$$\mathcal{A}(\zeta, \mu; p) = \left\| \begin{matrix} a(\zeta+0)\xi_p(\mu) + a(\zeta-0)(1-\xi_p(\mu)) & h_p(\mu)(a(\zeta+0) - a(\zeta-0)) \\ h_p(\mu)(b(\zeta+0) - b(\zeta-0)) & b(\zeta+0)(1-\xi_p(\mu)) + b(\zeta-0)\xi_p(\mu) \end{matrix} \right\|, \tag{7.1}$$

where $\xi_p(\mu)$ is the function defined by equality (5.1) and $h_p(\mu)$ is a branch of the root $\sqrt{\xi_p(\mu)(1-\xi_p(\mu))}$, is said to be the symbol of the operator $A = P\widetilde{T}_a + Q\widetilde{T}_b$. We equip the cylinder $X = \Gamma_0 \times [0,1]$ with the topology, where a neighborhood of every point (ζ_0, μ_0) is defined by one of the following equalities:

$$\begin{aligned} u(\zeta_0, 0) &= \big\{(\zeta,\mu):\ |\zeta - \zeta_0| < \delta,\ \zeta \prec \zeta_0,\ 0 \le \mu \le 1\big\} \\ &\quad \cup \big\{(\zeta_0, \mu):\ 0 \le \mu < \varepsilon\big\}, \\ u(\zeta_0, 1) &= \big\{(\zeta,\mu):\ |\zeta - \zeta_0| < \delta,\ \zeta \succ \zeta_0,\ 0 \le \mu \le 1\big\} \\ &\quad \cup \big\{(\zeta_0, \mu):\ \varepsilon < \mu \le 1\big\}, \\ u(\zeta_0, \mu_0) &= \big\{(\zeta_0, \mu):\ \mu_0 - \delta_1 < \mu < \mu_0 + \delta_2\big\} \quad (\mu_0 \ne 0, 1), \end{aligned} \tag{7.2}$$

where $0 < \delta_1 < \mu_0$, $0 < \delta_2 < 1 - \mu_0$, $0 < \varepsilon < 1$, and $0 < \delta < 1$. Here the relation $\zeta \prec \zeta_0$ means that the point ζ precedes the point ζ_0 on the circle Γ_0 [5]. If $a(\zeta) \in \Pi C$, then the function $a(\zeta, \mu; p)$ is continuous on the compact Hausdorff space X. It turns out that the function $\det \mathcal{A}$, where $\mathcal{A}(\zeta, \mu; p)$ is the matrix defined by equality (7.1), may not be continuous on the cylinder X. However, if $\det \mathcal{A}(\zeta, \mu; p) \ne 0$, then the function

$$g_A(\zeta, \mu) = \frac{\det \mathcal{A}(\zeta, \mu; p)}{\det(b(\zeta+0)b(\zeta-0))} \tag{7.3}$$

is continuous on the cylinder X. A proof of this statement is presented, for instance, in [20, Section 5].

---

[5] That is, $\zeta_0/\zeta = \exp(i\varphi)$, where $0 \le \varphi < \pi$.

Theorem 7.1. *Let* $a(\zeta), b(\zeta) \in \Pi C_{\langle p\rangle}^{n\times n}$. *The operator*

$$A = P\widetilde{T}_a + Q\widetilde{T}_b$$

is a $\Phi_\pm$*-operator in the space* $\widetilde{\ell_p^n}$ *if and only if the condition*

$$\inf |\det \mathcal{A}(\zeta, \mu; p)| \neq 0 \quad ((\zeta, \mu) \in X) \tag{7.4}$$

holds. If condition (7.4) *is fulfilled, then the operator* A *is a* Φ*-operator and*

$$\operatorname{Ind} A = -\operatorname{ind} \frac{\det \mathcal{A}(\zeta, \mu; p)}{\det(b(\zeta+0)b(\zeta-0))}. \tag{7.5}$$

Moreover, if $n = 1$*, then one of the numbers* $\dim\ker A$ *or* $\dim\operatorname{coker} A$ *is equal to zero.*

Proof. Let condition (7.4) be fulfilled. From this condition (if $\mu = 0$ and $\mu = 1$) it follows that $\det b(\zeta \pm 0) \neq 0$. Put $c(\zeta) = a(\zeta)b^{-1}(\zeta)$. From Theorem 4.1 it follows that $c(\zeta) \in \Pi C_{\langle p\rangle}^{n\times n}$. The operator A can be represented in the form

$$P\widetilde{T}_a + Q\widetilde{T}_b = (I + P\widetilde{T}_c Q)(P\widetilde{T}_c P + Q)\widetilde{T}_b.$$

The operators $I + P\widetilde{T}_c Q$ and $\widetilde{T}_b$ are invertible in $\widetilde{\ell_p^n}$, and the restriction of the operator $P\widetilde{T}_c P + Q$ to ℓ_p^n coincides with the operator T_c. In [20, Section 2] it is shown that

$$\det \mathcal{A}(\zeta, \mu; p) = \det c(\zeta, \mu; p)\det(b(\zeta+0)b(\zeta-0)). \tag{7.6}$$

From Theorem 5.2 it follows that the operator T_c is a Φ-operator in ℓ_p^n. Hence the operator A is a Φ-operator in $\widetilde{\ell_p^n}$. Equality (7.6) and the results of Section 5 imply formula (7.4) and the last statement of the theorem.

With the aid of relation (7.6) one can also prove the necessity of the hypotheses of the theorem. However, one should prove first that if the operator A is a Φ_+-operator or a Φ_--operator, then $\det b(\zeta \pm 0) \neq 0$ ($|\zeta| = 1$). First note that if one assumes that the operator A is a Φ_+-operator and the condition $\det b(\zeta \pm 0) \neq 0$ is violated, then (as in Theorem 5.1) one can choose an operator $P\widetilde{T}_a + Q\widetilde{T}_\beta$ with $\beta \in \Pi C_{\langle p\rangle}^{n\times n}$, which is a Φ_+-operator (or a Φ_--operator), and, moreover, the matrix function $\beta(\zeta)$ is continuous on some arc $\zeta'\zeta''$,

$$\det \beta(\zeta') \neq 0, \quad \det \beta(\zeta'') \neq 0, \quad \det \beta(\zeta_0) = 0,$$

where ζ_0 is some point on the arc $\zeta'\zeta''$. The matrix function $\beta(\zeta)$ can be represented in the form of the product $\beta(\zeta) = \beta_1(\zeta)\beta_2(\zeta)$, where $\beta_2 \in C_{\langle p\rangle}^{n\times n}$ and $\det \beta_2(\zeta_0) = 0$. Therefore the equality

$$P\widetilde{T}_a + Q\widetilde{T}_\beta = (P + Q\widetilde{T}_{\beta_2} Q)(P\widetilde{T}_a + Q\widetilde{T}_{\beta_1}) + T$$

holds, where $T \in \mathcal{J}(\widetilde{\ell_p^n})$. If the operator $P\widetilde{T}_a + Q\widetilde{T}_\beta$ is a Φ_--operator, then from the last inequality it follows that the operator $P + Q\widetilde{T}_{\beta_1} Q$ is also a Φ_--operator. This implies that the operator T_{β_2} is a Φ_--operator in the space ℓ_p^n. Since $\det \beta_2(\zeta_0) = 0$, this fact contradicts Theorem 4.2.

Let us show that the operator $A = P\widetilde{T}_a + Q\widetilde{T}_b$ is not a Φ_+-operator. Assume the contrary, then the operator $A^* = P\widetilde{T}_{a^*} + Q\widetilde{T}_{b^*}$ is a Φ_--operator in the space $\widetilde{\ell}_q^n$ $(q^{-1}+p^{-1} = 1)$. Consider the operator $\widetilde{A}$ defined in the space $\widetilde{\ell}_q^{3n}$ by the matrix

$$\widetilde{A} = \begin{pmatrix} I & 0 & P \\ 0 & I & Q \\ \widetilde{T}_{a^*} & \widetilde{T}_{b^*} & 0 \end{pmatrix}.$$

Since the equality

$$\widetilde{A} = \begin{pmatrix} I & 0 & 0 \\ 0 & I & 0 \\ \widetilde{T}_{a^*} & \widetilde{T}_{b^*} & I \end{pmatrix} \begin{pmatrix} I & 0 & 0 \\ 0 & I & 0 \\ 0 & 0 & -A^* \end{pmatrix} \begin{pmatrix} I & 0 & P \\ 0 & I & Q \\ 0 & 0 & I \end{pmatrix} \tag{7.7}$$

holds and outermost multiples in this equality are invertible operators in $\widetilde{\ell}_q^{3n}$, we deduce that the operator $\widetilde{A}$ is a Φ_--operator. The operator $\widetilde{A}$ is a paired operator (in contrast to the operator A^*, which is the transposed operator to a paired operator). Indeed, $\widetilde{A} = P\widetilde{T}_{\widetilde{a}} + Q\widetilde{T}_{\widetilde{b}}$, where

$$\widetilde{a} = \begin{pmatrix} E_n & 0 & E_n \\ 0 & E_n & 0 \\ a^* & b^* & 0 \end{pmatrix}, \quad \widetilde{b} = \begin{pmatrix} E_n & 0 & 0 \\ 0 & E_n & E_n \\ a^* & b^* & 0 \end{pmatrix},$$

and E_n is the identity matrix of order n. Taking into account that $\det \widetilde{b}(\zeta_0) = \det b^*(\zeta_0) = 0$, we arrive at the contradiction to what has been proved above. The theorem is proved. □

7.2. By Σ_p denote the smallest Banach subalgebra of the algebra $L(\widetilde{\ell}_p)$ that contains the set $\mathcal{G}_p$ of all operators of the form $P\widetilde{T}_a + Q\widetilde{T}_b$, where $a, b \in \Pi C_{\langle p \rangle}$, and by $\mathcal{K}_p$ denote the set of the operators of the form

$$A = \sum_{j=1}^{k} A_{j1} A_{j2} \dots A_{jr}, \tag{7.8}$$

where $A_{jl} \in \mathcal{G}_p$. On the set $\mathcal{G}_p$ we define the mapping

$$\pi : A \to \mathcal{A}(x) \quad (x = (\zeta, \mu) \in \Gamma_0 \times [0, 1]),$$

where $\mathcal{A}(x)$ is the symbol of the operator A. By the scheme proposed in [7], one can extend the mapping π to the homomorphism defined on the whole algebra Σ_p. Further one can prove that Σ_p is an algebra with Fredholm symbol. Without going into details, we will show only how to construct an operator permuting indices at an arbitrarily given point.

By A_z $(z \in \Gamma_0)$ denote the operator defined in the space $\widetilde{\ell}_p$ by the equality $A_z\{\xi_k\}_{-\infty}^{\infty} = \{z^{-2k}\xi_{-k}\}_{-\infty}^{\infty}$. Consider the operator M_z acting from $\mathcal{K}_p$ to $L(\widetilde{\ell}_p)$ by the rule $M_z A = -A_z PAQA_z$. Let $A \in \mathcal{K}_p$ and let $\mathcal{A}(x) = \|\alpha_{jk}(x)\|_{j,k=1}^2$ be its symbol. As in [20, Section 4] it can be shown that for the operator A there exists a compact operator T such that for the symbol $\|\beta_{jk}\|_{j,k=1}^2$ of the operator $B = M_z A - T \in \mathcal{K}_p$ the equality $\alpha_{12}(z,\mu) = \beta_{21}(z,\mu)$ holds and, moreover,

$$\sup_{z\in\Gamma_0, A\in\mathcal{K}_p} \frac{\|M_z A\|}{\|A\|} \leq 1. \tag{7.9}$$

Thus the results which have been proved above and the results of the paper [7] (see also [20]) imply[6] the following proposition.

Theorem 7.2. *Suppose $A \in \Sigma_p$ and its symbol has the form $\mathcal{A}(x) = \|\alpha_{jk}(x)\|_{j,k=1}^2$. The operator A is a Φ_+-operator or a Φ_--operator in $\widetilde{\ell}_p$ if and only if the condition*

$$\inf |\det \mathcal{A}(\zeta,\mu)| \neq 0 \quad (|\zeta| = 1,\ 0 \leq \mu \leq 1) \tag{7.10}$$

holds. If condition (7.10) is fulfilled, then the operator A is a Φ-operator and

$$\operatorname{Ind} A = -\operatorname{ind}\det \frac{\mathcal{A}(\zeta,\mu)}{a_{22}(\zeta,0)a_{22}(\zeta,1)}.$$

Note that the operator $A' = \widetilde{T}_a P + \widetilde{T}_b Q$ transposed to a paired operator belongs to the algebra Σ_p, and its symbol is

$$\mathcal{A}'(\zeta,\mu) = \left\| \begin{matrix} a(\zeta+0)\xi_p(\mu) + a(\zeta-0)(1-\xi_p(\mu)) & h_p(\mu)(b(\zeta+0)-b(\zeta-0)) \\ h_p(\mu)(a(\zeta+0)-a(\zeta-0)) & b(\zeta+0)(1-\xi_p(\mu)) + b(\zeta-0)\xi_p(\mu) \end{matrix} \right\|.$$

It is the transposed matrix to the symbol $\mathcal{A}(\zeta,\mu)$ of the operator $A = P\widetilde{T}_a + Q\widetilde{T}_b$.

Note also that the results of the last subsection can be generalized to the algebras $\Sigma_p^{n\times n}$.

References

[1] I.C. Gohberg and N.Ya. Krupnik, *On the algebra generated by Toeplitz matrices.* Funkts. Anal. Prilozh. **3** (1969), no. 2, 46–56 (in Russian). English translation: Funct. Anal. Appl. **3** (1969), 119–127. MR0250082 (40 #3323), Zbl 0199.19201.

[2] I.C. Gohberg and N.Ya. Krupnik, *On an algebra generated by the Toeplitz matrices in the spaces h_p.* Matem. Issled. **4** (1969), no. 3, 54–62 (in Russian). English translation: **this volume**. MR0399922 (53 #3763a), Zbl 0254.47045.

[3] R.V. Duduchava, *Discrete Wiener-Hopf equations composed of Fourier coefficients of piecewise Wiener functions.* Dokl. Akad. Nauk SSSR **207** (1972), 1273–1276 (in Russian). English translation: Soviet Math. Dokl. **13** (1972), 1703–1707. MR0313865 (47 #2418), Zbl 0268.47031.

[6] In Theorem 4.2 from [7] which is needed for the proof, it is required that the set of Φ-operators in the algebra $P\Sigma_p(L(\ell_p))$ is dense in $P\Sigma_p$. In our case (cf. the proof of the analogous Theorem 5.1 from [20]), it is sufficient to require that each operator in Σ_p with non-degenerated symbol can be approximated by operators $A^{(n)} \in \Sigma_p$ such that $\alpha_{11}^{(n)}(x) \neq 0$ on $\Gamma_0 \times [0,1]$. The possibility of such an approximation is easily verified.

[4] R.V. Duduchava, *Discrete Wiener-Hopf equations in l_p spaces with weight.* Soobshch. Akad. Nauk Gruz. SSR **67** (1972), 17–20 (in Russian). MR0306962 (46 #6083), Zbl 0249.47030.

[5] I.B. Simonenko, *A new general method of investigating linear operator equations of singular integral equation type. I.* Izv. Akad. Nauk SSSR Ser. Matem. **29** (1965), 567–586 (in Russian). MR0179630 (31 #3876), Zbl 0146.13101.

[6] I.B. Simonenko, *A new general method of investigating linear operator equations of singular integral equation type. II.* Izv. Akad. Nauk SSSR Ser. Matem. **29** (1965), 757–782 (in Russian). MR0188738 (32 #6174), Zbl 0146.13101.

[7] I.C. Gohberg and N.Ya. Krupnik, *Banach algebras generated by singular integral operators.* In: "Hilbert Space Operators Operator Algebras (Proc. Internat. Conf., Tihany, 1970)." Colloquia Math. Soc. Janos Bolyai **5** (1972), 239–264. MR0380519 (52 #1419), Zbl 0354.45008.

[8] I.B. Simonenko, *Some general questions in the theory of the Riemann boundary problem.* Izv. Akad. Nauk SSSR Ser. Matem. **32** (1968), 1138–1146 (in Russian). English translation: Math. USSR Izvestiya **2** (1968), 1091–1099. MR0235135 (38 #3447), Zbl 0186.13601.

[9] E.M. Semenov, *A new interpolation theorem.* Funkcional. Anal. Prilozh. **2** (1968), no. 2, 68-80 (in Russian). English translation: Funct. Anal. Appl. **2** (1968), no. 2, 158–168. MR0236694 (38 #4989), Zbl 0202.12805.

[10] I.B. Simonenko, *The Riemann boundary-value problem for n pairs of functions with measurable coefficients and its application to the study of singular integrals in L_p spaces with weights.* Izv. Akad. Nauk SSSR Ser. Matem. **28** (1964), 277–306 (in Russian). MR0162949 (29 #253), Zbl 0136.06901.

[11] S.K. Pichorides, *On the best values of the constants in the theorems of M. Riesz, Zygmund and Kolmogorov.* Studia Math. **44** (1972), 165–179. MR0312140 (47 #702), Zbl 0238.42007.

[12] I.C. Gohberg and N.Ya. Krupnik, *Norm of the Hilbert transformation in the L_p space.* Funkcional. Anal. Prilozhen. **2** (1968), no. 2, 91–92 (in Russian). English translation: Funct. Anal. Appl. **2** (1968), no. 2, 180–181. Zbl 0177.15503.

[13] I.C. Gohberg and N.Ya. Krupnik, *The spectrum of singular integral operators in L_p spaces.* Studia Math. **31** (1968), 347–362 (in Russian). English translation: **this volume**. MR0236774 (38 #5068), Zbl 0179.19701.

[14] S.B. Stechkin, *On bilinear forms.* Dokl. Akad. Nauk SSSR **71** (1950), 237–240 (in Russian). MR0033868 (11,504c), Zbl 0035.19703.

[15] I.I. Hirschman, Jr., *On multiplier transformations.* Duke Math. J. **26** (1959) , no. 2, 221–242. MR0104973 (21 #3721), Zbl 0085.09201.

[16] M.A. Karasnosel'skii, *On a theorem of M. Riesz.* Dokl. Akad. Nauk SSSR **131** (1960), 246–248 (in Russian). English translation: Soviet Math. Dokl. **1** (1960), 229–231. MR0119086 (22 #9852), Zbl 0097.10202.

[17] I.C. Gohberg and I.A. Feldman, *Convolution Equations and Projection Methods for their Solution.* Nauka, Moscow, 1971 (in Russian). English translation: Amer. Math. Soc., Providence, RI, 1974. German translation: Birkhäuser, Stuttgart, 1974 and Akademie-Verlag, Berlin, 1974. MR0355674 (50 #8148), Zbl 0214.38503, Zbl 0278.45007.

[18] I.C. Gohberg and N.Ya. Krupnik, *Algebra generated by one-dimensional singular integral operators with piecewise continuous coefficients.* Funkcional. Anal. Prilozhen. **4** (1970), no. 3, 26–36 (in Russian). English translation: Funct. Anal. Appl. **4** (1970), no. 3, 193–201. MR0270164 (42 #5057), Zbl 0225.45005.

[19] I.C. Gohberg, *Normal solvability and the index of a function of an operator.* Izv. Akad. Nauk Mold. SSR 1963, no. 11, Ser. Estestv. Tekh. Nauk (1964), 11–25 (in Russian). MR0223918 (36 #6965), Zbl 0152.33601.

[20] I.C. Gohberg and N.Ya. Krupnik, *Singular integral operators with piecewise continuous coefficients and their symbols.* Izv. Akad. Nauk SSSR Ser. Mat. **35** (1971), 940–964 (in Russian). English translation: Math. USSR Izvestiya **5** (1971), no. 4, 955–979. MR0291893 (45 #981), Zbl 0235.47025.

[21] I.C. Gohberg and N.Ya. Krupnik, *On the quotient norm of singular integral operators.* Matem. Issled. **4** (1969), no. 3, 136–139 (in Russian). English translation: Amer. Math. Soc. Transl. (2) **111** (1978), 117–119. MR0259671 (41 #4306), Zbl 0233.47035.

[22] L.A. Coburn, *Weyl's theorem for nonnormal operators.* Michigan Math. J. **13** (1966) no. 3, 285–288. MR0201969 (34 #1846), Zbl 0173.42904.

Operator Theory:
Advances and Applications, Vol. 206, 185–199

The Symbol of Singular Integral Operators on a Composed Contour

Israel Gohberg and Nahum Krupnik

Let Γ be a contour in the complex plane that consists of n simple closed curves $\Gamma_1, \dots, \Gamma_n$ having one common point t_0. We orient the contour Γ by choosing the counter-clockwise orientation on each curve Γ_j. Everywhere in what follows we will suppose that the contour has the following properties:

1) every arc of the contour Γ that does not contain the point t_0 satisfies the Lyapunov condition;
2) the curve Γ_j does not intersected the domain F_k $(k \neq j)$ bounded by the curve Γ_k;
3) no two curves of the contour are tangential at the point t_0.

In this paper singular integral operators of the form

$$(A\varphi)(t) = c(t)\varphi(t) + \frac{d(t)}{\pi i} \int_\Gamma \frac{\varphi(\tau)}{\tau - t}\, d\tau \tag{0.1}$$

are considered in the space $L_p(\Gamma, \varrho)$ with weight $\varrho(t) = |t - t_0|^\beta$ and $1 < p < \infty$. We will suppose that the coefficients $c(t)$ and $d(t)$ are continuous everywhere on Γ except for possibly at the point t_0. Moreover, it is supposed that each of the functions $c(t)$ and $d(t)$ has finite (in general, different) one sided limits as t tends to t_0 along each arc of the contour Γ.

Singular integral operators of the form (0.1) in the classes of piecewise Hölder continuous functions were considered in the famous monograph by N.I. Mushelishvili [1]. In particular, some sufficient conditions for the validity of Noether theorems were obtained there.

In this paper the (non-closed) algebra $\mathfrak{H}$ generated by all operators of the form (0.1) is studied. It is shown that this algebra is homomorphic to some algebra of

---

The paper was originally published as И.Ц. Гохберг, Н.Я. Крупник, О символе сингулярных интегральных операторов на сложном контуре, Труды симпозиума по механике сплошной среды и родственным проблемам анализа (Тбилиси, 1971), том 1, с. 46–59. Мецниереба, Тбилиси, 1973. Proceedings of the Symposium on Continuum Mechanics and Related Problems of Analysis (Tbilisi, 1971), Vol. 1, pp. 46–59. Mecniereba, Tbilisi, 1973. MR0385644 (52 #6504), Zbl 0292.47048.

matrix functions. The matrix-function A which is the image of an operator $A \in \mathfrak{H}$ under this homomorphism is called the symbol of the operator A. The symbol A is defined on the set $\Gamma \times [0,1]$, it depends on the space $L_p(\Gamma, \varrho)$ and is a matrix function of variable order. This order is equal to $2n$ at the points (t_0, μ) $(0 \le \mu \le 1)$ and to two at all other points.

A necessary and sufficient condition for an operator A to be a Φ-operator is that the determinant of the symbol A be different from zero. The index of an operator A is expressed in terms of its symbol.

The mentioned results are also obtained for the algebras generated by matrix singular integral operators of the form (0.1).

The paper consists of six sections.

Section 1 has an auxiliary nature. Properties of the operator of singular integration along Γ are studied there. In Sections 2 and 3 operators of the form (0.1) and algebras generated by these operators are studied under additional assumptions on the coefficients. In Section 4, necessary and sufficient conditions for an operator of the form (0.1) to be a Φ-operator are obtained with the aid of the above mentioned results. In Section 5 main results are established. Formulas for the symbol and the index are found.

In the last section, possible generalizations and relations to results from other papers are mentioned.

1. Auxiliary propositions

1.1. By Λ denote the set of all functions $a(t)$ $(t \in \Gamma)$ that are continuous everywhere on Γ except possibly at the point t_0 and such that finite (in general, different) limits exist as t tends to t_0 along each arc of the contour.

To each function $a(t) \in \Lambda$ assign the numbers a_j $(j = 1, \dots, 2n)$ being its limiting values at the points t_0. By a_{2k} denote the limit of the function $a(t)$ as t tends to t_0 along the arc Γ_k $(k = 1, \dots, n)$ inward to t_0 and by a_{2k-1} denote the limit of $a(t)$ as t tends to t_0 along the arc Γ_k outward to t_0. The set of the functions $a(t)$ in Λ such that $a_{2k-1} = a_{2k}$ $(k = 1, \dots, n)$ is denoted by Λ^+ and the set of the functions $a(t) \in \Lambda$ for which $a_{2k+1} = a_{2k}$ $(k = 1, \dots, n-1)$ and $a_{2n} = a_1$ is denoted by Λ^-. The intersection $\Lambda^+ \cap \Lambda^-$ coincides with the set $C(\Gamma)$ of all continuous functions on Γ.

By S denote the operator of singular integration along Γ:

$$(S\varphi)(t) = \frac{1}{\pi i} \int_\Gamma \frac{\varphi(\tau)}{\tau - t}\, d\tau.$$

This operator is bounded in $L_p(\Gamma, \varrho)$ (see [2]). Introduce the operators P and Q defined by the equalities

$$P = \frac{1}{2}(I + S), \quad Q = \frac{1}{2}(I - S).$$

In what follows by $\mathfrak{F} = \mathfrak{F}(L_p(\Gamma, \varrho))$ denote the set of all linear compact operators acting in $L_p(\Gamma, \varrho)$. It is known that if $a(t) \in C(\Gamma)$, then the operator $a(t)P - Pa(t)I$ belongs to $\mathfrak{F}$. This statement plays an important role in the theory of singular integral operators with continuous coefficients. In the case of functions $a(t) \in \Lambda^{\pm}$ the following proposition holds.

Lemma 1.1. *If $a(t) \in \Lambda^+$, then $PaP - aP \in \mathfrak{F}$.*

Proof. The function $a(t)$ can be represented in the form $a(t) = b(t)g(t)$, where $b(t) \in C(\Gamma)$ and the function $g(t)$ is constant on each curve Γ_k $(k = 1, \dots, n)$.

Let us show that $PgP = gP$. Let $\varphi(t)$ $(t \in \Gamma)$ be an arbitrary function satisfying the Hölder condition on each contour Γ_k $(k = 1, \dots, n)$. Then the function $\psi(t) = (P\varphi)(t)$ satisfies the Hölder condition on each arc of the contour Γ that does not contain the point t_0. Moreover, it admits an analytic continuation into each domain F_k $(k = 1, \dots, n)$ and has an integrable singularity in a neighborhood of the point t_0. The function $g(t)\psi(t)$ has the same properties. By using these properties of the function $g(t)\psi(t)$, it can be straightforwardly verified that for each point $t \in \Gamma$ different from t_0 the equality $(Sg\psi)(t) = (g\psi)(t)$ holds. From this equality it follows that $gP\varphi = PgP\varphi$, whence $PgP = gP$. Since $b(t) \in C(\Gamma)$, we have $bP - PbI \in \mathfrak{F}$. From here and the equality $PgP = gP$ it follows that $PaP - aP \in \mathfrak{F}$. The lemma is proved. □

1.2. In general, the operator $aP - PaI$ is not compact for $a(t) \in \Lambda^+$. This statement follows from the following example.

Let Γ be a contour consisting of curves Γ_1 and Γ_2 such that $[-1, 0] \subset \Gamma_1$ and $[0, 1] \subset \Gamma_2$. We take the characteristic function of the contour Γ_2 as the function $a(t)$ $(t \in \Gamma)$. Assume that $PaI - aP \in \mathfrak{F}$. From here it follows that $SaI - aS \in \mathfrak{F}$. Let $\chi(t)$ be the characteristic function of the segment $[-1, 0]$ and $A = \chi(t)(SaI - aS)$. Let us show that $A \in \mathfrak{F}$. This will lead to a contradiction.

Consider in the space $L_2(\Gamma)$ the sequence of normalized functions defined by the equalities

$$\varphi_n(t) = \begin{cases} \sqrt{n} & \text{for} \quad t \in [0, 1/n], \\ 0 & \text{for} \quad t \in \Gamma \setminus [0, 1/n], \end{cases} \qquad (n = 1, 2, \dots).$$

Let us show that it is not possible to extract a convergent subsequence from the sequence $\psi_n = A\varphi_n$. First note that the norm $\|\psi_n\|_p$ in the space $L_p(\Gamma)$ $(1 < p < 2)$ can be estimated as follows:

$$\|\psi_n\|_p^p \le n^{(p-2)/p} \int_{-\infty}^{0} \left| \ln \left| \frac{x-1}{x} \right| \right|^p dx,$$

whence $\|\psi_n\|_p \to 0$. From here it follows that if the sequence ψ_n contains a subsequence ψ_{k_n} that converges in $L_2(\Gamma)$, then $\|\psi_{k_n}\|_2 \to 0$. But the latter is impossible because

$$\|\psi_n\|_2^2 \ge \frac{1}{\pi^2} \int_{-1}^{0} \left| \ln \left| \frac{x-1}{x} \right| \right|^2 dx.$$

1.3.

Lemma 1.2. *If $a(t) \in \Lambda^-$, then $QaQ - aQ \in \mathfrak{F}$.*

Proof. Let $s_1, \dots, s_n$ be n points belonging to the domains $F_1, \dots, F_n$, respectively[1]. Consider the functions $f_j(t)$ defined by the equalities

$$f_j(t) = \alpha_j \left(\frac{t - s_j}{t - s_{j+1}} \right)^{\delta_j} \quad (j = 1, \dots, n;\ s_{n+1} = s_1).$$

Each function $f_j(t)$ is analytic in the extended complex plane with the cut joining the points s_j and s_{j+1} and lying in $F_j \cup F_{j+1} \cup \{t_0\}$. Without loss of generality we can assume that the one sided limits of the function $a(t)$ at the point t_0 are different from zero. Choose the numbers α_j and δ_j so that the limits of the function $f_j(t)$ as t tends to t_0 along Γ_j by the inward and outward arc are equal to a_{2j-1} and 1, respectively. By $f(t)$ denote the product $f(t) = f_1(t) \cdots f_n(t)$. The function $f(t)$ is holomorphic in the complement of $F_1 \cup \dots \cup F_n \cup \Gamma$ in the extended complex plane. It can be straightforwardly verified that $QfQ = fQ$. In view of the choice of the function $f(t)$, the product $g(t) = a(t) f^{-1}(t)$ belongs to $C(\Gamma)$. Since $QgI - gQ \in \mathfrak{F}$ and $QfQ = fQ$, we have $QaQ - aQ \in \mathfrak{F}$. The lemma is proved. □

Lemmas 1.1 and 1.2 immediately imply the following.

Lemma 1.3. *Let $a(t)$ and $b(t)$ be arbitrary functions belonging simultaneously to one of the sets Λ^+ or Λ^-. Then*

$$PaPbP - PabP \in \mathfrak{F}, \quad QaQbQ - QabQ \in \mathfrak{F}.$$

2. Singular operators with coefficients in Λ^+ and Λ^-

Let us agree on the following notation: $L_p^m(\Gamma, \varrho)$ is the direct sum of m copies of the space $L_p(\Gamma, \varrho)$; Λ_m^+ (resp. Λ_m^-) is the set of all matrix functions of the form $\|a_{jl}(t)\|_1^m$, where $a_{jl} \in \Lambda^+$ (resp. $a_{jl} \in \Lambda^-$);

$$S_m = \|\delta_{jl} S\|_1^m, \quad P_m = \frac{1}{2}(I + S_m), \quad Q_m = \frac{1}{2}(I - S_m).$$

Theorem 2.1. *Let $a(t), b(t) \in \Lambda_m^+(\Lambda_m^-)$. The operator $A = aP_m + bQ_m$ is a Φ_+-operator or a Φ_--operator in the space $L_p^m(\Gamma, \varrho)$ if and only if the condition*

$$\inf |\det a(t) b(t)| \neq 0 \quad (t \in \Gamma) \tag{2.1}$$

holds. If condition (2.1) *is fulfilled, then*

$$\operatorname{ind} A = \frac{1}{2\pi} \sum_{k=1}^{n} \{\arg \det b(t) a^{-1}(t)\}_{\Gamma_k}. \tag{2.2}$$

---

[1]The domain bounded by the curve Γ_j is denoted by F_j.

An auxiliary statement will be needed for the proof of this theorem. By $S_m^{(j)}$ denote the operator $\|\delta_{kl}S^{(j)}\|_1^m$, where $S_m^{(j)}$ is the operator of singular integration along Γ_j, and by $P_m^{(j)}$ and $Q_m^{(j)}$ denote the operators

$$P_m^{(j)} = \frac{1}{2}(I + S_m^{(j)}), \quad Q_m^{(j)} = \frac{1}{2}(I - S_m^{(j)}),$$

respectively.

Lemma 2.1. *Suppose $a(t)$ and $b(t)$ are bounded measurable matrix functions on Γ that coincide with the identity matrix of order m on $\Gamma \setminus \Gamma_j$. The operator $N_j = aP_m^{(j)} + bQ_m^{(j)}$ (resp. $N_j = P_m^{(j)}a + Q_m^{(j)}b$) is normally solvable in $L_p^m(\Gamma_j, \varrho)$ if and only if the operator $N = aP_m + bQ_m$ (resp. $N = P_m a + Q_m b$) is normally solvable in $L_p^m(\Gamma, \varrho)$. Moreover,*

$$\dim\ker N = \dim\ker N_j, \quad \dim\operatorname{coker} N = \dim\operatorname{coker} N_j.$$

The proof of this lemma is analogous to the proof of the corresponding lemma from [3].

Proof of Theorem 2.1. Let condition (2.1) be fulfilled. For the sake of definiteness assume that the matrix functions $a(t)$ and $b(t)$ belong to Λ_m^+. In this case, in view of Lemma 1.1, the operator A can be written in the form

$$A = P_m a P_m + Q_m b Q_m + P_m b Q_m + T,$$

where $T \in \mathfrak{F}$. It is easy to see that the operator

$$B = P_m a^{-1} P_m + Q_m b^{-1} Q_m - P_m a^{-1} P_m b Q_m b^{-1} Q_m$$

is a regularizer for the operator A. Therefore the operator A is a Φ-operator. Let us prove equality (2.2). To this end, write the operator A in the form

$$A = (P_m a P_m + Q_m b Q_m)(I - P_m a^{-1} P_m b Q_m) + T_1, \tag{2.3}$$

where $T_1 \in \mathfrak{F}$. The operator $I - P_m a^{-1} P_m b Q_m$ is invertible and $I + P_m a^{-1} P_m b Q_m$ is its inverse. Hence $\operatorname{ind} A = \operatorname{ind} A_1$, where $A_1 = P_m a P_m + Q_m b Q_m$.

Let $c(t)$ be a matrix function in Λ_m^+. By $c^{(j)}(t)$ $(j = 1, \dots, n)$ denote the matrix function

$$c^{(j)}(t) = \begin{cases} c(t) & \text{for} \quad t \in \Gamma_j, \\ E_m & \text{for} \quad t \in \Gamma \setminus \Gamma_j, \end{cases}$$

where E_m is the identity matrix of order m.

In view of Lemma 1.3, the operator A_1 can be represented in the form of the product

$$A_1 = (a^{(1)} P_m + Q_m)(P_m + Q_m b^{(1)}) \dots (a^{(n)} P_m + Q_m)(P_m + Q_m b^{(n)}) + T_2, \tag{2.4}$$

where $T_2 \in \mathfrak{F}$.

Lemma 2.1 allows us to reduce the problem of calculating the index of the operators $a^{(j)} P_m + Q_m$ and $P_m + Q_m b^{(j)}$ to the case when the contour consists of one closed curve Γ_j and the matrix functions $a^{(j)}(t)$ and $b^{(j)}(t)$ are continuous on Γ_j. As is known in this case

$$\operatorname{ind}(a^{(j)}P_m + Q_m) = \operatorname{ind}\det a^{(j)}(t), \quad \operatorname{ind}(P_m + Q_m b^{(j)}) = \operatorname{ind}\det b^{(j)}(t).$$

From here and equality (2.4) it follows that $\operatorname{ind} A_1 = \operatorname{ind}\det b(t)a^{-1}(t)$.

Let us prove the necessity of condition (2.1). Suppose the operator A is a Φ_+-operator (or a Φ_--operator). We choose a matrix function $\widetilde{a}(t) \in \Lambda_m^+$ sufficiently close to $a(t)$ (in the uniform norm) and such that 1) $\inf|\det\widetilde{a}(t)| \neq 0$ and 2) the operator $\widetilde{A} = \widetilde{a}P_m + bQ_m$ is a Φ_+-operator (resp. Φ_--operator). The operator $\widetilde{A}$ can be represented in the form (2.3). Therefore $P_m\widetilde{a}P_m + Q_m bQ_m$ is also a Φ_+-operator (resp. Φ_--operator). In its turn, the operator $P_m\widetilde{a}P_m + Q_m bQ_m$ can be represented in the form (2.4). Again, with the aid of Lemma 2.1 the problem is reduced to the case when the contour consists of one closed curve. From here it follows that $\inf|\det b^{(j)}(t)| \neq 0$ $(j = 1, \dots, n)$. Hence $\inf|\det b(t)| \neq 0$. Analogously, using the equality

$$A = (I + P_m bQ_m b^{-1}Q_m)(P_m aP_m + Q_m bQ_m) + T' \quad (T' \in \mathfrak{F})$$

instead of (2.3), we conclude that $\inf|\det a(t)| \neq 0$ $(t \in \Gamma)$. The theorem is proved. □

3. Algebra generated by singular operators with coefficients in Λ_m^+ and Λ_m^-

By $\mathfrak{L}(L_p^m(\Gamma, \varrho))$ denote the algebra of all bounded linear operators acting in the space $L_p^m(\Gamma, \varrho)$ and by $\mathfrak{A}_m^+$ denote the smallest Banach subalgebra of the algebra $\mathfrak{L}(L_p^m(\Gamma, \varrho))$ that contains all operators of the form $aP_m + bQ_m$, where $a, b \in \Lambda_m^+$. The algebra $\mathfrak{A}_m^+$ is the closure in $\mathfrak{L}(L_p^m(\Gamma, \varrho))$ of the set $\mathfrak{H}_m^+$ of the operators of the form

$$A = \sum_{j=1}^{k} A_{j1} \dots A_{js}, \tag{3.1}$$

where $A_{jl} = a_{jl}P_m + b_{jl}Q_m$ and $a_{jl}, b_{jl} \in \Lambda_m^+$.

From Lemmas 1.1–1.3 it follows that every operator A in $\mathfrak{H}_m^+$ of the form (3.1) can be represented in the form

$$A = P_m aP_m + Q_m bQ_m + P_m AQ_m + T, \tag{3.2}$$

where

$$a(t) = \sum_{j=1}^{k} a_{j1}(t) \dots a_{js}(t), \quad b(t) = \sum_{j=1}^{k} b_{j1}(t) \dots b_{js}(t)$$

and $T \in \mathfrak{F}$. The algebras $\mathfrak{A}_m^-$ and $\mathfrak{H}_m^-$ are defined analogously. Each operator of the form (3.1) with coefficients $a_{jl}(t)$ and $b_{jl}(t)$ in Λ_m^- admits a representation in the form

$$A = P_m aP_m + Q_m bQ_m + Q_m AP_m + T \quad (a, b \in \Lambda_m^-,\ T \in \mathfrak{F}). \tag{3.3}$$

Theorem 3.1. *Suppose an operator* $A \in \mathfrak{H}_m^{\pm}$ *has the form* (3.2) *or* (3.3) *and suppose that*

$$a(t) = \|\alpha_{jl}(t)\|_1^m, \quad b(t) = \|\beta_{jl}(t)\|_1^m.$$

Then the inequalities

$$\sup|\alpha_{jl}(t)| \leq |A|, \quad \sup|\beta_{jl}(t)| \leq |A| \tag{3.4}$$

hold, where

$$|A| = \inf\|A + T\| \quad (T \in \mathfrak{F}).$$

Proof. Indeed, repeating the proof of Theorem 2.1 one can obtain that the operator A is a Φ-operator if and only if the condition $\inf|\det a(t)b(t)| \neq 0$ $(t \in \Gamma)$ is fulfilled. First, consider the case $m = 1$. By Ω denote the complement to the set of all Φ-points of the operator A. The set Ω is the union of the range of the functions $a(t)$ and $b(t)$. Since $\max|\lambda| \leq |A|$ $(\lambda \in \Omega)$, we have $|a(t)| \leq |A|$ and $|b(t)| \leq |A|$. In the general case (when $m \geq 1$), the operator A is the operator matrix $A = \|F_{jl}\|_1^m$ with entries $F_{jl} \in \mathfrak{H}_1^{\pm}$. The operators F_{jl} can be represented in the form

$$F_{jl} = P\alpha_{jl}P + Q\beta_{jl}Q + PF_{jl}Q + T_{jl},$$

where $T_{jl} \in \mathfrak{F}$. In view of what has been proved above, we have $|\alpha_{jl}(t)| \leq |F_{jl}|$ and $|\beta_{jl}(t)| \leq |F_{jl}|$. Because $|F_{jl}| \leq |A|$, inequalities (3.4) are proved. □

Theorem 3.2. *Each operator A in $\mathfrak{A}_m^+$ can be represented in the form*

$$A = P_m a P_m + Q_m b Q_m + P_m A Q_m + T, \tag{3.5}$$

where $a(t), b(t) \in \Lambda_m^+$ and $T \in \mathfrak{F}$.

Matrix functions $a(t)$ and $b(t)$ are uniquely determined by the operator A.

The operator A is a $\Phi_{\pm}$-operator in $L_p^m(\Gamma, \varrho)$ if and only if the condition

$$\inf|\det a(t)b(t)| \neq 0 \quad (t \in \Gamma) \tag{3.6}$$

holds. If condition (3.6) *is fulfilled, then the operator A is a Φ-operator and*

$$\operatorname{ind} A = \frac{1}{2\pi}\sum_{j=1}^{n}\{\arg\det b(t)a^{-1}(t)\}_{\Gamma_j}. \tag{3.7}$$

Proof. Let $A \in \mathfrak{A}_m^+$ and A_n be a sequence of operators in $\mathfrak{H}_m^+$ that converges in norm to the operator A. The operators A_n can be represented in the form

$$A_n = P_m a_n P_m + Q_m b_n Q_m + P_m A_n Q_m + T_n,$$

where $a_n, b_n \in \Lambda_m^+$ and $T_n \in \mathfrak{F}$. From inequalities (3.4) it follows that the matrix functions $a_n(t)$ and $b_n(t)$ converge uniformly to the matrix functions $a(t)$ and $b(t)$, respectively. Moreover, the matrix functions $a(t)$ and $b(t)$ do not depend on a choice of the sequence $\{A_n\}$ converging to A. It is clear that $a(t)$ and $b(t)$ belong to Λ_m^+. Since the operators $P_m a_n P_m + Q_m b_n Q_m + P_m A_n Q_m$ tend in norm to the operator $P_m a P_m + Q_m b Q_m + P_m A Q_m$, we have $\lim\|T - T_n\| = 0$. Thus the operator A can be represented in the form (3.5). Conditions (3.6) and formula (3.7) are obtained in the same way as in the proof of Theorem 2.1. □

It is easy to see that Theorem 3.2 remains true if one replaces in its formulation $\mathfrak{A}_m^+$ by $\mathfrak{A}_m^-$, Λ_m^+ by Λ_m^-, and equality (3.5) by the equality

$$A = P_m a P_m + Q_m b Q_m + Q_m A P_m + T.$$

4. Singular integral operators with coefficients in Λ_m

By Λ_m denote the set of all matrix functions $c(t) = \|c_{jl}(t)\|_1^m$ with entries $c_{jl}(t)$ in Λ. Let $c(t) \in \Lambda_m$ and $c_1, \dots, c_{2n}$ be constant matrices that are the limits of the matrix function $c(t)$ as $t \to t_0$ [2]. By $\xi(\mu) = \xi_{p,\beta}(\mu)$ $(0 \le \mu \le 1)$ denote the function

$$\xi(\mu) = \begin{cases} \dfrac{\sin(\theta\mu)\exp(i\theta\mu)}{\sin\theta\exp(i\theta)} & \text{if } \theta \neq 0, \\ \mu & \text{if } \theta = 0, \end{cases} \tag{4.0}$$

where $\theta = \pi - 2\pi(1+\beta)/p$. Here β and p are the parameters from the definition of the space $L_p(\Gamma, \varrho)$.

If μ varies from 0 to 1, then the range of the function $\xi(\mu)$ is the circular arc (or the segment of the straight line) joining the points 0 and 1. From the points of this arc the segment $[0,1]$ is seen at the angle $\pi - |\theta|$. If $2(1+\beta) < p$, then $\operatorname{Im}\xi(\mu) \le 0$; if $2(1+\beta) > p$, then $\operatorname{Im}\xi(\mu) \ge 0$.

Theorem 4.1. *Let $a(t), b(t) \in \Lambda_m$. The operator $A = aP_m + bQ_m$ is a Φ_+-operator or a Φ_--operator in the space $L_p^m(\Gamma, \varrho)$ if and only if the following two conditions are fulfilled:*

$$\inf_{t\in\Gamma} |\det a(t)| > 0, \quad \inf_{t\in\Gamma} |\det b(t)| > 0, \tag{4.1}$$

$$\det\left(b_1^{-1}a_1a_2^{-1}b_2\dots a_{2n}^{-1}b_{2n}\xi(\mu) + (1-\xi(\mu))E_m\right) \neq 0, \tag{4.2}$$

where E_m is the identity matrix of order m.

If conditions (4.1) *and* (4.2) *are fulfilled, then the operator A is a Φ-operator and*

$$\begin{aligned} \operatorname{ind} A = &\sum_{j=1}^{m} \frac{1}{2\pi}\{\arg\det b(t)a^{-1}(t)\}_{t\in\Gamma_j} \\ &- \frac{1}{2\pi}\left\{\arg\det\left(b_1^{-1}a_1a_2^{-1}b_2\dots a_{2n}^{-1}b_{2n}\xi(\mu) + (1-\xi(\mu))E_m\right)\right\}_{\mu=0}^{1}. \end{aligned} \tag{4.3}$$

For the proof of this theorem we will need the following.

Lemma 4.1. *Let $c(t) \in \Lambda_m$ and $\det c_j \neq 0$ $(j = 1, \dots, 2n)$. Then the matrix function $c(t)$ can be represented in the form $c(t) = f(t)x(t)g(t)$, where $g(t) \in \Lambda_m^+$, $f(t) \in \Lambda_m^-$, and $x(t)$ is a matrix function having the following properties:* 1) *$x(t) = E_m$ for $t \in \Gamma \setminus \Gamma_1$;* 2) *$x_2 = E_m$; and* 3) *$x_1 = c_1c_2^{-1}c_3c_4^{-1}\dots c_{2n-1}c_{2n}^{-1}$.*

[2] Recall that the limit of the matrix function $c(t)$ as $t \to t_0$ along the arc Γ_k $(k = 1, \dots, n)$ inward to t_0 is denoted by c_{2k} and the limit of $c(t)$ as t tends to t_0 along Γ_k outward to t_0 is denoted by c_{2k-1}.

Proof. Let $x(t) \in \Lambda_m$ be an arbitrary nonsingular matrix function satisfying conditions 1)–3). By $g(t)$ denote a nonsingular matrix function in Λ_m^+ such that $g_{2k-1} = c_1 c_2^{-1} c_3 \dots c_{2k-2}^{-1} c_{2k-1}$ $(k = 1, \dots, n)$. Put $f(t) = c(t)g^{-1}(t)x^{-1}(t)$. Since $f_{2k+1} = f_{2k}$ $(k = 1, \dots, n-1)$ and $f_1 = f_{2n}$, we have $f(t) \in \Lambda_m^-$. The lemma is proved. □

Proof of Theorem 4.1. Assume conditions (4.1) and (4.2) are fulfilled. According to Lemma 4.1, the matrix function $c(t) = b^{-1}(t)a(t)$ can be represented in the form $c(t) = f(t)x(t)g(t)$, where $f \in \Lambda_m^-$, $g \in \Lambda_m^+$, and the matrix function $x(t)$ has properties 1)–3) of Lemma 4.1. Then in view of Lemmas 1.1–1.3, the operator $cP_m + Q_m$ can be represented in the form

$$cP_m + Q_m = f(xP_m + Q_m)(gP_m + f^{-1}Q_m) + T, \tag{4.4}$$

where $T \in \mathfrak{F}$. The operator $gP_m + f^{-1}Q_m$ is a Φ-operator because it has a regularizer $g^{-1}P_m + fQ_m$. Since $x(t) = E_m$ on $\Gamma \setminus \Gamma_1$, in view of Lemma 2.1 the operator $xP_m + Q_m$ is a Φ_+-operator (resp. Φ_--operator) in $L_p^m(\Gamma, \varrho)$ if and only if the operator $xP_m^{(1)} + Q_m^{(1)}$ is a Φ_+-operator (resp. Φ_--operator) in $L_p^m(\Gamma_1, \varrho)$. Since $\inf|\det x(t)| \neq 0$ and $\det(\xi(\mu)x_1 + (1-\xi(\mu))x_2) \neq 0$, we obtain[3] (see [4]) that the operator $xP_m^{(1)} + Q_m^{(1)}$ is a Φ-operator in the space $L_p^m(\Gamma_1, \varrho)$. From Lemma 2.1 it follows that the operator $xP_m + Q_m$ is a Φ-operator in $L_p^m(\Gamma, \varrho)$. Thus conditions (4.1) and (4.2) are sufficient for the operator $aP_m + bQ_m$ to be a Φ-operator in $L_p^m(\Gamma, \varrho)$. Now we turn to the proof of the necessity of the hypotheses of the theorem. First, let us show that if the operator $A = aP_m + bQ_m$ $(a, b \in \Lambda_m)$ is a Φ_+-operator or a Φ_--operator, then it is a Φ-operator. Indeed, let A_n be a sequence of singular integral operators converging in the norm to A. The operators A_n can be chosen so that their coefficients satisfy conditions (4.1) and (4.2). In view of what has just been proved, the operators A_n are Φ-operators. Taking into account the property of stability of the index of Φ-operators, we obtain that the operator A is a Φ-operator.

We prove the necessity of condition (4.1) by contradiction. Assume that the operator A is a Φ-operator and condition (4.1) does not hold. For instance, let $\inf|\det a(t)| = 0$. Then one can find an operator $B = \widetilde{a}P_m + bQ_m$ (sufficiently close to A) such that $\det \widetilde{a}(\widetilde{t}) = 0$ $(\widetilde{t} \neq t_0)$ and $\det \widetilde{a}(t') \neq 0$, $\det \widetilde{a}(t'') \neq 0$, where (t', t'') is some neighborhood of the point $\widetilde{t}$. The matrix function $\widetilde{a}(t)$ can be represented in the form $\widetilde{a}(t) = h(t)u(t)$, where $u(t) \in C_m(\Gamma)$ and $\det u(\widetilde{t}) = 0$. Then $B = (hP_m + bQ_m)(uP_m + Q_m) + T$ $(T \in \mathfrak{F})$.

Since the operator B is a Φ-operator, we see that the operator $uP_m + Q_m$ is a Φ_+-operator. The latter fact contradicts Theorem 2.1 because $u(t) \in \Lambda_m^+$ and $\det u(\widetilde{t}) = 0$. Let us prove condition (4.2). To that end we represent the operator $cP_m + Q_m$, where $c(t) = b^{-1}(t)a(t)$, in the form (4.4). Since $cP_m + Q_m$ is a Φ-operator, we deduce that $xP_m + Q_m$ also is a Φ-operator. In view of Lemma 2.1,

---

[3] In [4] it was assumed that Γ is a Lyapunov curve. However, none of the results of that paper and their proofs are changed if the contour is piecewise Lyapunov.

the operator $xP_m^{(1)} + Q_m^{(1)}$ is a Φ_--operator in $L_p^m(\Gamma_1, \varrho)$. From [4] it follows that $\det(\xi(\mu)x_1 + (1-\xi(\mu))x_2) \neq 0$, whence (3.2) is fulfilled.

It remains to prove formula (3.3). Since $\operatorname{ind} A = \operatorname{ind}(cP_m + Q_m)$, from equality (3.4) it follows that

$$\operatorname{ind} A = \operatorname{ind}(xP_m + Q_m) + \operatorname{ind}(gP_m + Q_m) + \operatorname{ind}(P_m + f^{-1}Q_m). \tag{4.5}$$

From [4, Theorem 1] and Lemma 2.1 it follows that

$$\operatorname{ind}(xP_m + Q_m) = \frac{1}{2\pi}\{\arg\det x(t)\}_{\Gamma_1} - \frac{1}{2\pi}\{\arg\det(\xi(\mu)x_1 + (1-\xi(\mu))x_2\}_{\mu=0}^1.$$

On the other hand, from Theorem 3.2 it follows that

$$\operatorname{ind}(gP_m + Q_m) = -\frac{1}{2\pi}\sum_{j=1}^n \{\arg\det x(t)\}_{\Gamma_j},$$

$$\operatorname{ind}(P_m + f^{-1}Q_m) = -\frac{1}{2\pi}\sum_{j=1}^n \{\arg\det f(t)\}_{\Gamma_j}.$$

Taking into account these equalities and (4.5), we get (4.3). The theorem is proved. □

5. Symbols of singular operators

5.1. The symbol of the operator $A = a(t)P_m + b(t)Q_m$ acting in the space $L_p^m(\Gamma, \varrho)$ is a matrix function $A(t, \mu)$ $(t \in \Gamma, 0 \le \mu \le 1)$ whose order depends on t. If $t \neq t_0$, then the symbol is the matrix function of order $2m$ defined by

$$A(t, \mu) = \left\| \begin{matrix} a(t) & 0 \\ 0 & b(t) \end{matrix} \right\|.$$

The symbol at the points (t_0, μ) $(0 \le \mu \le 1)$ is the matrix function

$$A(t, \mu) = \|u_{jk}(\mu)\|_{j,k=1}^{2n}$$

of order $2mn$, where the matrix functions $u_{jk}(\mu)$ (of order m) are defined by the following equalities.

For $j < k$,

$$u_{jk}(\mu) = \begin{cases} (-1)^{j+1}(b_k - b_{k+1})\xi(\mu)^{\frac{k-j}{2n}}(1-\xi(\mu))^{1-\frac{k-j}{2n}} & \text{if } k \text{ is even,} \\ (-1)^{j+1}(a_{k+1} - a_k)\xi(\mu)^{\frac{k-j}{2n}}(1-\xi(\mu))^{1-\frac{k-j}{2n}} & \text{if } k \text{ is odd;} \end{cases}$$

for $j > k$,

$$u_{jk}(\mu) = \begin{cases} (-1)^{j}(b_k - b_{k+1})\xi(\mu)^{1-\frac{j-k}{2n}}(1-\xi(\mu))^{\frac{j-k}{2n}} & \text{if } k \text{ is even,} \\ (-1)^{j}(a_{k+1} - a_k)\xi(\mu)^{1-\frac{j-k}{2n}}(1-\xi(\mu))^{1-\frac{j-k}{2n}} & \text{if } k \text{ is odd;} \end{cases}$$

and for $j = k$,

$$u_{kk}(\mu) = \begin{cases} \xi(\mu)a_k + (1 - \xi(\mu))a_{k+1} & \text{if } k \text{ is odd,} \\ \xi(\mu)b_k + (1 - \xi(\mu))b_{k+1} & \text{if } k \text{ is even.} \end{cases}$$

In these equalities the matrix b_1 is also denoted by b_{2n+1} and $\xi(\mu) = \xi_{p,\beta}(\mu)$ is the function defined in Section 4.

Theorem 5.1. *The operator $A = aP_m + bQ_m$ $(a, b \in \Lambda_m)$ is a Φ_+-operator or a Φ_--operator if and only if the condition*

$$\det A(t, \mu) \neq 0 \quad (t \in \Gamma,\ 0 \leq \mu \leq 1) \tag{5.1}$$

holds. If condition (5.1) *is fulfilled, then the operator A is a Φ-operator and*

$$\operatorname{ind} A = \frac{1}{2\pi} \sum_{j=1}^{n} \{\arg \det b(t)a^{-1}(t)\}_{t \in \Gamma_j} - \frac{1}{2\pi} \{\arg \det A(t_0, \mu)\}_{\mu=0}^{1}. \tag{5.2}$$

For the proof of this theorem we will need the following.

Lemma 5.1. *Let $a(t), b(t) \in \Lambda_m$ and the conditions $\det a_j b_j \neq 0$ $(j = 1, \dots, 2n)$ hold. Then*

$$\begin{aligned} \det A(t_0, \mu) &= \det(b_1 a_2 b_3 a_4 \dots b_{2n-1} a_{2n}) \\ &\quad \times \det\big((1 - \xi(\mu))E_m + \xi(\mu) b_1^{-1} a_1 a_2^{-1} b_2 \dots a_{2n}^{-1} b_{2n}\big). \end{aligned} \tag{5.3}$$

Proof. Let α_j and β_j $(j = 1, \dots, 2n)$ be some nonsingular matrices of order m. It is not difficult to verify that

$$\begin{aligned} &\det \begin{pmatrix} \alpha_1 & 0 & 0 & \dots & 0 & \beta_1 \\ \alpha_2 & \beta_2 & 0 & \dots & 0 & 0 \\ 0 & \beta_3 & \alpha_3 & \dots & 0 & 0 \\ \dots & \dots & \dots & \dots & \dots & \dots \\ 0 & 0 & 0 & \dots & \alpha_{2n} & \beta_{2n} \end{pmatrix} \\ &= \det(\beta_1 \alpha_2 \beta_3 \alpha_4 \dots \beta_{2n-1} \alpha_{2n}) \\ &\quad \times \det(E_m - \beta_1^{-1} \alpha_1 \alpha_2^{-1} \beta_2 \dots \beta_{2n-1}^{-1} \alpha_{2n-1} \alpha_{2n}^{-1} \beta_{2n}). \end{aligned} \tag{5.4}$$

Consider the matrices

$$M(\mu) = \begin{pmatrix} \gamma a_1 & 0 & 0 & \dots & 0 & -\delta b_1 \\ \delta a_2 & \gamma b_2 & 0 & \dots & 0 & 0 \\ 0 & \delta b_3 & \gamma a_3 & \dots & 0 & 0 \\ \dots & \dots & \dots & \dots & \dots & \dots \\ 0 & 0 & 0 & \dots & \delta a_{2n} & \gamma b_{2n} \end{pmatrix},$$

$$N(\mu) = \begin{pmatrix} \gamma E_m & 0 & 0 & \dots & 0 & -\delta E_m \\ \delta E_m & \gamma E_m & 0 & \dots & 0 & 0 \\ 0 & \delta E_m & \gamma E_m & \dots & 0 & 0 \\ \dots & \dots & \dots & \dots & \dots & \dots \\ 0 & 0 & 0 & \dots & \delta E_m & \gamma E_m \end{pmatrix},$$

where $\gamma = \gamma(\xi) = \xi^{1/2n}$ and $\delta = \delta(\xi) = (1 - \xi)^{1/2n}$.

From equality (5.4) it follows that

$$\det M(\mu) = \det(b_1a_2b_3a_4\dots b_{2n-1}a_{2n})\det((1-\xi)E_m + \xi b_1^{-1}a_1a_2^{-1}b_2\dots a_{2n}^{-1}b_{2n})$$

and $\det N(\mu) = 1$.

One can check straightforwardly that $N(\mu)A(t_0,\mu) = M(\mu)$, which implies the statement of the lemma. □

Proof of Theorem 5.1. Let $A(t,\mu) \neq 0$. Successively substituting $\mu = 0$ and $\mu = 1$ into this condition for $t = t_0$, we obtain that $\det a_jb_j \neq 0$ $(j = 1,\dots,2n)$. From equalities (5.3) and (5.1) it follows that for the operator A all the hypotheses of Theorem 4.1 are fulfilled. Hence the operator A is a Φ-operator. Formula (5.2) is a corollary of equalities (4.3), (5.1), and (5.3).

The necessity of conditions (5.1) is proved analogously. Indeed, let A be a Φ_+-operator (or a Φ_--operator). From Theorem 4.1 and equalities (5.1), (5.3) it follows that $\det A(t,\mu) \neq 0$. □

5.2. By $\mathfrak{H}_m$ denote the (non-closed) algebra of all operators of the form

$$A = \sum_{j=1}^{k} A_{j1}A_{j2}\dots A_{jr}, \qquad (5.5)$$

where $A_{jl} = a_{jl}P_m + b_{jl}Q_m$ and $a_{jl}, b_{jl} \in \Lambda_m$.

The matrix function

$$A(t,\mu) = \sum_{j=1}^{k} A_{j1}(t,\mu)A_{j2}(t,\mu)\dots A_{jr}(t,\mu)$$

where $A_{jl}(t,\mu)$ is the symbol of an operator A_{jl}, is called the symbol of an operator $A \in \mathfrak{H}_m$ defined by equality (5.5).

Theorem 5.2. *Let $A \in \mathfrak{H}_m$. If the operator A is compact, then $A(t,\mu) \equiv 0$. The symbol $A(t,\mu)$ of the operator A does not depend on a manner of representation of the operator A in the form (5.2). The mapping $A \to A(t,\mu)$ is a homomorphism of the algebra $\mathfrak{H}_m$ onto the algebra of symbols of all operators in $\mathfrak{H}_m$. The operator $A \in \mathfrak{H}_m$ is a Φ_+-operator or a Φ_--operator if and only if the condition*

$$\det A(t,\mu) \neq 0 \quad (t \in \Gamma, 0 \leq \mu \leq 1)$$

holds. If this condition is fulfilled, then the operator A is a Φ-operator and its index is calculated by the formula

$$\operatorname{ind} A = \frac{1}{2\pi}\left(\sum_{j=1}^{n}\{\arg\det b(t)a^{-1}(t)\}_{t\in\Gamma_j} - \{\arg\det A(t_0,\mu)\}_{\mu=0}^{1}\right),$$

where

$$a(t) = \sum_{j=1}^{k} a_{j1}a_{j2}\dots a_{jr}, \quad b(t) = \sum_{j=1}^{k} b_{j1}b_{j2}\dots b_{jr}.$$

The proof of this theorem in the general case ($n \geq 1$) is developed by the same scheme as it is developed in the case $n = 1$ (see [5, 6, 7]). Note that Theorem 5.1 is used substantially in the proof of this theorem.

6. Concluding remarks

6.1. In this section by Γ denote an oriented contour consisting of a finite number of closed and open piecewise Lyapunov curves having a finite number of intersection points. We can think of such a contour as consisting of a set of simple closed Lyapunov curves $\gamma_1, \dots, \gamma_n$ that do not have common points except for the endpoints of open arcs. The endpoints of one or several arcs are called nodes (see [1]). The points of the contour different from the nodes are called ordinary. We assume that none of two lines are tangential to each other at the nodes.

By Λ denote the set of the functions that are continuous at each ordinary point of the contour Γ and have finite (in general, distinct) limits as t tends to the nodes along each line.

Let t_j ($j = 1, \dots, q$) be some node joining r arcs $\ell_1^{(j)}, \dots, \ell_s^{(j)}, \ell_{s+1}^{(j)}, \dots, \ell_r^{(j)}$, the first s of which are the beginnings of open curves and the other $r - s$ are the ends of open curves. For each function $c(t)$, by $c_1^{(j)}, \dots, c_r^{(j)}$ denote the limits of the function $c(t)$ as t tends to t_j along the arcs $\ell_1^{(j)}, \dots, \ell_r^{(j)}$, respectively. If $\inf |c| > 0$, then by $c^{(j)}$ denote the number $c_1^{(j)} \dots c_s^{(j)} (c_{s+1}^{(j)} \dots c_r^{(j)})^{-1}$.

We will consider operators $aP + bQ$ in the space $L_p(\Gamma, \varrho)$ where $1 < p < \infty$;

$$\varrho(t) = \prod_{k=1}^{m} |t - t_k|^{\beta_k};$$

$t_1, \dots, t_m$ are pairwise distinct points of the contour Γ that include all the nodes; $\beta_1, \dots, \beta_m$ are real numbers satisfying the condition $-1 < \beta_k < p - 1$ ($k = 1, \dots, m$). To each node t_j assign the function $\xi_j(\mu)$ ($0 \leq \mu \leq 1$) defined by equality (4.0) with the corresponding number β_j in place of the number β.

Theorem 6.1. *Let $a(t), b(t) \in \Lambda$. The operator $A = aP + bQ$ is a Φ_+-operator or a Φ_--operator in $L_p(\Gamma, \varrho)$ if and only if the following conditions:*

$$\inf_{t\in\Gamma} |a(t)| > 0, \quad \inf_{t\in\Gamma} |b(t)| > 0, \tag{6.1}$$

$$\xi_j(\mu) c^{(j)} + 1 - \xi_j(\mu) \neq 0 \quad (j = 1, \dots, q;\ 0 \leq \mu \leq 1) \tag{6.2}$$

hold, where $c(t) = a(t)/b(t)$.

Suppose conditions (6.1)–(6.2) are fulfilled and

$$2\pi\kappa = \sum_{j=1}^{n} \{\arg c(t)\}_{t\in\gamma_j} + \sum_{j=1}^{q} \big\{ \arg(c^{(j)}\xi_j(\mu) + 1 - \xi_j(\mu)) \big\}_{\mu=0}^{1}.$$

Then for $\kappa \geq 0$ the operator A is left-invertible and $\dim \operatorname{coker} A = \kappa$*; for $\kappa \leq 0$ the operator A is right-invertible and* $\dim \ker A = -\kappa$.

The proof is carried out as follows. We first consider the case when the contour consists only of several open arcs having one common point t_0 and the operator A has the form $cP + Q$, where $c(t)$ is a function that is equal to 1 at all starting and terminating points of the open curves of the contour Γ different form t_0. The contour Γ is complemented to the contour $\widetilde{\Gamma}$ satisfying the conditions of Section 1 and the function $c(t)$ is extended to $\widetilde{\Gamma}$ by the equality $c(t) = 1$ for $t \in \widetilde{\Gamma} \setminus \Gamma$. After that, by the usual procedure (see [3]) Theorem 6.1 for the considered case is deduced from Theorem 4.1. In the general case the function $c(t)$ is represented as the product $c(t) = c_0(t)c_1(t) \cdots c_q(t)$, where the function $c_j(t)$ $(j = 1, \dots, q)$ is different from 1 only in some neighborhood of the node t_j that does not contain other nodes different from t_j and the function $c_0(t)$ is continuous on Γ. In the same way as in [3], one can show that $cP+Q = (c_0P+Q)(c_1P+Q)\dots(c_qP+Q)+T$. This allows us to reduce the problem in the general case to the case considered above.

6.2. As in Section 5, one can define the symbol of the operator $A = aP + bQ$ and the symbol of a more general operator

$$A = \sum_{k=1}^{r} A_{k1}A_{k2}\dots A_{kr}, \tag{6.3}$$

where $A_{kj} = a_{kj}P_m + b_{kj}Q_m$ are singular integral operators with matrix piecewise continuous coefficients acting in the space $L_p^m(\Gamma, \varrho)$. One can show that Theorem 5.2 is valid for the operators of the form (6.3). We will not provide details here.

References

[1] N.I. Mushelishvili, *Singular Integral Equations.*

1st Russian edition, OGIZ, Moscow, Leningrad, 1946, MR0020708 (8,586b).

English translation of 1st Russian edition: Noordhoff, Groningen, 1953, MR0058845 (15,434e), Zbl 0051.33203.

Reprinted by Wolters-Noordhoff Publishing, Groningen, 1972, MR0355494 (50 #7968), by Noordhoff International Publishing, Leyden, 1977, MR0438058 (55 #10978), by Dover Publications, 1992.

2nd Russian edition, revised, Fizmatgiz, Moscow, 1962. MR0193453 (33 #1673). Zbl 0103.07502.

German translation of 2nd Russian edition: *Singuläre Integralgleichungen.* Akademie-Verlag, Berlin, 1965. Zbl 0123.29701.

3rd Russian edition, corrected and augmented, Nauka, Moscow, 1968. MR0355495 (50 #7969), Zbl 0174.16202.

[2] E.G. Gordadze, *The Riemann-Privalov problem in the case of a non-smooth boundary curve.* Trudy Tbilis. Mat. Inst. Razmadze **33** (1967), 25–31 (in Russian). MR0243084 (39 #4408), Zbl 0205.09201.

[3] I.C. Gohberg and N.Ya. Krupnik, *On the spectrum of singular integral operators in L_p spaces with weight.* Dokl. Akad. Nauk SSSR **185** (1969), 745–748 (in Russian). English translation: Soviet Math. Dokl. **10** (1969), 406–410. MR0248565 (40 #1817), Zbl 0188.18301.

[4] I.C. Gohberg and N.Ya. Krupnik, *Systems of singular integral equations in L_p spaces with weight.* Dokl. Akad. Nauk SSSR **186** (1969) 998–1001 (in Russian). English translation: Soviet Math. Dokl. **10** (1969), 688–691. MR0248566 (40 #1818), Zbl 0188.18302.

[5] I.C. Gohberg and N.Ya. Krupnik, *Algebra generated by one-dimensional singular integral operators with piecewise continuous coefficients.* Funkcional. Anal. Prilozhen. **4** (1970), no. 3, 26–36 (in Russian). English translation: Funct. Anal. Appl. **4** (1970), no. 3, 193–201. MR0270164 (42 #5057), Zbl 0225.45005.

[6] I.C. Gohberg and N.Ya. Krupnik, *Singular integral equations with continuous coefficients on a composed contour.* Matem. Issled. **5** (1970), no. 2(16), 89–103 (in Russian). English translation: **this volume**. MR0447996 (56 #6306), Zbl 0223.45005.

[7] I.C. Gohberg and N.Ya. Krupnik, *Singular integral operators with piecewise continuous coefficients and their symbols.* Izv. Akad. Nauk SSSR Ser. Mat. **35** (1971), 940–964 (in Russian). English translation: Math. USSR Izvestiya **5** (1971), no. 4, 955–979. MR0291893 (45 #981), Zbl 0235.47025.

Operator Theory:
Advances and Applications, Vol. 206, 201–211

One-dimensional Singular Integral Operators with Shift

Israel Gohberg and Nahum Krupnik

Introduction

Let Γ be a closed or open oriented Lyapunov arc and $\omega(t)$ be a bijective mapping of Γ onto itself. An operator of the form

$$A = a(t)I + b(t)S + (c(t) + d(t)S)W \tag{1}$$

is usually called a *one-dimensional singular integral operator with shift* $\omega(t)$. Here $a(t)$, $b(t)$, $c(t)$, and $d(t)$ are bounded measurable functions on Γ, S is the operator of singular integration along Γ given by

$$(S_\Gamma\varphi)(t) = \frac{1}{\pi i}\int_\Gamma \frac{\varphi(\tau)}{\tau - t}\,d\tau \quad (t \in \Gamma),$$

and W is the shift operator defined by

$$(W\varphi)(t) = \varphi(\omega(t)).$$

Consider the simplest case of a shift, when $W^2 = I$, that is, the case when $\omega(\omega(t)) = t$. Besides that, we will assume that the function $\omega(t)$ has the derivative $\omega'(t)$ satisfying the Hölder condition with exponent α $(0 < \alpha < 1)$ and that $\omega(t) \not\equiv t$.

The operator A will be considered in the space $L_p(\Gamma, \varrho)$ with weight[1]

$$\varrho(t) = \prod_{k=1}^{n} |t - t_k|^{\beta_k},$$

where $t_k \in \Gamma$, $1 < p < \infty$, and $-1 < \beta_k < p - 1$. The operator A is a bounded linear operator in this space. Usually (see [1, 2, 3]), the operator A of the form (1)

The paper was originally published as И.Ц. Гохберг, Н.Я. Крупник, Об одномерных сингулярных интегральных операторах со сдвигом, Изв. Акад. Наук Армян. ССР, Сер. Матем. **8** (1973), № 1, 3–12. MR0333840 (48 #12162), Zbl 0255.47058.

[1] Note that obtained results remain true also for a wider class of spaces.

is considered simultaneously with the operator A_W defined in the space $L_p^2(\Gamma, \varrho) = L_p(\Gamma, \varrho) \times L_p(\Gamma, \varrho)$ by the matrix

$$A_W = \left\| \begin{array}{cc} a(t)I + b(t)S & c(t)I + d(t)S \\ W(c(t)I + d(t)S)W & W(a(t)I + d(t)S)W \end{array} \right\|.$$

It is easy to check the identities

$$W(a(t)I + b(t)S)W = a(\omega(t))I + \varepsilon b(\omega(t))S + T,$$

where T is a compact operator, and $\varepsilon = 1$ if the mapping $t = \omega(\tau)$ does not change the orientation of the contour Γ and $\varepsilon = -1$ otherwise. From here it follows that the operator A_W is equal modulo a compact operator to the operator $\widetilde{A}_W$ defined in the space $L_p^2(\Gamma, \varrho)$ by the equality

$$\widetilde{A}_W = \left\| \begin{array}{cc} a(t) & c(t) \\ c(\omega(t)) & a(\omega(t)) \end{array} \right\| + \left\| \begin{array}{cc} b(t) & d(t) \\ \varepsilon d(\omega(t)) & \varepsilon b(\omega(t)) \end{array} \right\| \left\| \begin{array}{cc} S & 0 \\ 0 & S \end{array} \right\|. \tag{2}$$

The operator $\widetilde{A}_W$ is a singular integral operator (without shift) with matrix coefficients. For such operators, necessary and sufficient conditions under which they are Φ-operators are known in the cases of continuous and piecewise continuous coefficients (see [4, 5, 6]).

In the papers [1, 2, 3] (see also [8, 7] and the references given in [1]), in particular, the following statement is obtained.

Theorem 1. *Let $a(t)$, $b(t)$, $c(t)$, and $d(t)$ be continuous functions. The operator A defined by equality* (1) *is a Φ-operator in the space $L_p(\Gamma, \varrho)$ if and only if the operator $\widetilde{A}_W$ is a Φ-operator in the the space $L_p^2(\Gamma, \varrho)$. If A is a Φ-operator, then*

$$\operatorname{Ind} A = \frac{1}{2} \operatorname{Ind} \widetilde{A}_W.$$

The formulated theorem remains true for arbitrary bounded measurable coefficients $a(t)$, $b(t)$, $c(t)$, and $d(t)$ in the case when the mapping $\omega : \Gamma \to \Gamma$ does not change the orientation of the contour. The proof of this statement follows from the following three facts.

1) The identity[2]

$$\begin{aligned} &\left\| \begin{array}{cc} I & W \\ I & -W \end{array} \right\| \left\| \begin{array}{cc} X & Y \\ WYW & WXW \end{array} \right\| \left\| \begin{array}{cc} I & I \\ W & -W \end{array} \right\| \\ &= 2 \left\| \begin{array}{cc} X + YW & 0 \\ 0 & X - YW \end{array} \right\|, \end{aligned} \tag{3}$$

holds for arbitrary bounded linear operators X, Y and an involution W acting on some Banach space.

[2]Note that this identity allows us to obtain explicit formulas for the inversion of singular integral operators with shift in a series of cases.

2) The function $\omega(t) - t$ is different from zero everywhere on Γ.
3) For the operators $X = aI + bS$, $Y = cI + dS$, and $M = (\omega(t) - t)I$ the identity

$$(X - YW)M = M(X + YW) + T$$

holds, where T is a compact operator.

From assertions 1)–3) it follows also that the operator A is a Φ_+-operator (or a Φ_--operator) in the space $L_p(\Gamma, \varrho)$ if and only if the operator $\widetilde{A}_W$ is a Φ_+-operator (resp. Φ_--operator) in the space $L_p^2(\Gamma, \varrho)$.

Note that all these results remain also valid in the case when the coefficients $a(t)$, $b(t)$, $c(t)$, and $d(t)$ are matrices whose entries are arbitrary bounded measurable functions.

In the case when the function $\omega : \Gamma \to \Gamma$ changes the orientation of the contour Γ and the coefficients $a(t)$, $b(t)$, $c(t)$, and $d(t)$ are continuous, Theorem 1 follows from identity (3) and the compactness of the operator $(X - YW)N - N(X + YW)$, where $N = (\omega(t) - t)I + \lambda S$ and λ is an arbitrary complex number.

From identity (3) it follows also that if the operator $\widetilde{A}_W$ is a Φ-operator (resp. Φ_+-operator, Φ_--operator), then the operator A (with arbitrary bounded measurable coefficients) is a Φ-operator (resp. Φ_+-operator, Φ_--operator), too. However, the converse statement is not true anymore. In Section 3, an example of a Φ-operator A of the form (1) with piecewise continuous continuous coefficients is given, for which the operator $\widetilde{A}_W$ is not a Φ-operator.

In the present paper, one model class of singular integral operators with shift is investigated in detail in the case when the shift changes the orientation and the coefficients have finite limits from the left and from the right at each point. Conditions guaranteeing that such operators are Φ-operators are obtained. The algebra generated by such operators is studied. Formulas for the symbol and the index are obtained.

Generalizations of the results of this paper to more general classes of singular integral equations and wider classes of shifts will be given elsewhere.

1. Auxiliary statements

By $L(B)$ denote the Banach algebra of all bounded linear operators acting in a Banach space B and by $J(B)$ denote the two-sided ideal of the algebra $L(B)$ that consists of all compact operators.

Let us agree also on the following notation: $L_{p,\beta} = L_p([0,1], \varrho)$, where $\varrho(t) = t^\beta$, $-1 < \beta < p-1$, $1 < p < \infty$; ΠC $(= \Pi C(a,b))$ is the set of all functions defined on the segment $[a,b]$ that have finite limits from the left and from the right at each interior point and continuous at the endpoints a and b.

In what follows, $L^n_{p,\beta}$ $(= L^n_{p,\beta}(0,1))$ will denote the direct sum of n spaces $L_{p,\beta}$, and $\Pi C^{(n)}(a,b)$ will denote the algebra of all square matrices of order n with entries in $\Pi C(a,b)$.

By $\Sigma^{(n)}_{p,\beta}$ $(= \Sigma^{(n)}_{p,\beta}(0,1))$ denote the smallest subalgebra of the Banach algebra $L(L^n_{p,\beta})$ that contains all operators of the form $aI + bS$, where $a, b \in \Pi C^{(n)}(0,1)$ and S is the operator of singular integration on the segment $[0,1]$ in $L^n_{p,\beta}$.

As is shown in [9], the algebra $\Sigma^{(n)}_{p,\beta}$ is homomorphic to some algebra of matrices of order $2n$ whose entries are bounded functions on the square $[0,1]\times[0,1]$. Let us denote this homomorphism by π. By $\mathbf{A}_{p,\beta}(t,\mu)$ $(0 \le t, \mu \le 1)$ denote the matrix πA, where $A \in \Sigma^{(n)}_{p,\beta}$. The matrix function $\mathbf{A}_{p,\beta}(t,\mu)$ is called the symbol of the operator A in the space $L^n_{p,\beta}$. Besides that, following [9], we will write the matrix $\mathbf{A}_{p,\beta}(t,\mu)$ in the form of the block matrix of order two

$$\mathbf{A}_{p,\beta}(t,\mu) = \|a^{p,\beta}_{jk}(t,\mu)\|^2_{j,k=1}.$$

In [9, Theorem 5.1] it is proved that an operator $A \in \Sigma^{(n)}_{p,\beta}$ is a Φ_+-operator or a Φ_--operator in the space $L^n_{p,\beta}$ if and only if the function $\det \mathbf{A}_{p,\beta}(t,\mu)$ is different from zero on the square $[0,1]\times[0,1]$. If this condition is satisfied, then the operator A is a Φ-operator and its index is calculated by the formula[3]

$$\operatorname{Ind} A = -\frac{1}{2\pi}\left\{\arg \frac{\det \mathbf{A}_{p,\beta}(t,\mu)}{\det a^{p,\beta}_{22}(t,0)a^{p,\beta}_{22}(t,1)}\right\}_{0\le t,\mu\le 1}.$$

Besides that, in [9] it is shown that $J(L^n_{p,\beta}) \subset \Sigma^{(n)}_{p,\beta}$ and that if an operator $A \in \Sigma^{(n)}_{p,\beta}$ admits a regularization, then its regularizer also belongs to the algebra $\Sigma^{(n)}_{p,\beta}$. In particular, if an operator $A \in \Sigma^{(n)}_{p,\beta}$ is invertible, then $A^{-1} \in \Sigma^{(n)}_{p,\beta}$.

The symbol $\mathbf{A}_{p,\beta}(t,\mu)$ of an operator of the form $A = a(t)I + b(t)S$ is defined at the points (t,μ) $(0 < t < 1,\ 0 \le \mu \le 1)$ by the equality

$$\mathbf{A}_{p,\beta}(t,\mu) = \left\| \begin{matrix} \xi(\mu)x(t+0) + (1-\xi(\mu))x(t-0) & h(\mu)(y(t+0)-y(t-0)) \\ h(\mu)(x(t+0)-x(t-0)) & \xi(\mu)y(t-0) + (1-\xi(\mu))y(t+0) \end{matrix} \right\|, \tag{4}$$

where $x(t) = a(t) + b(t)$, $y(t) = a(t) - b(t)$, $\theta = \pi - 2\pi/p$,

$$\xi(\mu) = \begin{cases} \dfrac{\sin(\theta\mu)\exp(i\theta\mu)}{\sin\theta\exp(i\theta)} & \text{for} \quad \theta \ne 0, \\ \mu & \text{for} \quad \theta = 0, \end{cases} \tag{5}$$

and $h(\mu)$ is a fixed branch of the root $\sqrt{\xi(\mu)(1-\xi(\mu))}$.

The symbol is defined at the points $(0,\mu)$ and $(1,\mu)$ $(0 \le \mu \le 1)$ by

$$\mathbf{A}_{p,\beta}(0,\mu) = \left\| \begin{matrix} a(0) + (2\xi_\beta(\mu) - 1)b(0) & 0 \\ 0 & a(0) - b(0) \end{matrix} \right\|$$

[3]For the explanation of this formula, see [9], pp. 972–973 or p. 957 of the Russian original.

and

$$\mathbf{A}_{p,\beta}(1,\mu) = \left\| \begin{array}{cc} a(1) - (2\xi(\mu)-1)b(1) & 0 \\ 0 & a(1)-b(1) \end{array} \right\|,$$

where the function $\xi_\beta(\mu)$ is obtained from the function $\xi(\mu)$ by replacing the number θ in the right-hand side of (5) by $\pi - 2\pi(1+\beta)/p$.

Consider the operator R defined by the equality

$$(R\varphi)(t) = \frac{1}{\pi i}\int_0^1 \sqrt{\frac{t}{\tau}}\frac{\varphi(\tau)}{\tau - t}\,d\tau \quad (0 \le t \le 1). \tag{6}$$

From [10, Theorem 2.2] it follows that the operator R is bounded in the space $L_{p,\beta}$ if and only if the condition $2(1+\beta) < p$ is satisfied.

We will need the following statement.

Lemma 1. *Let numbers β and p $(-1 < \beta < p-1,\ 1 < p < \infty)$ satisfy the condition $2(1+\beta) < p$. Then the operator R belongs to the algebra $\Sigma^{(1)}_{p,\beta}$, and its symbol has the form*

$$\mathbf{R}_{p,\beta}(t,\mu) = \left\| \begin{array}{cc} r_{11}^{p,\beta}(t,\mu) & 0 \\ 0 & -1 \end{array} \right\|, \tag{7}$$

where

$$r_{11}^{p,\beta}(t,\mu) = \begin{cases} (2\xi_\beta(\mu)-1)^{-1} & \text{for } t = 0, \\ 1 & \text{for } 0 < t < 1, \\ 2\xi(\mu) - 1 & \text{for } t = 1. \end{cases}$$

Proof. Consider the operator

$$B = aI + SbI$$

in the space $L_{p,\beta}$, where $a(t) = \sin(\pi t/2)$ and $b(t) = i\cos(\pi t/2)$. From the definition of the symbol it follows straightforwardly that

$$\det \mathbf{B}_{p,\beta}(t,\mu) = \begin{cases} 1 & \text{for } 0 < t \le 1, \\ 2\xi_\beta(\mu) - 1 & \text{for } t = 0. \end{cases}$$

Since $2(1+\beta) < p$, we have $\pi - 2\pi(1+\beta)/p \neq 0$, whence $2\xi_\beta(\mu) - 1 \neq 0$. Thus, the operator B is a Φ-operator in the space $L_{p,\beta}$. Let us show that the index of B is equal to zero. Put

$$f_{p,\beta}(t,\mu) = \frac{\det \mathbf{B}_{p,\beta}(t,\mu)}{b_{22}(t,0)b_{22}(t,1)}.$$

It is not difficult to see that

$$f_{p,\beta}(t,\mu) = \begin{cases} 1 - 2\xi_\beta(\mu) & \text{for } t = 0, \\ -e^{-i\pi t} & \text{for } 0 < t \le 1. \end{cases}$$

The range of the function $f_{p,\beta}(t,\mu)$ consists of two circular arcs joining the points -1 and 1. Both these arcs are located in the upper half-plane. Therefore,

$$\{\arg f_{p,\beta}(t,\mu)\}_{0 \le t,\mu \le 1} = 0.$$

From formula (3) it follows that $\operatorname{Ind} B = 0$.

As is known [5, Theorem 1], in this case the operator B is invertible in $L_{p,\beta}$.

To find the inverse operator, we use formula (98,11) from Mushelishvili's monograph [11, 2nd Russian edition]. From this formula it follows that[4]

$$B^{-1} = aI - Z^{-1}SbZ,$$

where $Z = g(t)\sqrt{t}I$ and, moreover, the functions $g(t)$ and $1/g(t)$ are continuous on $[0,1]$. Since $B \in \Sigma^{(1)}_{p,\beta}$, we have $B^{-1} \in \Sigma^{(1)}_{p,\beta}$. In view of the equality

$$B^{-1} = aI + g^{-1}RbgI,$$

the operator $R_1 = RbI$ belongs to the algebra $\Sigma^{(1)}_{p,\beta}$.

Consider the operator $R_2 = RbI$, where $c(t) = \sqrt{t}$. Since the operator

$$R_2 - ScI = cS - ScI$$

is compact in $L_{p,\beta}$, we have $R_2 \in \Sigma^{(1)}_{p,\beta}$. Thus, $B(b+c)I \in \Sigma^{(1)}_{p,\beta}$. Since the function $b(t)+c(t)$ is not equal to zero on the segment $[0,1]$, we conclude that $B \in \Sigma^{(1)}_{p,\beta}$. Let $\mathbf{S}_{p,\beta}(t,\mu)$ and $\mathbf{C}_{p,\beta}(t,\mu)$ be the symbols of the operators S and $C = c(t)I$, respectively. Since the operator $(R-S)cI$ is compact, we have

$$(\mathbf{R}_{p,\beta}(t,\mu) - \mathbf{S}_{p,\beta}(t,\mu))\mathbf{C}_{p,\beta}(t,\mu) \equiv 0.$$

Since

$$\mathbf{C}_{p,\beta}(t,\mu) = \left\| \begin{matrix} \sqrt{t} & 0 \\ 0 & \sqrt{t} \end{matrix} \right\|,$$

we see that for all $t \neq 0$ the equality $\mathbf{R}_{p,\beta}(t,\mu) = \mathbf{S}_{p,\beta}(t,\mu)$ holds. From the equality

$$(aI + bS)(aI - g^{-1}RbgI) = I$$

it follows that the product $\mathbf{S}_{p,\beta}(0,\mu)\mathbf{R}_{p,\beta}(0,\mu)$ is the identity matrix. Therefore $\mathbf{R}_{p,\beta}(0,\mu) = \mathbf{S}^{-1}_{p,\beta}(0,\mu)$. Thus,

$$\mathbf{R}_{p,\beta}(t,\mu) = \begin{cases} \mathbf{S}_{p,\beta}(t,\mu) & \text{for} \quad t \neq 0, \\ \mathbf{S}^{-1}_{p,\beta}(0,\mu) & \text{for} \quad t = 0, \end{cases}$$

and this implies (7). The lemma is proved. □

2. Main statement

Let us denote by $\Sigma^{(n)}_p(-1,1;W)$ the smallest subalgebra of the Banach algebra $L(L^n_p(-1,1))$ $(1 < p < \infty)$ that contains all operators of the form

$$A = aI + bS + (cI + dS)W, \tag{8}$$

[4]From [11] it follows that the operator $C = aI + Z^{-1}SbZ$ has the property $CB\varphi = BC\varphi$ for all functions $\varphi(t)$ satisfying the Hölder condition on $[0,1]$. Since, besides that, the operator C is bounded in $L_{p,\beta}$, we have $B^{-1} = C$.

where the coefficients $a, b, c, d \in \Pi C^{(n)}(-1,1)$, the shift operator W is defined by the equality $(W\varphi)(t) = \varphi(-t)$, and S is the operator of singular integration on the segment $[-1,1]$.

Let $\sigma : L_p^n(-1,1) \to L_p^{2n}(0,1)$ be the mapping defined by the equality $(\sigma\varphi)(t) = (\varphi(t), \varphi(-t))$ $(0 \leq t \leq 1)$. Then for every operator $X \in L(L_p^n(-1,1))$ the operator $\sigma X \sigma^{-1}$ can be represented as the matrix

$$\sigma X \sigma^{-1} = \left\| \begin{matrix} X_{11} & X_{12} \\ X_{21} & X_{22} \end{matrix} \right\|$$

with entries $X_{jk} \in L(L_p^n(0,1))$ $(j,k = 1,2)$. In particular, for the operator A of the form (8) we have

$$\sigma A \sigma^{-1} = gI + hS_0 + fM_0, \tag{9}$$

where

$$g(t) = \left\| \begin{matrix} a(t) & c(t) \\ c(-t) & d(-t) \end{matrix} \right\|, \quad h(t) = \left\| \begin{matrix} b(t) & d(t) \\ -d(-t) & -b(-t) \end{matrix} \right\|,$$

$$f(t) = \left\| \begin{matrix} -d(t) & -b(t) \\ b(-t) & d(-t) \end{matrix} \right\|$$

and

$$(S_0\varphi)(t) = \frac{1}{\pi i} \int_0^1 \frac{\varphi(\tau)}{\tau - t}\, d\tau, \quad (M_0\varphi)(t) = \frac{1}{\pi i} \int_0^1 \frac{\varphi(\tau)}{\tau + t}\, d\tau \quad (0 \leq t \leq 1)$$

are operators acting in the space $L_p^{2n}(0,1)$.

Consider the operator ν defined by the equality $(\nu\varphi)(t) = \varphi(t^{1/2})$. Obviously, the operator ν is a bounded linear and invertible operator acting from the space $L_p^{2n}(0,1)$ to $L_{p,-1/2}^{2n}(0,1)$. It is easy to see that $\nu a(t) \nu^{-1} = a(t^{1/2})I$,

$$\nu S_0 \nu^{-1} = \frac{1}{2}(S_0 + M_1), \quad \nu M_0 \nu^{-1} = \frac{1}{2}(S_0 - M_1),$$

where the operator M_1 is defined in the space $L_{p,-1/2}^{2n}(0,1)$ by the equality

$$(M_1\varphi)(t) = \frac{1}{\pi i} \int_0^1 \sqrt{\frac{t}{\tau}} \frac{\varphi(\tau)}{\tau - t}\, d\tau \quad (0 \leq t \leq 1).$$

According to Lemma 1, the operator M_1 belongs to the algebra $\Sigma_{p,-1/2}^{(2n)}(0,1)$. Therefore, we have the embedding

$$\nu\sigma\Sigma_p^{(n)}(-1,1;W)\sigma^{-1}\nu^{-1} \subset \Sigma_{p,-1/2}^{(2n)}(0,1).$$

Let X be an operator in the algebra $\Sigma_p^{(n)}(-1,1;W)$. By $\mathbf{X}_p(t,\mu)$ $(0 \leq t, \mu \leq 1)$ denote the symbol of the operator $\nu\sigma X \sigma^{-1}\nu^{-1} \in \Sigma_{p,-1/2}^{(2n)}(0,1)$. We will say that the matrix function $\mathbf{X}_p(t,\mu)$ $(0 \leq t, \mu \leq 1)$ of order $4n$ is the symbol of the operator X. From the obtained rules and Lemma 1 in particular it follows that the symbol of the operator (8) belonging to the algebra $\Sigma_p^{(n)}(-1,1;W)$ is defined by the equality

$$\mathbf{A}_p(t^2,\mu) = \left\| \begin{matrix} \xi(\mu)x(t+0)+(1-\xi(\mu))x(t-0) & h(\mu)(y(t+0)-y(t-0)) \\ h(\mu)(x(t+0)-x(t-0)) & \xi(\mu)y(t-0)+(1-\xi(\mu))y(t+0) \end{matrix} \right\|$$

in the interval $0<t<1$, where

$$x(t) = \left\| \begin{matrix} a(t)+b(t) & c(t)+d(t) \\ c(-t)-d(-t) & a(-t)-b(-t) \end{matrix} \right\|,$$

$$y(t) = \left\| \begin{matrix} a(t)-b(t) & c(t)-d(t) \\ c(-t)+d(-t) & a(-t)+b(-t) \end{matrix} \right\|,$$

and $\xi(\mu)$, $h(\mu)$ are functions defined in Section 1.

For $t=1$,

$$\mathbf{A}_p(1,\mu) = \left\| \begin{matrix} \alpha(\mu) & 0 \\ 0 & \beta(\mu) \end{matrix} \right\|,$$

where

$$\alpha(\mu) = \left\| \begin{matrix} a(1) & c(1) \\ c(-1) & a(-1) \end{matrix} \right\| + (2\xi(\mu)-1) \left\| \begin{matrix} b(1) & d(1) \\ -d(-1) & -b(-1) \end{matrix} \right\|,$$

$$\beta(\mu) = \left\| \begin{matrix} a(1)-b(1) & c(1)-d(1) \\ c(-1)+d(-1) & a(-1)+b(-1) \end{matrix} \right\|$$

and for $t=0$,

$$\mathbf{A}_p(0,\mu) = \left\| \begin{matrix} \gamma(\mu) & 0 \\ 0 & \delta(\mu) \end{matrix} \right\|,$$

where

$$\gamma(\mu) = \left\| \begin{matrix} a(+0) & c(+0) \\ c(-0) & a(-0) \end{matrix} \right\| + \frac{\eta(\mu)}{2} \left\| \begin{matrix} b(+0)-d(+0) & d(+0)-b(+0) \\ b(-0)-d(-0) & d(-0)-b(-0) \end{matrix} \right\|$$

$$+ \frac{1}{2\eta(\mu)} \left\| \begin{matrix} b(+0)+d(+0) & b(+0)+d(+0) \\ -b(-0)-d(-0) & -b(-0)-d(-0) \end{matrix} \right\|,$$

$\eta(\mu) = 2\xi_{-1/2}(\mu)-1$, and

$$\delta(\mu) = \left\| \begin{matrix} a(+0)-b(+0) & c(+0)-d(+0) \\ c(-0)+d(-0) & a(-0)+b(-0) \end{matrix} \right\|.$$

We will write the symbol $\mathbf{X}_p(t,\mu)$ of an operator $X \in \Sigma_p^{(n)}(-1,1;W)$ in the form $\|x_{jk}(t,\mu)\|_{j,k=1}^2$, where $x_{jk}(t,\mu)$ are matrix functions of order $2n$.

In view of a property of symbols of operators in the algebra $\Sigma_{p,\beta}^{(2n)}(0,1)$ (see [8]) and the arguments presented above, we have the following.

Theorem 2. *An operator $A \in \Sigma_p^{(n)}(-1,1;W)$ is a Φ_+-operator or a Φ_--operator in $L_p^n(-1,1)$ if and only if the condition*

$$\det \mathbf{A}_p(t,\mu) \neq 0 \quad (0 \le t \le 1,\ 0 \le \mu \le 1)$$

holds. If this condition is fulfilled, then the operator A is a Φ-operator and

$$\operatorname{Ind} A = -\frac{1}{2\pi}\left\{\arg \frac{\det \mathbf{A}_p(t,\mu)}{\det a_{22}(t,0)a_{22}(t,1)}\right\}_{0\le t\le 1, 0\le\mu\le 1}.$$

3. Example

In this section an example of a singular integral operator with a shift (changing the orientation of the contour) is presented. This example shows that if coefficients of the operator have discontinuity points, then Theorem 1 does not hold.

Let $A = I + \alpha\chi(t)SW$, where $\alpha = \text{const}$, $\chi(t)$ $(-1 \le t \le 1)$ be the characteristic function of the segment $[0,1]$, S be the operator of singular integration along the segment $[-1,1]$, and W be the shift operator given by $(W\varphi)(t) = \varphi(-t)$. The operator A belongs to the algebra $\Sigma_p^{(1)}(-1,1)$. Its symbol in the space $L_2(-1,1)$ has the form

$$\mathbf{A}_2(t,\mu) = \begin{Vmatrix} a_{11}(t,\mu) & 0 \\ 0 & a_{22}(t,\mu) \end{Vmatrix},$$

where

$$a_{22}(t,\mu) = \begin{Vmatrix} 1 & -\alpha \\ 0 & 1 \end{Vmatrix}$$

and

$$a_{11}(t,\mu) = \begin{cases} \begin{Vmatrix} 1 & \alpha \\ 0 & 1 \end{Vmatrix} & \text{for} \quad 0 < t < 1, \\ \begin{Vmatrix} 1 & \alpha(2\mu-1) \\ 0 & 1 \end{Vmatrix} & \text{for} \quad t = 1, \\ \begin{Vmatrix} 1 + i\alpha\sin\pi\mu & -\alpha\cos\pi\mu \\ 0 & 1 \end{Vmatrix} & \text{for} \quad t = 0. \end{cases}$$

Therefore, $\det \mathbf{A}_2(t,\mu) = 1$ for $t \neq 0$ and $\det \mathbf{A}_2(0,\mu) = 1 + i\alpha \sin \pi\mu$.

If we put $\alpha = -i$, then $\det \mathbf{A}_2(t,\mu) \neq 0$ $(0 \le t, \mu \le 1)$. Hence the operator $I - i\chi(t)SW$ is a Φ-operator in $L_2(-1,1)$. On the other hand, if we put $\alpha = i$, then $\det \mathbf{A}_2(0,1/2) = 0$. Therefore, the operator $I + i\chi(t)SW$ is not a Φ-operator in the space $L_2(-1,1)$.

Let $B = I - i\chi(t)SW$ and $C = I + i\chi(t)SW$. Since B is a Φ-operator in $L_2(-1,1)$ and C is not a Φ-operator in this space, we have in view of equality (3) that the operator B_W (as well as the operator $\widetilde{B}_W$) is not a Φ-operator in $L_2^2(-1,1)$. Thus, for the operator B Theorem 1 is not true.

Note that in this example the operators B_W and $\widetilde{B}_W$ coincide. Notice also that it is possible to construct examples of singular integral operators with shift (changing the orientation of the contour) on a closed contour with piecewise continuous coefficients, for which Theorem 1 does not hold.

References

[1] G.S. Litvinchuk, *Noether thoery of a system of singular integral equations with Carleman translation and complex conjugate unknowns.* Izv. Akad. Nauk SSSR, Ser. Mat. **31** (1967), 563–586 (in Russian). English translation: Math. USSR Izvestiya **1** (1967), no. 3, 545–567. MR0213836 (35 #4693), Zbl 0154.13601.

G. S Litvinchuk, *Correction to my paper "Noether thoery of a system of singular integral equations with Carleman translation and complex conjugate unknowns".* Izv. Akad. Nauk SSSR, Ser. Mat. **32** (1968), 1414–1417 (in Russian). English translation: Math. USSR Izvestiya **2** (1968), no. 6, 1361–1365. MR0236636 (38 #4931), Zbl 0169.45203.

[2] E.I. Zverovich and G.S. Litvinchuk, *Boundary-value problems with a shift for analytic functions and singular functional equations.* Uspehi Mat. Nauk **23** (1968), no. 3(141), 67–121 (in Russian). English translation: Russian Math. Surv. **23** (1968), no. 3, 67–124. MR0229839 (37 #5405), Zbl 0177.10802.

[3] A.B. Antonevich, *On the index of a pseudodifferential operator with a finite group of shifts.* Dokl. Akad. Nauk SSSR **190** (1970), 751–752 (in Russian). English translation: Soviet Math. Dokl. **11** (1970), 168–170. MR0266246 (42 #1153), Zbl 0198.48003.

[4] I.C. Gohberg, *A factorization problem in normed rings, functions of isometric and symmetric operators and singular integral equations.* Uspehi Mat. Nauk **19** (1964), no. 1(115), 71–124 (in Russian). English translation: Russ. Math. Surv. **19** (1964) 63–114. MR0163184 (29 #487), Zbl 0124.07103.

[5] I.C. Gohberg and N.Ya. Krupnik, *On the spectrum of singular integral operators in L_p spaces with weight.* Dokl. Akad. Nauk SSSR **185** (1969), 745–748 (in Russian). English translation: Soviet Math. Dokl. **10** (1969), 406–410. MR0248565 (40 #1817), Zbl 0188.18301.

[6] I.C. Gohberg and N.Ya. Krupnik, *Systems of singular integral equations in L_p spaces with weight.* Dokl. Akad. Nauk SSSR **186** (1969) 998–1001 (in Russian). English translation: Soviet Math. Dokl. **10** (1969), 688–691. MR0248566 (40 #1818), Zbl 0188.18302.

[7] N.K. Karapetiants and S.G. Samko, *Singular integral operators with shift on an open contour.* Dokl. Akad. Nauk SSSR **204** (1972), 536–539 (in Russian). English translation: Soviet Math. Dokl. **13** (1972), 691–696. MR0308700 (46 #7814), Zbl 0286.45004.

[8] N.K. Karapetiants and S.G. Samko, *On a new approach to the investigation of singular integral equations with shift.* Dokl. Akad. Nauk SSSR **202** (1972), 273–276 (in Russian). English translation: Soviet Math. Dokl. **13** (1972), 79–83. MR0300159 (45 #9207), Zbl 0243.45004.

[9] I.C. Gohberg and N.Ya. Krupnik, *Singular integral operators with piecewise continuous coefficients and their symbols.* Izv. Akad. Nauk SSSR Ser. Mat. **35** (1971), 940–964 (in Russian). English translation: Math. USSR Izvestiya **5** (1971), no. 4, 955–979. MR0291893 (45 #981), Zbl 0235.47025.

[10] I.C. Gohberg and N.Ya. Krupnik, *Singular integral equations with unbounded coefficients.* Matem. Issled. **5** (1970), no. 3(17), 46–57 (in Russian). English translation: **this volume**. MR0291892 (45 #980), Zbl 0228.47037.

[11] N.I. Mushelishvili, *Singular Integral Equations.*

1st Russian edition, OGIZ, Moscow, Leningrad, 1946, MR0020708 (8,586b).

English translation of 1st Russian edition: Noordhoff, Groningen, 1953, MR0058845 (15,434e), Zbl 0051.33203.

Reprinted by Wolters-Noordhoff Publishing, Groningen, 1972, MR0355494 (50 #7968), by Noordhoff International Publishing, Leyden, 1977, MR0438058 (55 #10978), by Dover Publications, 1992.

2nd Russian edition, revised, Fizmatgiz, Moscow, 1962. MR0193453 (33 #1673), Zbl 0103.07502.

German translation of 2nd Russian edition: *Singuläre Integralgleichungen.* Akademie-Verlag, Berlin, 1965. Zbl 0123.29701.

3rd Russian edition, corrected and augmented, Nauka, Moscow, 1968. MR0355495 (50 #7969), Zbl 0174.16202.

Operator Theory:
Advances and Applications, Vol. 206, 213–217

Algebras of Singular Integral Operators with Shift

Israel Gohberg and Nahum Krupnik

In this note the results of [1] are generalized to the case of an arbitrary simple closed Lyapunov contour and an arbitrary Carleman shift changing the orientation of the contour.

1. Let $\mathcal{L}_1$ and $\mathcal{L}_2$ be Banach spaces and $L(\mathcal{L}_1, \mathcal{L}_2)$ be the Banach space of all bounded linear operators acting from $\mathcal{L}_1$ to $\mathcal{L}_2$. The algebra $L(\mathcal{L}, \mathcal{L})$ is denoted by $L(\mathcal{L})$. We will say that two algebras $\mathcal{A}_1 \subset L(\mathcal{L}_1)$ and $\mathcal{A}_2 \subset L(\mathcal{L}_2)$ are *equivalent* if there exists an invertible operator $M \in L(\mathcal{L}_1, \mathcal{L}_2)$ such that the set of operators of the form MAM^{-1}, where $A \in \mathcal{A}_1$, coincides with the algebra $\mathcal{A}_2$.

Let Γ be a contour in the complex plane. Let us introduce the following notation: S_Γ is the operator of singular integration along Γ given by

$$(S_\Gamma \varphi)(t) = \frac{1}{\pi i} \int_\Gamma \frac{\varphi(\tau)}{\tau - t} d\tau \quad (t \in \Gamma);$$

$PC(\Gamma)$ is the set of all functions continuous on Γ except for at most a finite set of points where they have discontinuities of the first kind; $L_p(\Gamma, \varrho)$ is the space L_p on Γ with a weight ϱ; $\Sigma(p, \Gamma, \varrho)$ is the smallest Banach subalgebra of the algebra $L(L_p(\Gamma, \varrho))$ that contains all operators aI ($a \in PC(\Gamma)$) and the operator S_Γ; $\Sigma(p, \Gamma, \varrho; B)$ is the smallest Banach subalgebra of the algebra $L(L_p(\Gamma, \varrho))$ that contains the algebra $\Sigma(p, \Gamma, \varrho)$ and the operator B. Let Σ be some subalgebra of the algebra $L(\mathcal{L})$. Denote by Σ_n the subalgebra of the algebra $L(\mathcal{L}^n)$ consisting of all operators of the form $\|A_{jk}\|_{j,k=1}^n$, where $A_{jk} \in \Sigma$.

In the following we will suppose that Γ is a simple closed oriented Lyapunov contour and $\nu : \Gamma \to \Gamma$ is a mapping changing the orientation of Γ and having the following properties:

$$\nu'(t) \in H(\Gamma), \quad \nu(\nu(t)) \equiv t \quad (t \in \Gamma), \tag{1}$$

The paper was originally published as И.Ц. Гохберг, Н.Я. Крупник, Об алгебрах сингулярных интегральных операторов со сдвигом, Матем. Исслед. **8** (1973), № 2(28), 170–175. MR0341191 (49 #5941), Zbl 0306.47022.

where $H(\Gamma)$ is the set of functions satisfying the Hölder condition on Γ. By V we denote the shift operator defined by

$$(V\varphi)(t) = \varphi(\nu(t)). \tag{2}$$

We will suppose[1] that $\varrho(t) = |t - t_0|^\alpha |t - \tau_0|^\beta$, where t_0, τ_0 are fixed points of the mapping ν and α, β are real numbers satisfying $-1 < \alpha < 1$, $-1 < \beta < 1$. Under these assumptions the following result holds.

Theorem 1. *The algebra $\Sigma_n(p, \Gamma, \varrho; V)$ is equivalent to a subalgebra of the algebra $\Sigma_{2n}(p, \Gamma_0^+, \varrho_0)$, where*

$$\Gamma_0^+ = \{\zeta : |\zeta| = 1, \operatorname{Im} z \geq 0\}, \quad \varrho_0(\zeta) = |\zeta - 1|^{(\beta-1)/2} |\zeta + 1|^{(p-1+\alpha)/2},$$

and $1 < p < \infty$.

The proof of this theorem is based on the following lemma.

Lemma 1. *Let ν be a mapping satisfying conditions (1) and let $\Gamma_0 = \{\zeta : |\zeta| = 1\}$ be the unit circle. Then there exists a mapping $\gamma : \Gamma_0 \to \Gamma$ such that*

$$\gamma'(\zeta) \in H(\Gamma_0), \quad \gamma'(\zeta) \neq 0 \quad (\zeta \in \Gamma_0) \tag{3}$$

and

$$(\gamma^{-1} \circ \nu \circ \gamma)(\zeta) = \zeta^{-1}, \tag{4}$$

where $\nu \circ \gamma$ denotes the composition of the mappings γ and ν.

Proof. It is not difficult to see that the mapping ν has exactly two fixed points on Γ. We denote them by t_0 and τ_0. Since Γ is a Lyapunov contour, there exists a mapping $\eta : \Gamma_0 \to \Gamma$ such that $\eta'(\zeta) \in H(\Gamma_0)$ and $\eta'(\zeta) \neq 0$ for $\zeta \in \Gamma_0$. Besides, we can choose η in such a way that $\eta(1) = t_0$. Put $\lambda = \eta^{-1} \circ \nu \circ \eta$. It is easy to see that the mapping $\eta : \Gamma_0 \to \Gamma_0$ changes the orientation of the unit circle Γ_0. Moreover, $\lambda(\lambda(\zeta)) = \zeta$ for $|\zeta| = 1$, $\lambda(1) = 1$, and $\lambda(\zeta_0) = \zeta_0$, where $\zeta_0 = \eta^{-1}(t_0)$. The points 1 and ζ_0 divide the circle Γ_0 into the arcs Γ_1 and Γ_2. Moreover, $\lambda : \Gamma_1 \to \Gamma_2$ and $\lambda : \Gamma_2 \to \Gamma_1$. Let $\mu : \Gamma_1 \to \Gamma_2$ be a mapping satisfying

$$\mu(\zeta_0) = -1, \quad \mu(1) = 1, \quad \mu(\zeta) \neq 0 \quad (\zeta \in \Gamma_1), \quad \mu'(\zeta) \in H(\Gamma_1).$$

We extend μ to the whole contour Γ by setting $\mu(\zeta) = 1/\mu(\lambda(\zeta))$ for $\zeta \in \Gamma_2$. It is easy to see that for all $\zeta \in \Gamma$ one has $\mu(\zeta) = 1/\mu(\lambda(\zeta))$.

Let

$$\mu'(\zeta_0 + 0) = \lim_{\zeta \in \Gamma_1, \zeta \to \zeta_0} \mu'(\zeta), \quad \mu'(\zeta_0 - 0) = \lim_{\zeta \in \Gamma_2, \zeta \to \zeta_0} \mu'(\zeta).$$

Since

$$\mu'(\zeta_0 - 0) = -\frac{\mu'(\zeta_0 + 0)\lambda'(\zeta_0)}{(\mu(\zeta_0))^2}, \quad \lambda'(\zeta_0) = -1, \quad \mu(\zeta_0) = -1,$$

we obtain that $\mu'(\zeta_0 - 0) = \mu'(\zeta_0 + 0)$. Analogously one can check that $\mu'(1 - 0) = \mu'(1 + 0)$. Thus $\mu' \in H(\Gamma_0)$.

---

[1]One can consider even more general weights ϱ for which the space $L_p(\Gamma, \varrho)$ is invariant with respect to the operator V.

Let $\gamma = \eta \circ \mu^{-1}$. Without difficulty one can verify that γ satisfies conditions (3) and (4). The lemma is proved. □

Proof of Theorem 1. Without loss of generality we can suppose that $n = 1$. In view of Lemma 1, there exists a mapping $\gamma : \Gamma_0 \to \Gamma$ satisfying conditions (3) and (4). The mapping γ^{-1} maps the fixed points of the mapping ν to the points $\zeta = \pm 1$. For definiteness, let $\gamma^{-1}(t_0) = -1$ and $\gamma^{-1}(\tau_0) = 1$. Consider the operators

$$(V_0\varphi)(\zeta) = \varphi(\zeta^{-1}), \quad (M_1\varphi)(\zeta) = \varphi(\gamma(\zeta)) \quad (\varphi \in L_p(\Gamma, \varrho),\ \zeta \in \Gamma_0).$$

The operator M_1 is an invertible operator belonging to $L(L_p(\Gamma, \varrho), L_p(\Gamma_0, \widetilde{\varrho}))$, where $\widetilde{\varrho}(\zeta) = |\zeta + 1|^\alpha |\zeta - 1|^\beta$. It is not difficult to verify (see, e.g., [2, Chap. 1, Section 4]) that $M_1 S_\Gamma M_1^{-1} = S_0 + T$, where $S_0 = S_{\Gamma_0}$ and T is a compact operator. It is known (see [3, Section 5]) that $T \in \Sigma(p, \Gamma_0, \widetilde{\varrho})$. Since

$$M_1 a M_1^{-1} = a(\gamma(\zeta))I \quad (a \in PC(\Gamma)), \qquad M_1^{-1} V M_1 = V_0,$$

we have $M_1 A M_1^{-1} \in \Sigma(p, \Gamma_0, \widetilde{\varrho}; V_0)$ for every operator $A \in \Sigma(p, \Gamma, \varrho; V)$. Analogously one can prove that $M_1^{-1} B M_1 \in \Sigma(p, \Gamma, \varrho; V)$ for every operator $B \in \Sigma(p, \Gamma_0, \widetilde{\varrho}; V_0)$. Thus the algebras $\Sigma(p, \Gamma, \varrho; V)$ and $\Sigma(p, \Gamma_0, \widetilde{\varrho}; V_0)$ are equivalent.

In the same way, using the operator M_2 defined by

$$(M_2\varphi)(x) = \frac{2i}{i+x}\varphi\left(\frac{i-x}{i+x}\right) \quad (\varphi \in L_p(\Gamma, \widetilde{\varrho}),\ x \in \mathbb{R}),$$

one can obtain the equivalence of the algebras $\Sigma(p, \Gamma_0, \widetilde{\varrho}; V_0)$ and $\Sigma(p, \mathbb{R}, h; U)$, where $h(x) = |i + x|^\delta |x|^\beta$ ($\delta = p - \alpha - \beta - 2$) and $(U\varphi)(x) = \varphi(-x)$.

Let $\mathbb{R}^+$ denote the positive half-line. Consider the operator M_3 acting boundedly from $L_p(\mathbb{R}, h)$ to $L_p^2(\mathbb{R}^+, h)$ by the rule $(M_3\varphi)(x) = (\varphi(x), \varphi(-x))$ for $x > 0$. One can check straightforwardly that

$$\begin{aligned} M_3 S_\mathbb{R} M_3^{-1} &= \begin{Vmatrix} S & -N \\ N & -S \end{Vmatrix}, \\ M_3 a M_3^{-1} &= \begin{Vmatrix} aI & 0 \\ 0 & \widetilde{a}I \end{Vmatrix}, \\ M_3 U M_3^{-1} &= \begin{Vmatrix} 0 & I \\ I & 0 \end{Vmatrix}, \end{aligned} \tag{5}$$

where $S = S_{\mathbb{R}^+}$, $\widetilde{a}(x) = a(-x)$ for $x > 0$, and

$$(N\varphi)(x) = \frac{1}{\pi i}\int_0^\infty \frac{\varphi(y)}{y+x}\,dy \quad (x > 0).$$

Finally, consider the operator $M_4 \in L(L_p(\mathbb{R}^+, h), L_p(\mathbb{R}^+, \widetilde{h}))$, where

$$\widetilde{h}(x) = x^{(\beta-1)/2}|i + x|^{\delta/2},$$

defined by $(M_4\varphi)(x) = \varphi(\sqrt{x})$. It is not difficult to see that

$$M_4 S M_4^{-1} = \frac{1}{2}(S + R), \quad M_4 N M_4^{-1} = \frac{1}{2}(S - R),$$

where

$$(R\varphi)(x) = \frac{1}{\pi i}\int_0^\infty \sqrt{\frac{x}{y}}\frac{\varphi(y)}{y-x}\,dy \quad (x > 0).$$

Let us show that the operator R belongs to the algebra $\Sigma(p, \mathbb{R}^+, \widetilde{h})$. The operator $R = M_4(S - N)M_4^{-1}$ is bounded in the space $L_p(\mathbb{R}^+, \widetilde{h})$. Since $RS\varphi = SR\varphi = \varphi$ on a dense set (see, e.g., [2, Chap. 9, Section 7]), we have $R = S^{-1}$. Therefore, by [3, Lemma 6.1], $R \in \Sigma(p, \mathbb{R}^+, \widetilde{h})$. Put

$$M_5 = \left\| \begin{matrix} M_4 & 0 \\ 0 & M_4 \end{matrix} \right\| M_3.$$

In view of what has just been proved, $M_5AM_5^{-1} \in \Sigma_2(p, \mathbb{R}^+, \widetilde{h})$ for all operators $A \in \Sigma(p, \mathbb{R}, h; U)$. Thus, the algebra $\Sigma(p, \mathbb{R}, h; U)$ is equivalent to the subalgebra

$$\widetilde{\Sigma} = \{M_5AM_5^{-1} : A \in \Sigma(p, \mathbb{R}, h; U)\}$$

of the algebra $\Sigma_2(p, \mathbb{R}^+, \widetilde{h})$.

Put

$$\widetilde{M_2} = \left\| \begin{matrix} M_2 & 0 \\ 0 & M_2 \end{matrix} \right\|.$$

By what has been proved above, the algebra $\Sigma(p, \Gamma, \varrho; V)$ is equivalent to the subalgebra $\Sigma = \{\widetilde{M_2}^{-1}A\widetilde{M_2} : A \in \widetilde{\Sigma}\}$ of the algebra $\Sigma_2(p, \Gamma_0^+, \varrho_0)$. The theorem is proved. □

2. A Banach subalgebra Σ of the algebra $L(\mathcal{L})$ is said to be an algebra with Fredholm symbol (see [4]) if there exists a homomorphism π that maps Σ onto an algebra of matrix functions and has the following property: an operator $A \in \Sigma$ is a Φ-operator if and only if the function $\det \pi(A)$ is bounded away from zero. From [3, Theorem 5.1] it follows that $\Sigma_k(p, \Gamma_0^+, \varrho_0)$ is a Banach algebra with Fredholm symbol. Theorem 1 implies the following.

Corollary 1. *The algebra $\Sigma_n(p, \Gamma, \varrho; V)$ satisfying the conditions of the previous section is an algebra with Fredholm symbol.*

If $A \in \Sigma_n(p, \Gamma, \varrho; V)$, then the operator $\widetilde{A} = \widetilde{M}A\widetilde{M}^{-1}$, where $\widetilde{M} = \|\delta_{jk}M\|_{j,k=1}^n$ and $M = \widetilde{M_2}^{-1}M_5M_2M_1$, belongs to the algebra $\Sigma_{2n}(p, \Gamma_0^+, \varrho)$. It is natural to define the symbol of an operator A as the symbol of the operator $\widetilde{A}$ in the algebra $\Sigma_{2n}(p, \Gamma_0^+, \varrho)$.

From [3, Theorem 5.1] and the equality $\widetilde{A} = \widetilde{M}A\widetilde{M}^{-1}$ we get the following.

Theorem 2. *An operator $A \in \Sigma_n(p, \Gamma, \varrho; V)$ is a Φ-operator or a $\Phi_\pm$-operator in the space $L_p^n(\Gamma, \varrho)$ if and only if the determinant of its symbol is bounded away from zero. If this condition is satisfied, then the operator A is a Φ-operator.*

Note that one can calculate the index of a Φ-operator $A \in \Sigma_n(p, \Gamma, \varrho; V)$ by formula (5.3) of [3].

References

[1] I.C. Gohberg and N.Ya. Krupnik, *One-dimensional singular integral operators with shift.* Izv. Akad. Nauk Arm. SSR, Ser. Matem. **8** (1973), no. 1, 3–12 (in Russian). English translation: **this volume**. MR0333840 (48 #12162), Zbl 0255.47058.

[2] I.C. Gohberg and N.Ya. Krupnik, *Introduction to the Theory of One-Dimensional Singular Integral Operators.* Shtiinca, Kishinev, 1973 (in Russian). MR0405177 (53 #8971), Zbl 0271.47017.

German translation: *Einführung in die Theorie der Eindimensionalen Singulären Integraloperatoren.* Birkhäuser Verlag, Basel, Boston, 1979. MR0545507 (81d:45010), Zbl 0413.47040.

English translation (from the German edition of 1979) was revised and separated into two volumes:

One-Dimensional Linear Singular Integral Equations. I. Introduction. Operator Theory: Advances and Applications, **53**. Birkhäuser Verlag, Basel, 1992. MR1138208 (93c:47061), Zbl 0743.45004.

One-Dimensional Linear Singular Integral Equations. II. General Theory and Applications. Operator Theory: Advances and Applications, **54**. Birkhäuser Verlag, Basel, 1992. MR1182987 (93k:47059), Zbl 0781.47038.

[3] I.C. Gohberg and N.Ya. Krupnik, *Singular integral operators with piecewise continuous coefficients and their symbols.* Izv. Akad. Nauk SSSR Ser. Mat. **35** (1971), 940–964 (in Russian). English translation: Math. USSR Izvestiya **5** (1971), no. 4, 955–979. MR0291893 (45 #981), Zbl 0235.47025.

[4] I.C. Gohberg and N.Ya. Krupnik, *Banach algebras generated by singular integral operators.* In: "Hilbert Space Operators Operator Algebras (Proc. Internat. Conf., Tihany, 1970)." Colloquia Math. Soc. Janos Bolyai **5** (1972), 239–264. MR0380519 (52 #1419), Zbl 0354.45008.

MIX
Papier aus verantwortungsvollen Quellen
Paper from responsible sources
FSC® C105338

If you have any concerns about our products,
you can contact us on
ProductSafety@springernature.com

In case Publisher is established outside the EU,
the EU authorized representative is:
Springer Nature Customer Service Center GmbH
Europaplatz 3, 69115 Heidelberg, Germany

Printed by Libri Plureos GmbH
in Hamburg, Germany